Object Lessons and the Formation of Knowledge

OBJECT LESSONS
& the Formation of Knowledge

THE UNIVERSITY OF MICHIGAN MUSEUMS, LIBRARIES, & COLLECTIONS 1817–2017

Kerstin Barndt and
Carla M. Sinopoli, Editors

University of Michigan Press
Ann Arbor

Published in the United States of America by the
University of Michigan Press
Manufactured in China
♾ Printed on acid-free paper

2020 2019 2018 2017 4 3 2 1

A CIP catalog record for this book is available from the British Library.

Library of Congress Cataloging-in-Publication Data

Names: Barndt, Kerstin, editor of compilation. | Sinopoli, Carla M., editor of compilation.
Title: Object lessons and the formation of knowledge : the University of Michigan museums, libraries, and collections, 1817–2017 / Kerstin Barndt and Carla M. Sinopoli, editors.
Description: Ann Arbor : University of Michigan Press, 2017. |
Includes bibliographical references and index.
Identifiers: LCCN 2017008990 | ISBN 9780472130276 (hardback) | ISBN 9780472122646 (e-book)
Subjects: LCSH: University of Michigan—Museums—History. | University of Michigan—Libraries—History. | University of Michigan—Archives—History. | Museums—Collection management—Michigan—Ann Arbor—History. | Collection management (Libraries)—Michigan—Ann Arbor—History. | Archives—Michigan—Ann Arbor—Management—History. | College museums—Michigan—Ann Arbor—History. | Academic libraries—Michigan—Ann Arbor—History. | Learning and scholarship—Michigan—History. | Michigan—Intellectual life. | BISAC: HISTORY / United States / State & Local / Midwest (IA, IL, IN, KS, MI, MN, MO, ND, NE, OH, SD, WI). EDUCATION / History. | SOCIAL SCIENCE / Anthropology / Cultural. | EDUCATION / Higher. | SOCIAL SCIENCE / Archaeology. | SCIENCE / Study & Teaching. | LANGUAGE ARTS & DISCIPLINES / Library & Information Science / General. Classification: LCC AM101.M65 O35 2017 | DDC 069.09774/35—dc23
LC record available at https://lccn.loc.gov/2017008990

Contents

Foreword

The University of Michigan (U-M) has sought to advance knowledge and understanding since our founding in 1817 as one of the nation's first public universities.

Our mission is to serve the people of Michigan and the world through preeminence in creating, communicating, preserving, and applying knowledge, art, and academic values, and in developing leaders and citizens who will challenge the present and enrich the future.

The university's extraordinary museum, library, and art collections reflect our unique public origin. Like the university, they exist to better society.

For decades, they have thrived as a vital component of our character, with impact that is virtually immeasurable. They serve as centers for top scholarship, enhance the education of our students, foster cultural appreciation, and provide a platform for meaningful interaction between the university and people in our communities.

We are proud that our collections are as old as the state itself.

When the state of Michigan was formed in 1837, legislative action established the first collections at the university. The state's House of Representatives set up the organization and government of the university and provided for the Board of Regents to create a philosophical apparatus, a library, and a "Cabinet of Natural History."

These acts made the University of Michigan, in essence, the first public museum in the brand new state.

The ambition from the beginning was for the University of Michigan's collections to be a national resource for public engagement and the pursuit of knowledge.

Professor of botany and zoology Asa Gray was given $5,000 and purchased 3,400 books during a trip to Europe for the "philosophical apparatus." The fledgling Geological Survey of Michigan made the university a repository for its work. Early contributions from the survey included geological, mineralogical, botanical, and zoological specimens found during explorations of the state.

The first university president, Henry Tappan, made the collections a focus of his leadership. In his inauguration speech in 1852, he laid out his vision, which would shape the modern American university.

"We need first of all to provide the full material of learning," he said. "Our library must be enlarged so as to furnish all the helps of learning which are to be found in books; we need an observatory and a complete philosophical apparatus; our collections in natural history need to be extended; and collections in the fine arts to be begun and carried to a point to furnish the necessary models for artistic works."

In the decades since, U-M's public mission has been the bedrock principle on which our collections have been enhanced and curated.

Today our collections include more than twenty million specimens, housed in museums, libraries, and special exhibits across our campus. They capture the amazing breadth of our university, spanning eons of natural history and some of the most transformative periods in American life, and contribute to the excellent education we provide for our students and research opportunities for scholars around the world.

Public collections this large and accessible are rare and of tremendous value to society. They allow us to meet our special obligation, as a public institution, to extend the reach of our teaching and research across the full breadth of our society.

More than four hundred thousand people

enjoy our collections each year. Our visitors include schoolchildren from the region, U-M students and faculty, and campus and international researchers.

The University of Michigan's collections give life to the remarkable history of our world and provide incredible opportunities for the scholars who will create the exciting new knowledge of tomorrow.

As we move forward into our third century, our collections will continue to serve as an enduring and treasured resource for the academy and our community, state and nation.

Mark S. Schlissel
President
University of Michigan

INTRODUCTION

Kerstin Barndt and Carla M. Sinopoli

When Professor Alexander Winchell, chair of geology, zoology, and botany and museum curator on the Ann Arbor campus, filled out a questionnaire for the federal Bureau of Education about the "Museum of the University of Michigan" in 1873, he noted 1838 as its date of inception (fig. 1). The same year also marked the beginning of the University Library with the acquisition of John James Audubon's *Birds of America*, a work of immense artistic and research value that would provide a lasting bond between library and museum. The inauguration of both institutions, however, was only possible due to the weight that the university's charter placed on the building of collections. The original 1817 university charter was visionary in this sense as it explicitly empowered presidents and professors to establish not only colleges and schools but also "libraries, museums, atheneums [sic], botanic gardens."[1] This commitment was further strengthened in 1837 when the university was established in Ann Arbor and the newly constituted Board of Regents approved the creation of a Cabinet of Natural History.[2]

The 1873 questionnaire grants us insights into the nature of the university's first collections: natural history, ethnology, geology, medicine, and art. While the Board of Regents and professors of various departments were responsible for the collections, the Museum of the University of Michigan, as it was then called, provided an institutional home for the university's diverse and growing collections.

As Winchell was compiling the questionnaire, recent law school graduate Joseph Beal Steere, who had always had a passion for natural history, was in the middle of his ambitious five-year collecting expedition to South America and Asia, from where he shipped boxes filled with thousands of objects from the realms of nature and culture—bird skins, mammals, ferns, pottery, ethnographic objects, and musical instruments—to campus. Due to Steere's and others' collecting efforts, President James B. Angell could boast in his 1873 annual report about the growth of the collections: "we have 30,733 entries and 105,499 specimens, and, as is believed, the museum of only one of our institutions of learning surpasses ours in the number."[3] In the Gallery of Fine Arts, *Nydia, the blind Flower Girl of Pompeii*, a sculpture by Randolph Rogers, proved to be the most popular exhibit; other displays drew crowds as well, more than ten thousand per year: "The rooms of the Museum are regularly flooded with visitors from all around the country."[4]

Viewed from the current landscape of over twenty on-campus collections, libraries, and museums, Winchell's 1873 snapshot invites us to consider the common historical ground from which the early collections expanded: a young state university in the making with a pioneering spirit that founded one of the earliest and most comprehensive *public* university museums and libraries in the United States. In revisiting this history, the contributors to this volume address the defining transformation from the cross-disciplinary University Museum of the nineteenth century to the multiple and specialized research museums and libraries of the twenty-first century.

Over the last two hundred years, the University of Michigan has amassed research collections of enormous breadth and scope. These collections have served the university and its scholars as essential infrastructure for basic and applied research, teaching, and academic reputation. Few other North American university collections compare in size to Michigan's. The university's

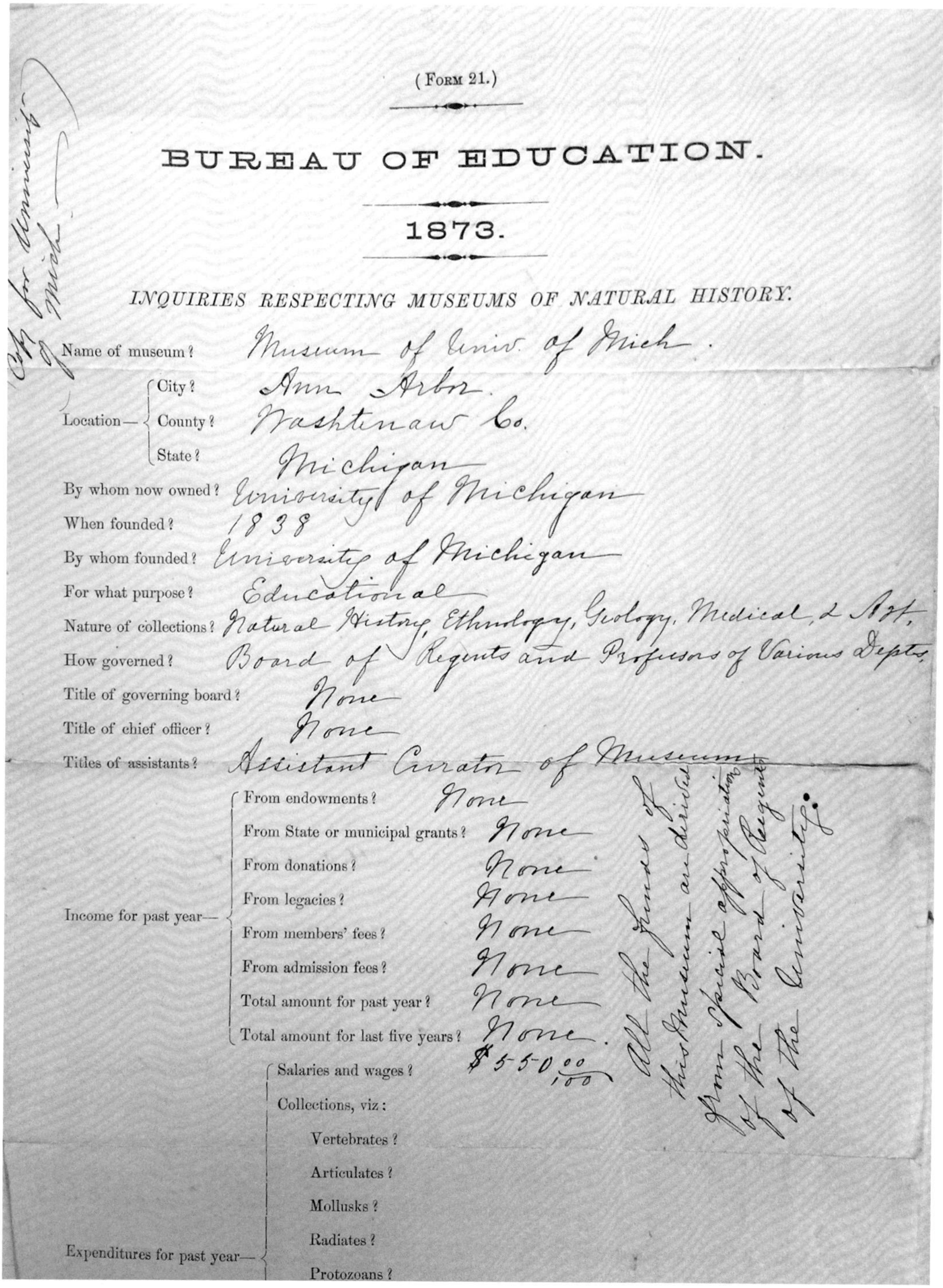

(Form 21.)

BUREAU OF EDUCATION.

1873.

INQUIRIES RESPECTING MUSEUMS OF NATURAL HISTORY.

Name of museum? Museum of Univ. of Mich.

Location— City? Ann Arbor. County? Washtenaw Co. State? Michigan

By whom now owned? University of Michigan

When founded? 1838

By whom founded? University of Michigan

For what purpose? Educational

Nature of collections? Natural History, Ethnology, Geology, Medical, & Art,

How governed? Board of Regents and Professors of Various Depts.

Title of governing board? None

Title of chief officer? None

Titles of assistants? Assistant Curator of Museum

Income for past year— From endowments? None
From State or municipal grants? None
From donations? None
From legacies? None
From members' fees? None
From admission fees? None
Total amount for past year? None
Total amount for last five years? None.

All the funds of this Museum are derived from special appropriation of the Board of Regents of the University.

Expenditures for past year— Salaries and wages? $550 00/100
Collections, viz:
Vertebrates?
Articulates?
Mollusks?
Radiates?
Protozoans?

Fig. 1. Federal Inquiry about University Museums, 1873; filed by Alexander Winchell. Bentley Historical Library.

Fig. 2. University of Michigan Botanical Garden and Old General Library, 1900, postcard. The Botanical Garden had its home on Central Campus before it was moved off campus in 1906. Opened in 1883, the Library also housed the Art Gallery. Bentley Historical Library.

treasures encompass world-class zoological and botanical collections with over sixteen million specimens that afford the study of global biodiversity; fossils in the Museum of Paleontology reach back millions of years and yield important data for the survey of life on earth; the archaeological collections assembled in the Kelsey Museum of Archaeology and the Museum of Anthropological Archaeology store knowledge about the long history of human culture and everyday life from around the globe; the galleries of the Museum of Art display paintings, prints, and sculpture spanning two millennia and four continents; and the University Library provides access to over twelve million volumes of scripts, papyri, books, and folios.

We take the university's bicentennial celebration as an important threshold to consider the role that collections have played in the history of the university and to recognize the centrality of museums and libraries to the university's teaching and research mission. Despite the importance and breadth of the University of Michigan collections, no publication exists that both summarizes them and considers their significance. *Object Lessons* takes important steps to address this lacuna and shows how on-campus collections and museums have provided an essential pillar for the university's ascent to one of the world's leading research universities. Further, while bound locally to the history of the university and the state of Michigan, the collections have also connected the university to other academic institutions via collegial networks and specimen exchanges; to major museums, including the Smithsonian Institution in Washington, DC, which early on recognized the University Museum as a partner institution; and to the world at large through global research expeditions.

The essays assembled here consider collections with astonishing continuity, well cared for over the course of almost two hundred years. They also consider lost collections and paths not taken. With *Object Lessons*, we engage multiple layers of knowledge and time embedded in the university's collections: the history of contact and colonialism that binds Michigan's European settlers and its indigenous populations to one another; traces of deep geological time and ecological crisis; and the history of science, culture, and the arts.

In this introductory chapter, we call attention to some of the lessons learned through our collective explorations of the history of the University of Michigan collections. We also briefly note some of the many stories not told in this book, left out not because they don't merit telling but due to constraints of organization and space. Our goal in this volume, however, is not merely to look backwards. We also address the continued and, indeed, growing relevance of the collections to today's university and the challenges they face. We consider both their ongoing potential for innovative scholarship and the new ways that the university's collections are being studied and disseminated in the age of big data and rapidly changing digital technologies. We conclude our introduction with a brief overview of the structure of *Object Lessons*.

CONNECTED HISTORIES

Today the University of Michigan's museums, archives, and collections consist of many distinct, administratively separate organizations, with varied reporting structures and administrative homes.[5] The catalog in the third part of this book presents on twelve museums, including the university's living collections housed in the botanical garden and arboretum, five libraries and special collections, and four important art collections, each with its own rich history. As we have noted, this is not how the university's collections began. In the early decades of the university, there were two divisions: the library and the museum. The first held books and the second objects. Now, of course, these boundaries have also blurred: our libraries and archives contain photographs, DVDs, maps, papyri, artworks, and artifacts; and our museums count documents, books, images, digital art, and sound recordings among their collections.

When Winchell wrote his response to the Bureau of Education survey in 1873, he accurately described a single university museum, located in the recently remodeled Mason Hall.[6] It exhibited geological and biological specimens alongside casts of classical sculptures and objects made by Michigan's indigenous Anishinaabe communities. By the early 1880s, when the first dedicated museum building was opened (figs. 2, 3), the art and classical archaeology collections did not accompany the natural history collections into the new space. Instead, they were soon displayed in a gallery in the new university library, launching a trajectory of division that would ultimately lead to the many museums and collections that exist today.

Given this history of "descent with modification," it is not surprising that the majority of our campus museums claim their origins in the initial 1837 Cabinet of Natural History. Nor is it inaccurate. Moreover, long after that founding moment, numerous ties have continued to connect the university's many museums and collections. Some of these ties are, as we elaborate below, the result of the actions of specific individuals. Professor of Latin language and literature Francis Kelsey, for example, discussed by John Griffiths Pedley in this volume, built the papyrology collections that are now part of the Special Collections Library and the archaeological collections that are housed in the museum named in his honor. He also offered the first anthropology course taught at the university, and played an important role in the creation of the Museum of Anthropology.

Other bonds that link collections across today's many museums and archives result from the ambitious scope of scientific research prior to the disciplinary differentiations that accelerated in the late nineteenth century. This broadly conceived idea of natural history still characterized nineteenth-century scientific expeditions. Early U-M naturalist Joseph Beal Steere studied and collected the entire natural—and cultural—world and brought large and wide-ranging collections back to campus.

Over time, as disciplinary boundaries became more clearly drawn and the museums and the

Fig. 3. Museum of Natural History with plaster cast of Apollo Belvedere at its center in the North Wing (former Mason Hall) of University Hall, c. 1875, photo print. Bentley Historical Library.

scientists who worked in them became ever more specialized, these early collections were divided. Thus, today, the University Herbarium, Museum of Zoology, and Museum of Paleontology each hold materials from the 1837 Geological Survey Collection discussed by Mackenzie Caple and Brian Williams in the first part of the volume. Many of the geological specimens themselves, however, that were long curated by the Department of Geology (now Earth and Environmental Sciences) have recently been transferred to the A. E. Seaman Mineral Museum at Michigan Technological University.[7] Similarly, spectacular Asian ceramics, textiles, and other objects, sometimes collected by the same individual, are found in both the Museum of Art and the Museum of Anthropological Archaeology.

In addition to the connections among collections within the university, numerous ties link them with larger national institutions and global networks. We have already noted the exchanges between the University Museum and the Smithsonian Institution that began in the early 1860s. And although the university lay on the frontier of the young United States and the even younger state of Michigan at its founding, it is nonetheless possible to trace how University of Michigan geologists and biologists were drawn into the tempestuous debates on evolution brought on by Charles Darwin's 1859 publication of *On the Origin of Species*. In her essay on the collections housed in Mason Hall and the first University Museum building, Kerstin Barndt discusses the relationship that bound religion to natural history in the early nineteenth century. She investigates how this relation changed throughout the century allowing the University Museum building that opened in 1881 to embrace and popularize the theory of evolution through exhibits and public programming. The focus on evolution was then carried into the University Museums Building, which opened in 1928 and whose central exhibition space has centered on the Hall of Evolution ever since (see Sinopoli and essay on Museum of Natural History, this volume).

Other connections thrust University of Michigan scientists into global politics, including the US history of colonialism and imperialism, in unexpected ways. Steere's exposure to the biological diversity in the Philippines during the 1870s so intrigued him that he returned to the archipelago a decade later, this time with several undergraduate students of zoology. One of these students was Dean C. Worcester (see Mark Rice this volume). Worcester also became fascinated by the Philippines and eventually advanced to secretary of the interior of the US colony after the Spanish-American War. Worcester's ties to the university contributed to the extraordinary relations our campus has had with the Philippines for more than a century, and to the remarkable Philippine collections in the Special Collections Library, Bentley Library, Museum of Zoology, Herbarium, and Museum of Anthropological Archaeology.

The chapters in this book work to reconstitute some of these lost connections, reminding us of shared origins and common stories. They also allow us to explore how the division of collections both echoed and played a role in the division of the academy into ever more specialized and, often, more isolated disciplines and departments. And they allow us to consider how, in today's era of growing commitment to inter- and transdisciplinary teaching and scholarship, and to hands-on, experiential, and engaged learning, the university's collections again serve as a medium to reexamine seemingly disparate materials and shape new research agendas.

NETWORKS: INDIVIDUALS, COLLECTIONS, AND INSTITUTIONS

The transformation of Michigan's original multidisciplinary University Museum into differentiated research museums with associated disciplines confirms established accounts regarding the history of science. Around 1900, human, social, and natural sciences established new protocols, methodologies, and disciplinary borders. Museum collections played an important part in this process, especially in the first decades of the twentieth century.[8] Museum historian Samuel Alberti, moreover, insists on the sociopolitical context in which disciplinary changes occurred. For museum disciplines "were enacted not only in the material culture of the collections, but also through personnel and administrative structures."[9] In this vein, we approach the networks through which collections and museums evolved and changed on the Ann Arbor campus—networks and relations between objects, curators, regents, administrators, and scholars as well as local and state politics.

The second part of *Object Lessons*—"Collectors, Archivists, and Curators"—presents thirteen men and women who laid the foundations for diverse university's collections. Many other people could have been featured and alternative stories could have been told. We chose to focus largely on key individuals in the early building of collections

Fig. 4. Collection Facility Research Museums Center, 2016. Photo by Carla Sinopoli.

and on men and women whose work spans their breadth. Connecting collecting subjects to the objects of collections, these biographies reflect on the life cycle of particular museums. They thus bring biographies and museums into close proximity, examining their relationships.[10]

Besides being builders of collections, archivists, and curators, museum-based researchers also were distinguished outside of their museums and archives. As scholars of national reputation, they shaped their disciplines through the founding of professional organizations, the building of top-ranked graduate programs, and the hiring of promising young scholars (often their own students in the early decades) to succeed them. To name but a few of the leadership positions held by a small number of the individuals discussed in this volume, we note that Carl Guthe, first director of the Museum of Anthropology, helped found the Society for American Archaeology; Museum of Paleontology director Ermine Cowles Case served as president of both the Michigan Academy of Science and the Paleontological Society; and Francis Kelsey served as president of the American Philological Association and the Archaeological Institute of America (see catalog entries and biographies for additional information).

In addition, many of the individuals discussed in this volume (and many whom we did not have room to include) remind us of the importance of visionary leadership—such as Alexander Ruthven, William Warner Bishop, Ermine Cowles Case, Francis Kelsey, Harris Harley Bartlett, Carl E. Guthe, and Lewis George Vander Velde, among many others—to define and create institutions and disciplines that have enduring legacies. Nonacademics, such as industrialist and regent William L. Clements, art collector Margaret Watson

Parker, and radical activist Agnes Inglis, also made remarkable contributions to the collections: the first two in their donations to the Clements Library and Museum of Art, respectively, and the latter in her efforts to care for and expand the University Library's extraordinary Labadie Collection.

Given our focus, we have largely emphasized key historic and founding figures for the shaping of our collections. However, we acknowledge that, with rare exceptions, the collections and institutions featured continue to thrive as centers of innovative research and teaching, thanks to the generations of distinguished successors to these founding figures and not least the curators, archivists, collection managers, exhibition preparators, and other professional staff who work in them today, many of whom have contributed to this book as authors or important advisors.

Stories Not Told (Enough)

This is a large book that covers substantial ground. Nonetheless, there are some important topics that we have only touched upon within the scope of this project and that certainly deserve further research.

While *Object Lessons* explores the history that resulted in the University of Michigan's remarkable collections, our goal has not been to be merely celebratory. Instead, we have sought to cast a critical scholarly eye on this history. Two themes that have particularly interested us are how early collection building and museum displays at the university intersected with the creation of the state of Michigan as a political and territorial entity and concomitantly, how the latter marginalized, and worked to objectify (literally) Michigan's Native populations.

Unlike many states, Michigan does not have an official state museum. The simultaneous creation of the university and the state in 1837 resulted in U-M becoming the de facto state museum as the primary recipient of the Michigan Geological Survey specimens. In the early twentieth century, the Museum of Zoology became the repository for the collections of the renewed Biological Survey overseen by Alexander Ruthven (see Rachel Miller, this volume). Similarly, the Museum of Anthropology became the repository for collections and records of the Michigan Archaeological Society. The ways that these close ties between state-making and university-making shaped the university's collections and distinctive institutional structure are discussed in Caple and Williams's essay on the Geological Survey as well as in Barndt's article on the first university museum and buildings, but certainly call for further scholarly study.

The university's complex history of relations with Michigan's Native American residents, as it involves collections and more, also merits far greater discussion than provided in this volume. Some aspects of this history have garnered recent publication attention, if not yet scholarly analysis.[11] Thus, in the early twenty-first century, the Museum of Natural History removed its mid-twentieth-century Native American dioramas following a period of intense controversy and critique of their context (in a "natural" history museum) and content. And critiques of the Museum of Anthropology's archaeological collections of ancestral human remains and burial objects excavated from Anishinaabe burial sites across Michigan have been the focus of protests and fierce debates from the 1970s on. Repatriation processes are underway for these and other Native American collections at the university that fall under the federal Native American Grave Protection and Repatriation Act (NAGPRA).[12]

While we feature catalogue and biographical entries about the Clements Library, the Bentley Library, the Gerald R. Ford Presidential Library and Museum, and the University Library's Special Collections, we have not attempted to cover the full scope of the University Library. Doing justice to the riches kept under the umbrella of the University Library would have doubled the length of this volume, and certainly merits its own book. Our expertise, as museum studies scholars and practitioners, is toward museums. We hope we have given the nonmuseum collections their due, but fear that our biases show in the hard decisions that had to be made about what and who to cover.

Looking Forward

Even as we look back over the history of the university's collections on the occasion of the bicentennial, we also hope that *Object Lessons* will be considered an important first step and inspire future enquiries into the riches of U-M's collections and museums. University collections are immense assets that remain relevant to present and future research, knowledge production, and object-based pedagogy. Exhibitions in campus museums, moreover, attract tens of thousands of visitors per year from local, regional, and global communities. Tightening budgets and spatial constraints have indisputably squeezed university museums in the United States and elsewhere, including our own. At the same time, the museums have begun to work more closely together, to build international databases and to share concepts as to how the cross-disciplinary classroom may engage with museum collections.[13] At the Universities of Oxford, Berkeley, Göttingen, Humboldt, and Harvard, to name just a few, recent research and museum projects, symposia, and anniversaries have sparked renewed interest in the diversity of research collections found on each campus and their place in the university's respective mission.[14] In Ann Arbor, the university museums have kept up with these international developments in the midst of dramatic change, expansion, and reorganization.

This period began several years ago with the construction of new wings on the university's Museum of Art and Kelsey Museum of Archaeology and will culminate in 2019 with the opening of a thoroughly reconceptualized Museum of Natural History in the new Biological Sciences Building. The recently renovated and expanded Clements Library has made one of the premier archives of American history more accessible to researchers and the public. Other important changes are occurring in the university's research museums, as more than twenty million objects and specimens from the Museums of Paleontology, Zoology, and Anthropological Archaeology are being moved from central campus to join the University Herbarium in a new storage and research facility—the Research Museums Center—on Varsity Drive (fig. 4).

At the same time that the physical accessibility of these collections to the central campus community is lessening, virtual accessibility to the collections has dramatically increased through collaborative digitization efforts that will bring the biological collections of the Museums of Zoology and Paleontology, the University Herbarium, Matthaei Botanical Gardens, and Nichols Arboretum into a single database structure. Similarly, the Kelsey Museum of Archaeology, Museum of Anthropological Archaeology, and Museum of Art will also be using a common database system. And information on all of the collections will be searchable online.

Object Lessons puts these changes into perspective as it provides a long view of expansion, revisions, and reorganizations. Momentous though they now appear, these current transformations form part of a larger dynamic that has bound together research, collecting, and teaching at the University of Michigan over the past two centuries and will continue to do so in the future.

OBJECT LESSONS IN THREE PARTS

As already alluded to, our book is divided into three main parts: a collection of essays that frame the book, a series of biographical entries, and a catalogue of the university's museums, collections, and special libraries.

The essays in the first part, "Museum Disciplines and Histories," provide the historical and conceptual foundation for *Object Lessons*, exploring central periods of change with an eye on how academic practices, pedagogical regimes, and disciplines have shaped collections and vice versa.

Chapters by Francis X. Blouin Jr. and Mackenzie Caple and Brian Williams take us back to the founding decades of the university and collections. Blouin describes the distinctive vision of Henry Philip Tappan, the university's first president, to create an "apparatus of learning" on campus and considers the role of collections in Tappan's Prussian-inspired model for U-M as a new kind of public research university. Caple and Williams examine the role that the First Geological Survey played in the foundation of the state and the University of Michigan's earliest museum collections.

In their respective chapters, Barndt and Sinopoli summarize the history of the university museums by focusing on the campus's first three museum buildings. Barndt begins with the first collections in Mason Hall (1841–80) and traces their move and reconceptualization to the first independent University Museum building (1881–1927). Sinopoli picks up this narrative in the early 1900s, when the collections moved again to the Ruthven University Museums Building (1928–present) to be transformed into the research and public museums we know today. Reconstructing the story of these three buildings allows us to trace the history of university collecting more generally—and natural history collecting in particular—from a single museum to multiple and differentiated research museums. This trajectory both parallels and plays a role in the development and separation of distinct academic disciplines over the last two centuries. It also anchors our exploration of the place that museums and collecting have occupied within the larger frameworks of the university's history and the history of Michigan.

Similarly, David Choberka focuses on the genesis of the campus art collections and the birth of the University of Michigan Museum of Art (UMMA), placed against the backdrop of university reforms and academic differentiation. Choberka discusses the growth of the art collections from the acquisition of the university's first art works in 1854 to the opening of the Frankel Family Wing in 2009. He finds surprising echoes between the collection's anchoring function within the university's first classical and liberal arts curriculum and the museum's expanding vision and activities in engaging the campus community today.

While Choberka's essay explores a museum that is growing and thriving, Peter M. McIsaac traces the history of campus museums that no longer exist: the historically important medical collections of medical models, anatomical and pathological specimens, and slides through which generations of students acquired their medical knowledge and faculty relied on in their research. These museums (along with a pharmacy museum, chemistry museum, and the first dentistry museum) were dissolved in the mid-twentieth century as medical training changed and specimen collections lost their pedagogical value. Only small remnants of what once were large and important collections of specimens and instruments now remain on campus.

As described above, the second part of the book zooms in on individual actors: museum and library directors, curators, explorers, and founders who built not only collections but also global networks of scholarship, research, and exchange.

Many museum researchers and archivists who followed the footsteps of the founding generation featured in the second part are included in the catalog of collections that compose the third part of *Object Lessons*. Here, readers find brief but comprehensive accounts of today's on-campus museums, libraries, and collections, describing their breadth and depth, development over time, and present usages. Information on opening hours, addresses, and contact information for collections that are not open to the general public are also included. While we hope that *Object Lessons* inspires further research into the linked histories of scholarship, teaching, and collecting in higher education, we would be delighted if it also sparked expanded interest among the general public in the riches on display across the university's many public exhibitions and museums.

Between the second and third parts of *Object Lessons*, photographs by internationally renowned artist Richard Barnes show museum spaces and objects in transition. As the research collections move from the Ruthven Museums Building with its long history into their new fully climate controlled home on Varsity Drive, layers of time seem to drift apart under the mounted birds' curious gaze. In her contribution to our volume, Amanda

Krugliak reads Barnes's striking photographic series as a meditation on time, loss, and transformation. "Barnes' photographs chart the velocity of change, the sense of displacement we experience as the momentum shifts, first meandering, and then speeding up. They measure the recession of our own stories into the distance."[15] With his photographic essay, Barnes brings an aesthetic temporality to this volume to accompany us as we move further into the twenty-first century of scientific collecting and curating.

Notes

1. University Act, August 26, 1817, signed by Acting-Governor William Woodbridge and Judges A. B. Woodward and John Griffin of the Michigan Territory. Cited in Burke Hinsdale, *History of the University of Michigan*, 1906, 9. The original handwritten copy of the Act is preserved in the University Library.

2. *Proceeding of the Board of Regents* University of Michigan. 1837–1864 (PBR), June 1837, 2; and November 1837, 17.

3. PBR, 1870–1876, October 1873, 310.

4. Alexander Winchell, "Report of Operations in the Museum of the University of Michigan for the Year Ending September 21, 1869," PBR 1864–1869, 356–66.

5. Indeed, the six museums in the College of LSA fall under four different divisions: Natural Sciences (Paleontology, Herbarium, Zoology), Humanities (Kelsey Museum of Archaeology), Social Sciences (Museum of Anthropological Archaeology), and Undergraduate Education (Museum of Natural History), and report to four different associate deans (and for the Herbarium and Museum of Zoology, a department chair).

6. In 1872, Mason Hall was integrated into the newly built University Hall with a large auditorium at its center. This remodeling opened new museum space in Mason Hall for the natural history and art collections on display. See Barndt, this volume.

7. *The University Record*, May 13, 2015. https://record.umich.edu/articles/u-m-partners-michigan-tech-historic-mineral-collection

8. For eloquent accounts of the role of museum collections played for the foundation of academic disciplines at institutions in the United States and Great Britain, see Steven Conn, *Museums and the American Intellectual Life, 1876–1926*, Chicago: University of Chicago Press, 1998, and Samuel J. M. M. Alberti, *Nature and Culture. Objects, disciplines and the Manchester Museum*, Manchester: Manchester University Press, 2009, 31–81.

9. Alberti, *Nature and Culture*, 3.

10. See *Museums and Biographies. Stories, Objects, Identities*, ed. Kate Hill, Woodbridge: The Boydell Press, 2012.

11. Though see, Amy Lonetree, *Decolonizing Museums: Representing Native America in National and Tribal Museums*. Chapel Hill: The University of North Carolina Press, 2012, especially chapter 4, and Raymond Silverman and Carla M. Sinopoli, "Besieged! Contemporary Political, Cultural, and Economic Challenges to Museums in the Academy as Seen from Ann Arbor," *University Museums and Collections Journal (UMACJ)*, vol. 4 (2011), 19–31.

12. Native American Grave Protection and Repatriation Act of 1990; see http://nagpra.umich.edu/about-nagpra for information on U-M NAGPRA activities.

13. In April 2000, representatives of collections and museums of several European universities found the network Universeum. The objective of the network is to share knowledge and experiences among its members and to undertake joint projects with the aim of enhancing access to the collections at all levels. See http://www.universeum.it; the ICOM committee UMAC (University Museums and Collections) of which the U-M research museums are members, is also maintaining a worldwide database of university museums and collections: http://publicus.culture.hu-berlin.de/collections

14. At the University of Oxford, the Pitt Rivers Museum reopened after substantial renovation in 2009. It showcases the university's collections of anthropology and archaeology and finds itself only a block away from the Ashmolean Museum, which was established in 1683 as one of the oldest (university) museums in the world. In 2014, Harvard University reopened its three art museums, the Fogg, Busch-Reisiger, and Arthur M. Sackler, under one expanded roof, and Julie Buckler and Yukio Lippit organized a conference interrogating the "University as Collector" at Harvard's Radcliff Institute for Advanced Study (April 2015). The collections of Berlin's Humboldt University are documented in *Teatrum Naturae and Artis—Theater der Natur und Kunst. Wunderkammern des Wissens*, eds. Horst Bredekamp, Jochen Brüning, Cornelia Weber, Berlin: Henschel, 2000. From 2019 on, the collections of Humboldt University will be displayed in a new museum building, the Humboldt Forum, alongside with three other major Berlin museums.

15. Amanda Krugliak, "Richard Barnes/Taking Inventory" this volume.

PART I

Museum Disciplines and Histories

DEFINING THE VISION

Henry Tappan and the Creation of an "Apparatus of Learning"

Francis X. Blouin Jr.

The University of Michigan is among a handful of universities in America that holds significant research collections assembled over the course of more than a century. These collections are of materials—biological specimens, artworks, books and manuscripts, archaeological artifacts. And it is upon those materials that much of learning was based in the mid-nineteenth century, as institutions of higher education transformed into the research universities we know today. Henry Phillip Tappan, the University of Michigan's first president, was among the first to articulate the importance of what he called the "apparatus of learning." For Tappan, these materials fit a broader vision of what he hoped the University of Michigan would become. Any exploration of his central importance in the development of the young institution in Ann Arbor really needs to step back and consider the process as well as the context for the investment in materials that this volume celebrates. The fundamental purpose of collections was to establish a connection. Materials would connect students and faculty to worlds of ideas and experiences far beyond the boundaries of Ann Arbor. Collections then were essential to ever-broadening ideas of what would constitute learning and knowledge.

As with nearly all collections of any sort, the assemblage of materials at the University of Michigan began somewhat idiosyncratically and then became increasingly formalized and structured. Collecting on the scale that occurred at the university can in retrospect be considered as falling into three distinct phases. The first is accumulation and connection—the importance of materials providing connection to the broader world. The second is rationalization—at this point, the collections become of a scale that their purposes need to be defined more specifically within the context of what the university aspires to be. The third is organization—when collections become so big that the classification, cataloguing, and preservation become an equally important complement to the activity of accumulation.

ACCUMULATION

The importance of collections and materials of learning at the University of Michigan can certainly be traced to its earliest days. The Catholepistemiad or University of Michigania was established in Detroit in 1817 by Gabriel Richard, John Monteith, and Augustus Woodward. It was hardly what we would think of as a cosmopolitan university. The "city" was incorporated in 1815 with a population of about 1,200 (exclusive of the army), though how that count was reached is not clear.[1] Little is known about the content and thrust of intellectual life at the newly established university, beyond Woodward's ambitious delineation of a curriculum. What little we do know can be derived by inference from the collections accumulated by Richard and Monteith, which still exist in large part at the University of Michigan and elsewhere.[2]

By the time of his death in 1832, Richard had accumulated a library that numbered in excess of forty-six hundred volumes. Its wide-ranging contents included the encyclopedia of Diderot, a Mohawk-English edition of the *Gospel According to Saint John*, and philosophical tracts by Voltaire and Rousseau as well as those of their contemporary critics. Richard was a member of the Sulpician order of Catholic priests, so not surprisingly there was a great deal of theology among the volumes. In any case, it was among the largest libraries in the American West and nearly two-thirds of the volumes survive. It was Richard's intent that the library would form the basis for the founding of a Sulpician college in Detroit. This never happened and the collection passed to the bishop of the Roman Catholic Archdiocese of Detroit. It was later divided between the University of Michigan (Bentley Library) and two seminaries of the archdiocese in Detroit (fig 5). Richard also established the first nongovernmental printing press in Detroit and published over fifty pamphlets and devotional items.

Monteith also accumulated a library, considerably smaller but not inconsequential, judging from the three hundred volumes that survive and are now at the Bentley Library. Though never considered "property" of the Catholepistemiad, one can only conclude that they informed the content and thrust of the intellectual life of the new institution. These volumes connected Richard and Monteith to the world of ideas of the Enlightenment in Europe as well as to the theological and philosophical debates of their own time. That they survive is testament to the importance of these ideas in the process of formation of the new university.[3]

In 1837, the university was reestablished in Ann Arbor. While certain lands held by the Catholepistemiad were passed on to the Ann Arbor institution and determined a connection between the two, the Richard and Monteith libraries did not transfer. Ann Arbor was a small community nestled in a small county and part of a new state in the American Union. One need only look at the famous painting of the university campus by Jasper Cropsey done in 1855 to see how thoroughly isolated it was (fig. 6).

During these first years in Ann Arbor, it was clear to the regents that collections would be a means to connect this institution to a wider world. Among its first acts in 1837 was to establish a "Committee on the Library, Philosophical Apparatus, and Cabinet of Natural History." Edward Mundy (who also served as lieutenant governor of the state), Elon Farnsworth, and John Porter were appointed as members of the committee.[4] The committee reported in November of that year that they had not made much progress, but they did have their eye on a collection of minerals (see Caple and Williams this volume, on the story of this aspect of the committee's work and the eventual appointment of Douglass Houghton).

Fig. 5. The Gabriel Richard library arrives on campus in 1951. F. Clever Bald, Director of the Michigan Historical Collection with Monsignor H.E. Donnelly and Reverend F.X. Canfield (left to right). Bentley Historical Library.

Accumulation at this point in the university's history moved in a variety of directions including the library. For faculty and students the library that was contemplated would provide connections to ideas and places well beyond the boundaries of Ann Arbor. At their March 1838 meeting, the regents approved the historic purchase of John James Audubon's four-volume ornithology series, *Birds of America*, and at their September 1838 meeting they asked Asa Gray to travel to Europe and purchase books for the library, approving a sum of \$1,500 for this purpose (later raised to \$5,000). The same resolution that appointed Asa Gray also articulated a framework for his pur-

Fig. 6. Jasper Francis Cropsey, University of Michigan, oil on canvas, 1855. Bentley Historical Library.

chases, emphasizing "Standard English Works in history, biography, voyages and travels, poetry, statistics, science. . . ."[5] A report of the library committee in 1840 praised Gray's work, noting that the library collection stood at 3,700 volumes "embracing the various departments of history, philosophy, classical literature, sciences and arts, jurisprudence, etc."[6] A large portion of the purchases secured by Gray were of works that "could not be obtained in America" and many of the editions were "scarce and rare in Europe."[7]

During the years in Ann Arbor prior to 1850, the regents regularly reported to the state superintendent of public instruction on the accumulation of materials including both library books and natural history specimens. This accumulation was clearly perceived as foundational to the new university. However, any comprehensive statement on how this pertained to a vision for the institution simply does not exist. There is an occasional comparison of the amount of materials to those acquired at Yale, with a regents' report for 1841 noting that the University of Michigan specimen collection was "much richer." This suggests that the regents had models in mind and had their eye on the strong institutions of the East Coast.[8]

The "act to provide for the organization and government of the University of Michigan" of 1837 states that "the objects of the university shall be to provide the inhabitants of the State with the means of acquiring a thorough knowledge of various branches of literature, science, and the arts."[9] In 1841, the regents stated that this "collegiate institution in a free state, and in the conducting of its interests should ever be upon liberal principles and irrespective of all sectarian predilections and prejudices."[10] There were also allusions to the importance that the institution promote "usefulness."[11] By midcentury, the university was a viable institution with an array of expertise represented on the faculty that included Latin and Greek, mathematics, mental and moral philosophy, medicine, pharmacy, zoology, chemistry, geology, and minerology.[12] Clearly the materials accumulated would be central to work in these academic specialties—but how was stated only in the broadest terms.

RATIONALIZATION

In 1850, the state of Michigan adopted a new constitution that among other things reorganized the governing structure of the university in Ann Arbor. It provided for an elected board of regents who would be responsible for appointing a president to oversee the operations of the institution. At a special set of meetings held in Detroit in the month of August 1852, the board unanimously elected Henry Philip Tappan, the first president of the University of Michigan under these new constitutional provisions (fig. 7).[13]

Tappan was a congregational minister and sometime professor at New York University.[14] He was recognized as an authority on higher education and in 1851 published *University Education*, a work largely focused on the benefits of the Prussian system. In it, Tappan proposed:

> A great institution would collect all that is now scattered and isolated among us, be the home of scholars, the nurse of scholarlike endeavors, the regulating and harmonizing center of thought and investigation. Our whole population would feel the plastic power of intellectual development and progress.
>
> Only with a strong faculty, intelligent students, ample buildings and all the apparatus of learning could one really experience a true university. Especially important was a full store of the material of science, literature, and the arts.[15]

Tappan's conception of what a university should be was breathtaking. It should embody all knowledge and point the way to new knowledge. And, while he had great admiration for the scholarly traditions at Oxford, Cambridge, and Paris, as well as for the oldest American colleges, Harvard and Yale, he thought they were not universities in his sense of the term, which was based on the Prussian model. In Tappan's view, he would come to Michigan and make it a university "worthy of the name."[16]

Tappan arrived in Ann Arbor in 1852 and delivered his inaugural speech as president on December 21 of that year. It was fifty-two pages long and represents what we would call today a "vision statement" of what the university could and should become. Tappan gazed far into the future and saw a university that would explore all branches of theoretical knowledge, all aspects of applied science, all manner of the arts, and all of professional education. By so doing it would not only educate students but would create new knowledge and thus advance a distinctly American civilization. Through his university he wanted to establish the new nation as a full participant in the learned conversation of all peoples. All this would be accessible to all citizens regardless of rank, position, or wealth. The only aristocracy for Tappan was the aristocracy of the intellect.[17]

Fig. 7. Portrait of Henry Tappan, 1860. Bentley Historical Library.

He opened his speech by boldly defining his concept of knowledge in two ways. First, in his words, "all knowledge must be of every thing standing around man, and objective to him." Tappan's second category is "knowledge of man himself—the being that knows." Using the term "man" in the generic sense, he went on to explain his categories. Knowledge that is "objective to man" is knowledge that an individual can observe, either by direct observation or by inference. Within this category he included relational analysis such geometry. He also listed scientific phenomena that existed in time and space, including physics, astronomy, and optics—what he called "an illuminated Geometry"; the study and classification of distinct substances, through chemistry, geology, and mineralogy; and the phenomenon of life itself—biology, zoology, and botany.[18]

Tappan's embrace of the natural and physical sciences put him and his university at the forefront of those universities that embraced science in a more systematic way. The Scientific Department of Yale College had been organized in 1846, adding an agricultural college in 1866. In 1847, the Lawrence Scientific School of Harvard College was created. Tappan would introduce a variety of formal scientific courses at the University of Michigan in the 1850s.

The goal of Tappan's first category of knowledge was toward two "commanding ideas"—utility and beauty. The dynamic economic growth of the time demonstrated the utility of advanced knowledge and more formal explorations of various branches of the physical and natural sciences were thus essential to the ambitions of the nation. And the beauty of new architecture, sculpture, painting, poetry, and music demonstrated a different aspect of his comprehensive definition of knowledge. Set within all of this was a serious consideration of individuals and society.[19]

Tappan's second category of knowledge related to "man himself—the being who knows." Tappan encouraged an understanding through science of "all that is going on" in the mind of man. Whereas the first category was the study of phenomena "outside" man, this second category was the study of man "within." Ontology, philosophy, psychology, aesthetics, logic, and morals were subjects that defined this dimension of knowledge. He then argued that this study of the individual must be joined with the "study of all that man has done on the great theater of the world." History, languages, literature, art, antiquities, law, religion, ethnology, and geography were components of this comprehensive understanding of people over time and space.[20]

To articulate such an ambitious and comprehensive conception of knowledge was one thing. To map a way to achieve this within the context of a single institution in the mid-nineteenth century in a small town in what was then the American West was no small challenge. Nonetheless, Tappan went on for pages setting out all that would be required. A distinct and important component of his vision for the university was the accumulation of material necessary for study and the generation of new knowledge—what he called the "material of learning."[21] Collections of all sorts were essential to connecting the young university to the vast world of knowledge and experience.

He urged that the university acquire "books, apparatus, specimens, and models sufficient to furnish all needed information, all needed facilities for scientific and learned investigation, and all needed illustrations. And the men who are to do the work, can alone determine what and how much are needed."[22] He came to Ann Arbor and saw that the university had erected substantial and useful buildings but had not achieved a sufficient accumulation of materials to put into those buildings. He said, "we have had gorgeous shells that seemed like mother of pearl, but there were no pearls within."[23]

For Tappan this was not simply a matter of accumulation. Rather, he went further in his inaugural and in subsequent reports to the regents to provide a rationalization for these materials set within a carefully articulated and vigorously promoted idea of the university as an intellectual center of the highest order and as a public treasury of knowledge and culture. He made three points in particular: first, that collections are an essential capital of the university; second, that collecting on the scale he envisioned was a state responsibility; and third, that collections will have a wide-ranging impact.

At the heart of Tappan's university was the faculty. He worked tirelessly to bring to Ann Arbor some of the most important intellectuals of the day, including Andrew Dickson White, Alexander Winchell, and Franz Brunnow. But he saw that the faculty, no matter how brilliant, could not be a repository for all knowledge. The work of the faculty had to be supplemented and sustained by a well-formulated library. He argued that "a great institution of learning without a Library affording the necessary aids of learning is simply an impossibility, unless we can bring together a large number of Professors each one a walking Library; but such Professors cannot be created and sustained without books."[24] Books were the central point of connection between the university and ideas and that connection occurred through the faculty. Books were then an essential capital of the university. For Tappan, "it is impossible to carry on the educational discipline of such an institution as ours without an ample supply of books in every branch of Science and Literature. Books, here, are not an amusement or a luxury; they are a prime necessity; they are the fixed capital of a University."[25]

Tappan's vision was expansive. Because materials of learning—whether books, apparatus, or objects—were all critical to the formation of knowledge as well as to instruction, he sought an ambitious program of acquisition—beyond anything any other institution might hope to realize. But to achieve this level of collecting would require a significant investment. The only institution with resources of the scale he had in mind was the state of Michigan. In looking at the landscape of higher education at the time, dominated by denominationally affiliated religious institutions, Tappan sensed the limitations on their resources. He would have no such limits for his vision of this state university. He would argue that

> there are institutions of magnitude requiring so large a number of Professors, Such extensive libraries and such a complete supply of apparatus and the material of learning, generally, that no private means, nor even the means rationally to be expected from any single religious denomination can suffice for their establishment. Besides their nature they can not be of any private or denominational interest; they stand related to the culture of the whole community; they are even cosmopolitan. Most clearly, therefore, Universities are the appropriate work of States.[26]

Throughout his years at the university, he would continually and vigorously argue for the level of state support to achieve all dimensions of his vision.

Collections were not static accumulations but rather were to be energizing and enriching of the community. Collections had a "cumulative" effect, an impact. In December 1857, Tappan reported on the status of collections. In this report we see not only the fact of accumulations but the dynamic interaction within the university between material and learning. At the start of this section of the report he cites his beloved observatory and its importance to the university and to research generally. Tappan's observatory, described elsewhere in this volume, represents the epitome of his vision—an institution that would bring faculty, materials (instruments) of learning, and students together to learn and to generate new knowledge. The observatory would be maintained in the service of the academic community of the university and for individuals beyond the boundaries of the campus for whom new knowledge would be useful and important.

In that 1857 report, Tappan notes that the library collection is growing: "affording facilities for consulting books, and in attracting the students to resort to the Library daily and habitually." He expresses a hope that the new Gallery of Fine Arts under the direction of Henry Simmons Frieze (see Choberka, this volume) "will prove the beginning of a movement destined to enrich the University with many treasures of Art." He sees the "limited" but growing collection of antiquities a very valuable aid in "illustrating the subjects of instruction assigned to them [i.e., faculty]." In the rich collections of specimens of mineralogy and geology and in anatomy, he sees the potential for a "noble Museum adequate to all the purposes of learned investigation." And finally, in the collections of apparatus of physics and civil engineering, he sees material for the important experiments and for field exercise.[27]

The breadth of material holdings would match the breadth of intellectual offerings in a way that would challenge and inform the most sophisticated minds. This was to be the high purpose of Tappan's university. As he said in 1856

A University is a collection of finished scholars in every Department of human knowledge, associated for the purpose of advancing and communicating knowledge. To accomplish these purposes they gather around them books on all subjects without any limit, specimens of art, specimens of natural history, apparatus for illustrating the laws of nature and prying into her secrets. . . .[28]

But for Tappan it was not just having these materials that was important. It was the integration of the material in the process of learning. Students and materials would be "living together, they aid and stimulate each other . . . they create an atmosphere filled with inspirations to thought, research, and culture."[29] The interrelationship between the materials of learning and learning itself was dynamic and constant.

Tappan's vision and his tireless work to achieve it was not a smooth road. He was generally dismissive of particular denominational interests represented on the Board of Regents. He was seen as not sufficiently attentive to aspects of student life on the campus, and he was held somewhat suspect by temperance interests in his insistence on having wine at dinner. On June 25, 1863, the regents unanimously voted to remove Tappan from the presidency. While denominational issues, temperance, and student life were expressed as a part of the problem, it is more likely that the regents and even some of the faculty could simply not keep up with his ambition and energy. There would be no compromise on the road to achieving his vision. Just a few years later, the regents rethought their decision and asked him to return. But that was not to be. Still, as James Turner argues, Tappan's vision and energies lived on mainly through Henry Frieze and later James B. Angell.[30]

CLASSIFICATION

The full story of the accumulation and rationalization of collections at the University of Michigan must include the important third dimension of classification. The number of volumes in the library, manuscripts in the archive, objects in the museums, and specimens in departments increased enormously in the hundred years following Tappan's departure. Accumulation and rationalization would not be enough. Access to the material required organization and cataloguing. This work coincided with the great movements in the classification of knowledge and the materials of it. Though beyond the scope of this article, it is important to remember that the story of collections at this university has associated with it the names of hundreds of individuals who over the course of many decades created catalogs and inventories that have been essential to the utility of these accumulations (fig. 8).[31]

In the process, these curators, librarians, and archivists engaged the most sophisticated debates over time on the nature and level of description required. To adhere to these evolving and exacting standards required careful reading of texts and objects, as well as a thorough grounding in taxonomies and specific descriptive vocabularies. Much of this work has now been transferred to digital form. The precise information and structure of these catalogs became the conceptual foundation for the design of online access systems. As the information was transferred, the impressive physicality of the main card catalog for the General Library and other handwritten or typed systems was lost. One could no longer see the total accumulations of descriptive data in tangi-

Fig. 8. Hatcher Graduate Library, Card Catalogue Room,1981. Bentley Historical Library.

ble form, assembled in drawers or binders. One could no longer see the varied typefaces on cards reflecting generations of individuals who prepared catalog data over decades. Nor could one see the varied handwriting of those who entered accession records into volumes in systematic ways. Now, instantly available on a handheld device, the accessibility and ease of access to descriptive information belies the extent of the effort to assemble the data within the framework of earlier technologies.

CONCLUSION

Henry Philip Tappan—in his commitment to having extensive holdings of the materials of learning on the campus, and in his commitment to the use of these materials by both the university community and a broad educated public—would clearly be pleased at the breadth and depth of the collections on the campus today. And, given his particular focus on the possibilities of a scientific approach to all knowledge, that all this could be accessible through digital technology would be another piece of evidence that the possibilities for his university to expand the boundaries of knowledge and the application of that knowledge were indeed endless.

Notes

1. "Extracts from the Journal of Rev. John Monteith Concerning the Introduction of the Gospel into Michigan" in D. M. Cooper, ed., *Pastor's New Year's Greeting*, Detroit: Memorial Presbyterian Church, vol. 12, 1893, 55.

2. Though most of the volumes of the Richard library survive, the contents of the entire library are known from an inventory that was prepared at the time of Richard's death in 1832. See Leonard A. Coombs and Francis X. Blouin Jr., eds., *Intellectual Life on the Michigan Frontier*, Ann Arbor: Bentley Historical Library, 1985. This volume has a complete catalog of extant volumes from the Richard library based on an inventory of his library that accompanied his will. The volume also has an inventory of the Monteith books that have survived. See in particular pages 22–26 of the volume that has a history of the collection written by Leonard Coombs.

3. See George McClure, "Piety and Paidea on the American Frontier," in Coombs and Blouin, *Intellectual Life*, 1–21.

4. See first meeting of the Board of Regents June 1837, 2–4. Note all page references to meetings of the Board of Regents are taken from *University of Michigan Regent's Proceedings with Appendixes and Index 1837–1864* (published by the University Ann Arbor, Oct. 1915). Hereafter cited as BRP.

5. Meeting September 1838, BRP, 56.

6. Meeting December 1840, BRP, 143.

7. Meeting December 1840, BRP, 143.

8. Meeting December 1841, BRP, 205.

9. Original of the 1837 act in Box 1 of the Michigan University Board of Regents collection, Bentley Historical Library.

10. Meeting December 1841, BRP, 211.

11. Meeting September 1845, BRP, 332.

12. See meetings: January 1845, BRP, 301; September 1845, BRP, 326; January 1850, BRP, 441.

13. Meeting August 1852, BRP, 521.

14. James Turner and Paul Bernard, "The Prussian Road to University? German Models and the University of Michigan 1837–c.1895," *Rackham Reports 1988–89*, Ann Arbor: Horace H. Rackham School of Graduate Studies, The University of Michigan, 1989, 14.

15. Henry P. Tappan D.D. *University Education*, New York: George Putnam, 1951, 89–90.

16. Turner and Bernard, "The Prussian Road," 14.

17. Henry P. Tappan D.D. "A Discourse delivered by Henry P. Tappan D. D. at Ann Arbor, Mich., on the occasion of his Inauguration as Chancellor of the University of Michigan, December 21, 1852," Detroit: Advertiser Power Presses, 1852.

18. Tappan, *Discourse*, 17–18.

19. Ibid., 18–19.

20. Ibid., 19–20.

21. Ibid., 23.

22. Ibid., 20.

23. Ibid.

24. Meeting December 1858, BRP, 813.

25. Meeting October 1854, BRP, 600.

26. Meeting October 1856, BRP, 650.

27. Meeting December 1857, BRP, 706.

28. Meeting December 1856, BRP, 654.

29. Meeting October 1856, BRP, 654.

30. Turner and Bernard, "The Prussian Road." This is the general thrust of Turner and Bernard's argument.

31. There is an enormous ephemeral literature on the cataloging and classification of materials of all sorts. Regarding the complexities involved in cataloging and classification from a modern perspective see Murtha Boca, ed., *Cataloging Cultural Objects: A Guide to Describing Cultural Works and their Images*, Chicago: American Library Association, 2006. For a perspective on classification in the nineteenth century, see Elizabeth Edwards, *The Camera as Historian Amateur Photographers and Historical Imagination 1885–1918*, Durham and London: Duke University Press, 2012, 117–20.

THE HOUGHTON GEOLOGICAL SURVEY AND THE FIRST UNIVERSITY COLLECTIONS

Mackenzie Caple and Brian Williams

A heavy and well-traveled ledger rests securely on a shelf in the University of Michigan's Bentley Historical Library (fig. 9). The leather covering bears evidence of age and frequent handling. The special tray case that houses the large ledger hints at the importance of the volume. The ledger in question contains the field notes of Douglass Houghton, dating from 1837 to 1841, with entries about geological formations and mineral composition, hand-drawn maps and sketches, and occasional observations about flora and fauna made during Houghton's geological survey of Michigan as the first state geologist. The significance of the ledger was readily apparent when it was received by university officials from Houghton's widow. A certified handwritten copy of the Board of Regents resolution regarding the ledger, made on March 26, 1858, is pasted inside the front cover of the ledger and reads:

> Resolved that the Librarian be authorized to receive and place in the Library the manuscript volume of Field Notes and Observations made by the late Professor Douglass Houghton, state geologist of Michigan, and to procure a suitable case to enclose the same under lock and key, and that the same shall be subject to examination only by the Regents, the President, and the several members of the respective Faculties, and such other persons as shall receive written permission from the President; but that the same shall not be removed from the Library without the direction of this Board.[1]

Given Houghton's role as the first state geologist and as the first appointed professor of geology and mineralogy (as well as chemistry and pharmacy) at the University of Michigan, the ledger is of considerable significance. Yet, for the museum and research collections at the University, the ledger is the touchstone, containing detailed records of several of the specimens collected for the natural history "cabinet." The specimens gathered by Houghton and others assisting with his survey were among the first in the university's collection, forming the genesis of a number of campus research collections.[2]

When Alexander Winchell was appointed professor of geology at U-M in 1855, he was quick to realize the importance of Houghton's field notes, noting that the "geological collection of the University is the fruit of the state geological survey" and that the field notes are "inseparable from the collections of the survey."[3] Access to Houghton's field notes, however, would be a source of contention for Winchell. While he accurately correlated the relationship of the notes to the geological specimens, he had great difficulty in gaining access to the ledger. Winchell even went so far as to write, in 1858, a fifteen-page "history of the controversy about Dr. Houghton's field notes" in which, bitterly at times, he recounted his unsuccessful efforts to get hold of the ledger. Winchell eventually gained access to the ledger and in 1859 successfully lobbied for a revival of the Michigan Geological Survey that had ended when Houghton died tragically in October 1845. Winchell himself then became state geologist; thousands of his fossil specimens, many of which are rare or type specimens, remain in the University of Michigan Museum of Paleontology.[4]

Creating a first-class cabinet of natural history was clearly an early priority for the university after it relocated from Detroit to Ann Arbor in 1837. At their November 1837 meeting, the regents established a "Committee on the Library, Philosophical

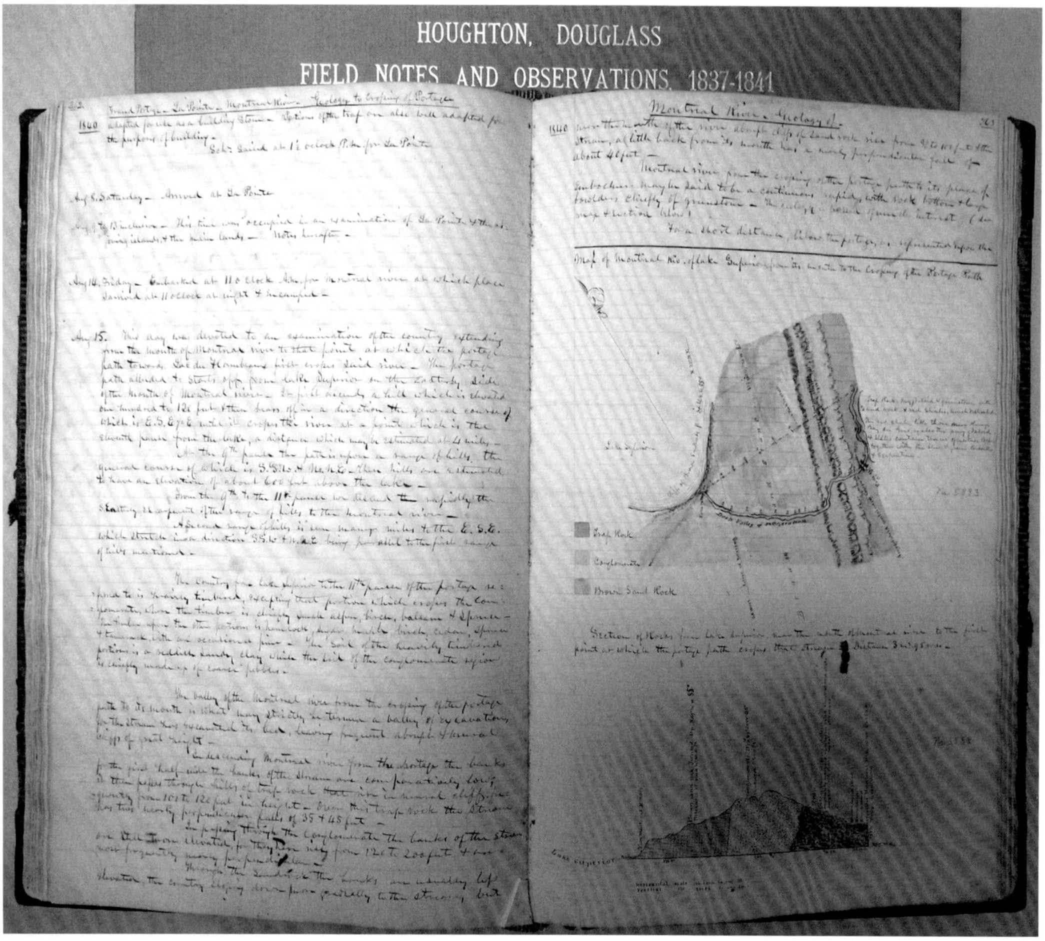

Fig. 9. Pages from Douglas Houghton's field notes from August 1840, detailing geology around the mouth of the Montreal River. Bentley Historical Library.

Apparatus, and Cabinet of Natural History."[5] In the field of natural history, it was resolved that the committee would take steps to "lay the foundation of a suitable collection of specimens for the University Cabinet" encompassing all areas of natural history.[6] They also resolved to suggest that the state geologist be directed to "secure the large mass of native copper on the shores of Lake Superior for the University Cabinet." The large copper boulder, popularly known as the "Ontonagan Boulder," was not just idle speculation (fig. 10). Henry Rowe Schoolcraft, one of the regents present at that meeting, had personally seen the massive copper boulder in 1820 when he was part of an expedition headed by Lewis Cass to explore the Michigan Territory and find the source of the Mississippi River. Houghton had also seen the Ontonagon boulder during his expeditions with Schoolcraft in the early 1830s and even chipped off specimens for the university's collection.[7] However, the 3,700 pound copper boulder would ultimately go elsewhere, eventually landing in Washington, DC.

When looking to fill the professorship of geology and mineralogy in 1839, the university committee appointed to the task affirmed the importance of providing funding to procure specimens, books, and apparatus, indicating that "every proper facility should be afforded to the professors" in order to establish their studies "on

Fig. 10. A copper specimen collected by Houghton in the Upper Peninsula. Photo by Christopher Stefano.

the most liberal and polished footings."[8] Further, they noted that the advancement of scientific investigation in all fields of natural history "is so closely linked with the process of academic and collegiate instruction" that course work in the sciences should begin as soon as possible. This was an important shift in thinking from the prevailing classical curriculum steeped in Latin and Greek. First and foremost, however, was establishing the scientific collections.

As for the university's collection, or cabinet, the first published catalog of the university, produced for the 1843–44 academic year, described the natural history collection as "a valuable cabinet of Minerals, consisting of between four and five thousand Specimens; and suits of Specimens illustrative of the Geology, Zoology and Botany of Michigan."[9] It is also worth noting that that first catalog was describing an institution consisting of just five faculty members (including Houghton) and fifty-three students (eleven juniors, nineteen sophomores, and twenty-three freshmen).

The bulk of the University's collection of minerals came from the purchase of Baron Louis Lederer's collection of foreign minerals, which was available for sale in New York in 1837.[10] Lederer was an Austrian consul general working in New York who had amassed a collection that was said to consist of "2600 specimens of foreign material, mainly said to be rare and not elsewhere obtained."[11] John Torrey, a professor and state botanist for New York, was brokering the deal, for which the sum of $4,500 was sought; he described the asking price as "rather high, but perhaps not above their real value, especially in a public institution." In January 1838, the regents passed a resolution authorizing purchase of the collection under the terms of a letter dated December 11, 1837.[12] Two bonds of $2,000 each were drawn up for the purchase, due in May 1840 and May 1841, plus interest.[13] Despite having no students enrolled at the time of the purchase, the university had made a major investment in building a cabinet of mineral and geological specimens and had already picked out its peers.

Discussing the Lederer collection in the 1841 annual report to Michigan's superintendent of public instruction, the regents claimed that "it was a much richer Cabinet than that of Yale which originally cost $20,000."[14] To this natural history cabinet would be added many geological, botanical, and zoological specimens directly from the State Geological Survey. In the university's annual report for 1840, the Lederer collection of minerals was touted as a collection not to be "excelled on this side of the Atlantic." Of the collections accumulated through the work of Houghton and his "exploring corps," the state legislature was praised for their foresight in providing for a geological survey "at one of their earliest sessions after the admission of the State into the union." The specimens yielded by the survey were described as the result of a thorough three-year reconnaissance of the entire state and that "much of the information thus brought to light is believed to be of practical importance aside from the value of the specimens as illustrations of the principles of science."[15]

Unfortunately, only about fifty of Houghton's mineralogical specimens remain in the university's collection (which now resides at the A. E. Seaman Mineral Museum at Michigan Technological University in Houghton, Michigan);[16] all others have been lost, mainly due to poor curation. Although those that are still known "are almost exclusively low-quality samples," they still carry great historical significance.[17] For example, the copper samples that Houghton brought back from the Upper Peninsula, along with his reports of vast ore deposits, kicked off Michigan's copper boom—which coincided nicely with the rise of the telegraph and thus was a huge boon for the young state's economy; by the middle of the century, the Upper Peninsula mines contributed over 80 percent of the native copper in the United States.[18]

A bill to authorize the first geological survey of the nascent state of Michigan was introduced just hours after Michigan achieved statehood on January 26, 1837, and became the twenty-sixth state in the union. Act 20, which provided for a geological survey of the state, was passed on February 23, 1837;[19] it required the appointment of a "competent person" to "make an accurate and complete geological survey" including "appropriate maps and diagrams" along with a "full and scientific description" of the state's "rocks, soils and minerals, and its botanical and geological productions, together with specimens of the same." The primacy given to the survey at the very moment of state founding was, in its essence, the advancement of a political claim to land and resources formerly controlled by Great Lakes indigenous peoples that had begun several decades earlier. Between 1807 and 1828, the Ojibwe (Chippewa), Potawatomi, and Odawa (Ottawa) signed a number of treaties ceding land to the United States.[20]

Fig. 11. Douglas Houghton portrait painted by Alva Bradish, 1850. Bentley Historical Library.

The specimens collected by the survey were to be deposited in the state library with any duplicate specimens placed in "literary and scientific institutions as the governor shall direct."[21] The survey of a sparsely populated state that was just getting organized passed largely because of Houghton, who was placed in charge of the survey. Houghton had come to Michigan in 1830 and in a few short years had made a very strong impression (fig. 11). Although he was born and raised in New York and received his education there, Houghton is best remembered in his adopted state of Michigan, where there is a city, a county, and a lake among other things named in his honor.

Houghton was born on September 21, 1809, in Troy, New York, the fourth of six children born to Jacob and Mary Lydia Douglass Houghton. A lawyer, Jacob moved his young family to Fredonia in western New York, where he set up practice and would also serve as judge, justice of the peace, and postmaster. Fredonia Academy was established in 1824, and Houghton and his siblings were among the earliest academy students when classes began in 1826. From a young age, Houghton was described as bright and inquisitive. Along with his more formal schooling, he conducted his own experiments in chemistry and physics and spent a great deal of time exploring the outdoors. He even began studying medicine as a sixteen-year-old through one of his father's friends.[22] In 1829, Houghton enrolled in the Rensselaer School (now the Rensselaer Polytechnic Institute), a scientific institute in Troy that had been established in 1824 by Stephen van Rensselaer.

Amos Eaton, a noted educator, botanist, and geologist, was a cofounder of Rensselaer and served as the school's first professor, teaching courses in chemistry, geology, surveying, and even philosophy. Eaton would have a significant impact on Houghton and, in turn, Michigan. Students studying under Eaton were required to demonstrate their mastery of a subject by presenting several lectures that provided proof of their understanding. Houghton was clearly proficient—earning his bachelor's degree in 1829, the same year he entered Rensselaer. He also earned an appointment as a faculty member when Eaton asked him to stay on as an assistant professor of chemistry and natural history. On the side, Houghton continued to study medicine.

In addition to being his academic teacher and mentor, Eaton likely inspired Houghton to create the Geological Survey of Michigan. Eaton was the first in the United States to conduct a comprehensive geological survey; in 1821, van Rensselaer paid Eaton to survey Rensselaer County in New York. The survey was certainly undertaken to benefit Rensselaer financially—Eaton described the types of soil (along with the crops that would grow best in each area) as well as making a thorough geological and lithographic survey. Houghton undoubtedly would have learned about this survey and seen how his scientific and geological interest could be used for economic gain, both personal and public.[23]

The circumstances that would bring Houghton to Michigan began in 1830 as an effort to enrich the citizens of Detroit. City leaders, concerned about the long and difficult winters, hoped to discourage the usual winter activities of drinking and carousing and decided that they wanted to bring in a resident lecturer to deliver talks about various scientific topics to pass the dull winter days in a more enlightening way. A letter was sent to Rensselaer seeking a recommendation for a lecturer. Lucious Lyon, a congressional delegate for the Michigan Territory, was tasked with following up with a visit to Rensselaer where Eaton promptly introduced the young and diminutive Houghton (he is described as 5 feet, 4 inches tall)[24] to Lyon. Houghton accepted this invitation and arrived in Detroit in November 1830. Despite his age, the populace took to him quickly—his series of twenty-six lectures all sold out, despite being standing-room only, and he soon became one of the town's most respected and recognized citizens.

In 1831, Houghton returned to Fredonia, New York, where he received his license to practice as a physician. However, later the same year he returned to Detroit, where his excellent reputation as a scientist and his friendship with Henry Rowe Schoolcraft led him to serve as botanist and physician on the expeditions led by Schoolcraft in 1831 and 1832.[25] He performed his botanical role exceptionally well, collecting many fine specimens of plants; several of these were sent to New York University's John Torrey (the foremost botanist on the continent at the time), who gave Houghton credit for discovering several new species.[26] The U-M Herbarium still has over a hundred of the botanical specimens Houghton collected on the expeditions of 1831 and 1832, all of which have been digitized. Unfortunately, Houghton never published a complete list of all the specimens he collected, and many of those that did survive were very poorly labeled, so it is not known how many originally existed and what has been lost.[27] As the expedition's physician, Houghton also provided dental services and vaccinations for Native Americans that the expedition treated along the way.

Although the Schoolcraft Expedition is popularly remembered as the expedition that reached the true source of the Mississippi for the first time,[28] it was not until 1832 that Schoolcraft even intended to reach the river's source. The purpose of the 1831 expedition was not watershed exploration—it was explicitly political. Governor Cass, at the request of the War Department, sent Schoolcraft into the Michigan Territory "to endeavor to put an end to the hostilities between the Chippewas and the Sioux."[29] Schoolcraft had a great deal of experience with the indigenous peoples of the Great Lakes, having married into a prominent family of the Anishanaabe. However, he had also been Sault Ste. Marie's Indian agent since 1822, and although he made treaties that benefitted his new family when possible, he also took advantage of them to gain notoriety via ethnographies and claims of discovery.[30] Houghton serving as a physician to the indigenous people the expedition encountered is thus complicated by their overall mission—the expedition was distinctly imperialist, with a definite tone of paternalism; given this light, it is impossible to view Houghton's services of vaccinations and tooth extractions merely as gestures of goodwill, no matter what Houghton himself thought about it. Additionally, Houghton's presence on the Schoolcraft Expedition is also notable because it was then that he first encountered the massive copper deposits in Michigan's Upper Peninsula, knowledge that surely played a great role in his persistence in lobbying for the creation of the state's geological survey several years later.

Soon after the conclusion of the Schoolcraft Expedition, Houghton decided to make Michigan his permanent home. Besides his scientific lectures, he helped establish a Protestant Episcopal church, founded a Young Men's Society, and ran a successful medical practice. In 1833, he married his childhood friend Harriet Stevens and brought her to Detroit. They moved into a new home at 19 Wayne Street.[31]

Houghton was nothing if not multidisciplinary. Besides his thriving medical practice and occasional work as a naturalist, he engaged in a good deal of real estate speculation of the land surrounding Detroit, recognizing that the area would boom, especially after Michigan gained statehood. He bought large tracts of land from the government and other landholders, then parceled it off and sold it at great profit after weathering some economic downturns. He insisted, though, that this was not profiteering or greed, but was rather motivated only by his great concern to care for and protect his family.

In 1837, following resolution of the border dispute with Ohio known as the Toledo War, Michigan became the twenty-sixth state. The compromise that ceded the Toledo strip to Ohio in exchange for Michigan gaining the Upper Peninsula was largely seen as a bad deal for Michigan. In addition, the prevailing image of Michigan among European Americans was of a "gloomy morass,"[32] a state made of swamps and sand hills, unfit for farming and suitable only for fur trappers and Native Americans. This view was quite understandable, considering that at the time the majority of European-American explorations of Michigan had been made along the shores of the Great Lakes—which largely *are* swamps and sand hills. Since overland travel was vastly slower and more difficult, few Europeans had ventured into Michigan's interior by the time statehood was achieved.

However, the relatively low number of European-American settlers does not mean that the state was a wilderness untouched by human activity. There was a substantial population of indigenous people, who managed the land in a number of ways, including planting corn and encouraging or setting the annual fires that maintained the park-like oak savanna ecosystem that made up much of the southeast part of the state.[33] In addition, there were already old influences from beyond the continent; during the 1837 survey, Bela Hubbard saw apple trees along the Saginaw and Tittabawasee rivers that "produced agreeable fruit, and some were apparently of a century's origin"; although Hubbard did not himself "speculate upon their origin," he notes that even "some of the earliest travellers allude to orchards"—an oddity indeed, since apples are native to Eurasia.[34]

Dispelling the popularly held belief that Michigan was an inhospitable territory was one of the goals of the Geological Survey of Michigan;[35] Governor Stevens T. Mason was doubtless eager to show the Upper Peninsula naysayers wrong. This, along with his personal friendship with and trust in Houghton, made him a staunch advocate for the survey. However, the idea for a state survey was met with some opposition in the new state legislature. Funds, after all, were very limited, and a comprehensive natural history survey did not, on the face of things, seem like a good use of public funds to many of the legislators. Although Houghton himself was at least partly motivated by scientific curiosity, the argument he put before the legislature was mainly economic since he had to convince them that the appropriations needed for the survey were a worthwhile use of the small budget.

The main argument for the survey was twofold: first, it would have a large but indirect economic impact because a greater knowledge of what the lands within the state contained would help convince settlers from the east to move to Michigan, and the increased population would drive up tax revenue as well as boost the economy as a whole. Second, the survey had the potential to have a large direct economic impact in the form of minerals. Houghton believed that Michigan contained large and valuable deposits of coal and various ores as well as a number of salt springs that could be profitably improved. Land containing such ore could be sold at a higher value than tracts without mining prospects, increasing the state's potential revenue from land sales. In addition, this also, no doubt, furthered the government's goal of eliminating tribal claims to land in the state; higher prices for parcels of land meant that they could offer higher compensation to the indigenous peoples who controlled large tracts of land, increasing the rate at which the government could divide up the land into smaller family plots. Even though many of these plots remained in the hands of indigenous peoples, the smaller, individualized holdings encouraged the Native Americans to "settle" and take up farming and more European lifestyles, rather than the traditional practice of moving from place to place with the seasons.

Houghton's charisma, persistence, and excellent scientific reputation combined to win over the support of the legislators, and in February 1837 Governor Mason approved the act organizing the State Geological Survey; it was one of the first actions of the legislature, introduced mere hours after Michigan's statehood became official. Houghton was soon thereafter appointed state geologist and thus head of the survey.

However, Houghton's struggles with the survey had only just begun. First, there was the issue of funding. The legislature appropriated $3,000 for the first year of work on the survey, a pitiably small amount even for that time. Houghton, though, was undeterred and quickly set about bringing together a group of assistants to staff each of the four departments of the survey (geology, zoology, botany, and topography). Abram Sager was appointed principal assistant and head of zoology and botany; C. C. Douglass was appointed "assistant to the Geologist";[36] Bela Hubbard became a general assistant; and William P. Smith was appointed subassistant, his principal duties being to prepare and preserve specimens. Hubbard, through a later reminiscence, gives us a rich description of Houghton in the field: "He was no carpet knight of science, and on his geological excursions never flinched from hard work and exposure. On these occasions he usually wore a suit of gray, the coat having large side-pockets, and hanging loosely upon his small frame. The hands and feet were very small, but the latter were incased [*sic*] in boots that came almost to his thighs. His shocking battered, bad hat was broad-brimmed and slouched, almost concealing his face, and his whole appearance was that of battered, weather-worn backwoodsman."[37]

The field season of the 1837 survey was rather brief due to the time required to find able and willing assistants and to make general plans, but it was profitable nonetheless. Sager and Smith were especially productive in the biological departments, collecting 250 bird specimens and "numerous other specimens of plant and animal life."[38] Unfortunately, only eighty specimens of birds attributed to Sager remain in the Museum of Zoology; the fate of the remainder is a mystery. In the geological department, Houghton, Hubbard, and Douglass had time to explore much of the Lower Peninsula only hastily, though they still succeeded in finding deposits that could be used for making glass, bricks, pottery, and lime and recorded sizable areas of coal, gypsum, and bog iron ore.[39] They were guided by Pierre Gruet, a man "of mixed French and Indian blood" who "seemed to be acquainted with every individual of either race that [the survey party] met."[40] Although both Houghton and Hubbard make only passing references to their guides,[41] without such a guide, who was informed by knowledge of multiple cultures, the surveyors would undoubtedly have been unable to navigate the peninsula's complex waterways or locate so many interesting and useful features. In addition to general mapping and surveying, Houghton also spent a great deal of time and effort in locating and studying salt springs, especially in Grand Rapids and along the Tittabawasee River, with the aim of selling the salt for state profit. Although he found none that had a salt content high enough to be immediately profitable, after comparing them carefully with similar springs in neighboring states, he believed that several of them could become highly profitable after they had been excavated to reach the brine at a lower depth, where fresh groundwater was not intermixed.

Although the harsh weather of Michigan winters prevented the survey team from undertaking field work during winter months, they were not idle. Houghton and his assistants spent that time organizing the information that they had collected into a more comprehensible state, in preparation for future publication as well as for the annual reports that were made to the legislature.[42]

The 1838 field season of the Geological Survey was more fruitful than the previous year. The monetary appropriation was larger, $12,000 instead of $3,000 (although some of it had to be used to pay for over-expenditures from the previous season), and preparations and plans were finalized before the weather cleared for the spring so there was much more time to spend traversing the state, collecting, and making observations.[43]

Several key people were also added to the staff of the survey. John Wright was hired as head botanist, with his assistant George Bull, relieving Sager of his botanical duties (Sager remained in charge of the zoological department), and S. W. Higgins filled the role of topographer, a position that had previously been vacant. Wright and Bull were extraordinarily successful in the botanical department, collecting between them around 750 different species of plants despite numerous difficulties. Notably, they collected a large number of rare and inconspicuous prairie and wetland plants (several of which have since been extirpated in the wild)—in contrast, earlier botanical specimens from Michigan had mainly been collected along the lake shores, since traveling by water was so much easier than traveling overland.[44]

Foremost among their impediments was the legislature's stipulation that they collect sixteen duplicates of each specimen. The stipulation required that the survey supply the state with "single and good specimens" but if "more similar specimens than one may be found, sixteen more, if possible, shall be procured."[45] The additional specimens were to be collected for the University of Michigan and were "to be distributed by the Regents amongst the University and its Branches"; at that time, the university was operating several branch institutions in the form of preparatory schools. With the rudimentary equipment and rough conditions they had to work with, preparing, preserving, and transporting so many specimens was a great burden. Unfortunately, if they actually achieved such a feat, nearly all the duplicates have since been lost. There are now only two reasonably complete sets of botanical specimens from the First Survey: the most complete resides at U-M Herbarium (MICH), while the second most complete set is at the Herbarium of the National Arboretum of the United States (NA), in Washington, DC. Fragmentary collections also exist at Gray Herbarium at Harvard University (GH) and New York Botanical Garden (NY), but they are few in number and poorly labeled. The fate of the rest is something of a mystery—they may have never existed at all, due to the difficulty of obtaining such a large number; they may have been lost in a fire that destroyed part of the U-M Herbarium in 1913; or they may have been poorly labeled and were discarded when institutions they had been sent to cleaned out their collections or ceased to exist.

The quality and quantity of the botanical collections is an excellent reflection of the character and knowledge of Houghton, Wright, and Bull. Wright was a very able botanist; he was also a physician with a successful practice in New York—which Houghton persuaded him to leave to join the survey. Houghton must have made a compelling argument in the name of science and exploration, for the survey could afford to pay very little and promised to be a great deal of work.[46]

Houghton soon began improvements on two springs the survey had identified on the Grand and Tittabawasee rivers that he thought could in time begin profitably producing salt. However, neither project went smoothly, with fires, mosquitoes, and other impediments greatly hindering progress.

The topography and mapping department focused mainly on the northern part of the Lower Peninsula (the southern part having been partly explored the previous year, it being the area closest to the survey's base in Detroit), especially the coast. They concluded that that part of the state would see little profit from ore, but they found much more of the land to be suitable for agriculture than was previously thought. The party extensively mapped lakes and rivers, which helped improve county borders and navigation, and also discovered a large seam of coal that stretched from Shiawassee to Jackson.

The regents kept their eye on Houghton beyond his roles as a collector of specimens for the university. In August 1839, a resolution was passed asking the committee on appointments to explore "the expediency of appointing Dr. D. Houghton, State Geologist, a professor in the University of Michigan."[47] The following October, the committee issued a lengthy report on the need to designate an individual to fill the professorship of geology and mineralogy. The urgency was based upon the belief that the construction of university buildings was

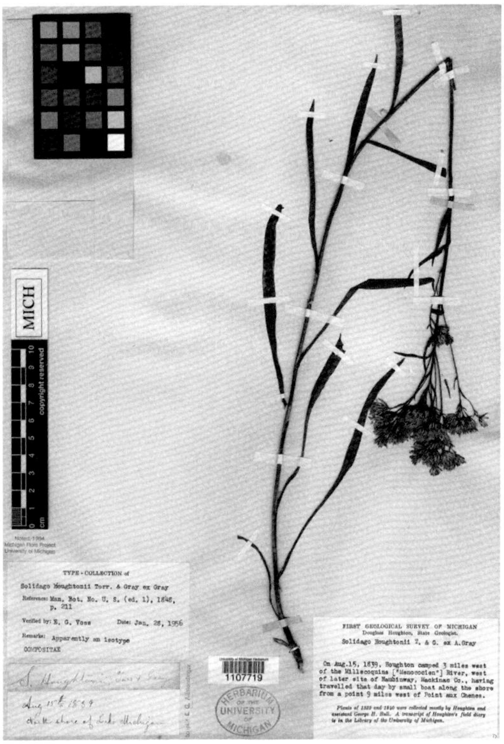

Fig. 12. A specimen of *Solidago Houghtonii* collected by Houghton in 1839. University of Michigan Herbarium.

"in such a state of forwardness" to give credence to the idea that "one of them will be ready to be appropriated to this object in the course of the Autumn."[48] Houghton had been sending back a prolific amount of specimens and the committee recognized that someone with the skill and expertise "to prepare the entire collection for exhibition and lecture" was sorely needed. The report was accepted along with a resolution that the professorship of geology and mineralogy be offered to Houghton and that he also be charged with instruction in the subjects of chemistry and pharmacology.[49] The appointment was to carry a salary of $1,500,[50] but because Houghton was engaged with and paid by the Geological Survey, no salary was paid to him by the university.[51] The regents also resolved that one of the buildings be "fitted up as soon as convenient under the advice and direction of Professor Houghton for the reception of the Library and specimens in Natural History."[52]

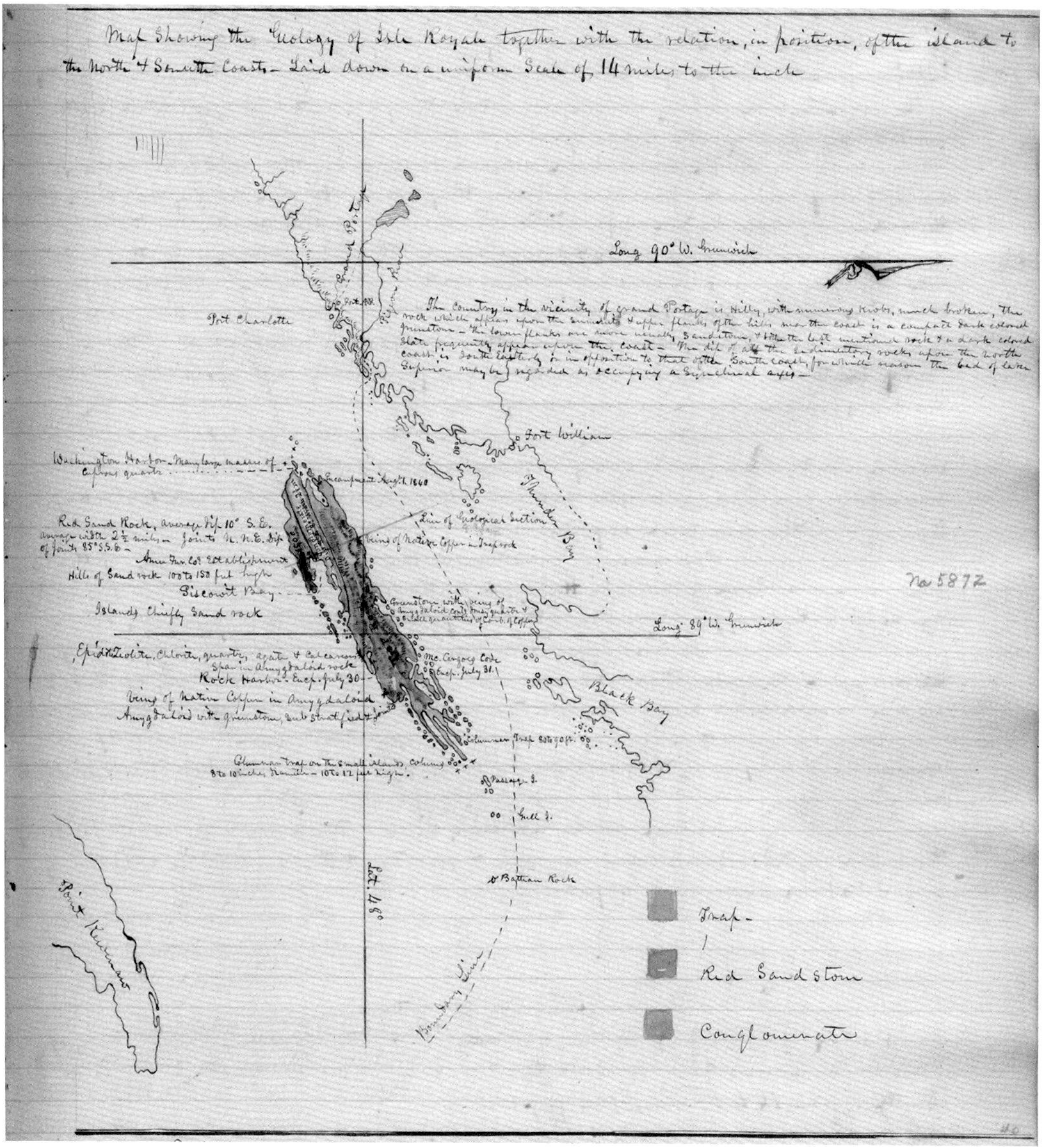

Fig. 13. Map of the geology of Isle Royale at a scale of 14 miles to the inch. From Houghton's field notes. Bentley Historical Library.

The year 1839 should have been the best one of the survey—there was a fairly generous appropriation of $12,000 from the state legislature and the corps of department heads and assistants was solid and well trained. It was instead a year that largely consisted of disappointments and setbacks. First, Houghton was frustrated because he had to spend a great deal of time personally overseeing the improvements to the two salt springs and so was unable to spend as much time as he wanted working on other aspects of the survey. Most significantly, though, Michigan suffered a financial panic in 1839, and the legislature decided that the survey was not as important as other parts of the state budget; although the geological and topographical portions of the survey continued (albeit with a reduced budget), in the fall of 1839 the botanical and zoological departments of the sur-

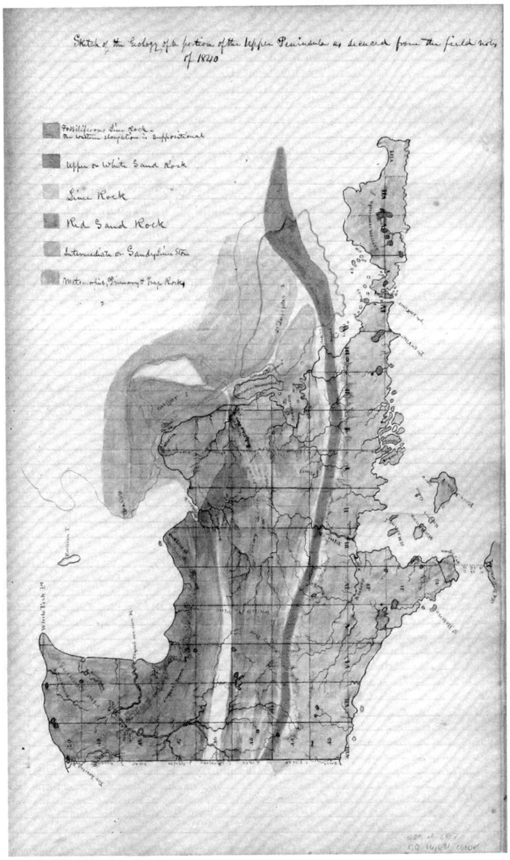

Fig. 14. 1840 sketch of the eastern Upper Peninsular indicating types of rocks. From Houghton's Field notes. Bentley Historical Library.

vey were completely defunded. Houghton saw this as a great tragedy, believing that a true scientific understanding of the state was not complete without a thorough exploration of its flora and fauna. Additionally, all members of the team suffered from exhaustion and exposure due to the harsh conditions under which they worked. Houghton, in particular, was plagued by rheumatism, a condition he described as "the result of excessive fatigue and exposure" that would be alleviated "were it not for the great number of hours I am compelled to work."[53] Nonetheless, in the field season of 1839, the survey was able to complete much of the mapping of the southern coast of the Upper Peninsula, and the improvements on the salt springs continued to move forward.[54] Houghton also collected a new species of goldenrod, endemic to Great Lakes shores, which was named *Solidago houghtonii* in his honor (fig. 12).[55]

The year 1840 was a productive year for the survey but fairly uneventful (figs. 13 and 14). The Lake Superior coast of the Upper Peninsula was thoroughly mapped, with Houghton spending extra time in the Keweenaw Peninsula; in the fall, his assistants returned to the Lower Peninsula, while Houghton stayed in the copper country until late in the season. In a July 22, 1840, letter written to Governor William Woodbridge, Houghton reported that he had "already shipped seventeen casks and boxes" containing "minerals and geological specimens illustrative of the country."[56] Houghton went on to note that he intended to collect more that season in order to be able to "exhibit the actual mineral character of the Upper Peninsula in such a manner that it cannot be mistaken."[57] He also explained that he was certain copper ore existed in large veins, but noted that it would take capital and labor to extract it, while disabusing people of the inaccurate belief that copper in its native state was readily abundant. Stories about massive float copper like the Ontonagon boulder served to distort the truth. The field notes indicate that Houghton visited the boulder on July 12, 1840, and "after much difficulty and labor" managed to chip off some pieces at the expense of "8 to 9 chisels."[58] Along with valuable minerals, Houghton also brought a birch-bark canoe from Anishinaabe in the Lake Superior region to the University Museum.[59] While there are no traces of this canoe in current collections, the boat must have resembled those taken by Houghton and his survey team, a testament to canoeing as an essential mode of transportation and a place of contact between surveyors, indigenous people, and traders. On June 8, 1840, Houghton recorded in his field notes: "Being desirous to ascend the Tahquamenon River I encamped half a mile above its mouth and having procured a young Indian guide, with two of my voyageurs in a bark canoe I commenced the ascent at 6 o'clock p.m."

Although the original act by the state legislature set the end date of the Geological Survey as March 1, 1841, Houghton and his assistant Higgins continued their work well beyond that. They concerned themselves less with grand discoveries and more with filling in details, consolidating reports, and drafting thorough and highly detailed maps. Though their activities in 1841 attracted much less public interest, there were still very significant results—Houghton's main achievement in 1841, for example, was solidifying Michigan's border with Wisconsin. Although this work was not strictly part of the survey, it was Houghton's work on the survey that gave him the expertise needed to finalize the line. Writing to his father in January 1841, Houghton noted that he would probably "be required to spend the coming season in Lake Superior country" dealing with the "difficulty" that was "occurring with respect to our boundary on the west, with Wisconsin."[60] He went on to note, "I suppose my services will be required there, which will divert my labor for a time from the Geology survey." Houghton was very concerned about the border settlement since it ran through Michigan's mineral district; he was worried that if an incompetent surveyor were to set the line, the state would lose some of its richest and most valuable resources. Therefore, he took on the role of Michigan's commissioner himself and ensured that Michigan did not lose valuable territory that it should rightly have.[61]

In early 1841, Houghton was still uncertain about relocating to Ann Arbor. Writing to his father, he stated that it was "now somewhat doubtful whether I will remove to Ann Arbor in the spring." Of his house, one of the four houses for professors being built on the campus, he added, "[m]y house there is finished, and it is quite a splendid mansion."[62] His residence was intended to be, at least temporarily, the home of the natural history specimens. In April 1840, the regents authorized the erection of partitions in the basement level of Houghton's residence "as may be necessary for his convenience."[63] At their January 1841 meeting, the regents made it official, explaining that one of the four recently built professors' houses was "temporarily appropriated to the reception of the Cabinet of Natural History which has been provided or is in the process of accumulation."[64] Writing to his father later in 1841, Houghton commented that "our university at Ann Arbor has commenced operations, and there is a tolerably strong class," but he did not yet think it was wise for him to move there.[65] He allowed that he might "think it advisable to do so in the coming spring," adding that he would be spending some time in Ann Arbor before winter set in to work on organizing the university cabinet. Houghton explained that the "collection of specimens, which is exceedingly choice, has now long been in boxes, and our people are desirous to have a portion at least so arranged as to permit examination."[66] A review of university expenditures around this time shows several bills submitted by Houghton for wood, hardware, and the work of a carpenter, suggesting that Houghton was also involved in overseeing the actual construction of some of the cases that would house the specimens.

By the time the regents sent their fourth annual report to the state superintendent of public instruction at the end of 1841, the natural history cabinet had moved from Houghton's professors' house to a room in one of the main university buildings. Houghton had also opened the Lederer collection in "cases according to their scientific classification." The report went onto to describe that beyond the foreign mineral collection, "the entire cabinet of the University comprehends a very extensive collection of rock specimens, fossils, American minerals, together with a Zoological collection of great value." The regents again boastfully claimed that their cabinet offered "greater facilities and inducements for study in Natural History than can be had in any Institution in this country."[67] That was a pretty heady claim for a university that just a few months earlier had commenced instruction to six freshmen and one sophomore.

In the following years, along with analyzing the data from the survey, Houghton found time to give a few lectures at the University of Michigan, though he never taught any formal classes as part of his professorship. In 1842, he was elected mayor of Detroit; although he did not campaign and did not desire the post, he served two terms and was by all accounts an able politician.[68] Of his election as mayor, Houghton wrote to his father that it was done while he was away from Detroit and without his knowledge, and that he "had determined to refuse to serve" but was talked into serving by friends. Feeling that it would be impossible to perform the mayoral duties "with satisfaction to myself," Houghton stated that "I will probably resign after two or three weeks."[69] He also convinced the United States Land Survey to partner with the Michigan Geological Survey—the earliest example of a state survey partnering with the federal government.[70] The two groups of surveyors coordinated their activities and shared data with each other, greatly increasing the efficiency and reducing the budgets of both endeavors. Houghton and Higgins, along with William Burt of the US Land Survey, managed to nearly complete all the surveying work in the state by 1845.

Tragically, Houghton was killed in a storm on Lake Superior off the Keweenaw Peninsula on October 13, 1845. With his death, the First Geological Survey of Michigan came to an end (though Hubbard and Burt continued the surveying effort).[71] Thankfully, many of his notes and observations remain, and although a comprehensive report of the First Survey was never published, the work was eventually expanded upon and published; Jackson, Poster, and Whitney, with the help of survey assistants Douglass, Hill, and Burt, "made use of his and their work to such an extent that it was impracticable to divide the credit."[72]

The volume of Houghton's field notes stored at the Bentley Library is a transcript of Houghton's work. It is essentially the "office copy" with notes from various sources formally compiled. In fact, Houghton's papers were widely scattered. Several volumes of his rough field notes, which turned up for auction in 1971, now reside at the Clarke Historical Library in Mount Pleasant, Michigan (a microfilm copy of those notes exists at the Bentley). Naturally, material that was with Houghton when his boat capsized and he perished was lost in Lake Superior. A few specific items were known to be lost. In the volume *Reports on the Mineral Region of Lake Superior* published in 1846, largely by Burt and Hubbard, in a listing of the district south of Lake Superior, township and ranges are given along with the notation, "The notes and maps of town 58 north, ranges 29 and 32 west, and town 59 north, range 29 west, were in possession of Dr. Houghton, and were lost with him."[73]

Although the First Geological Survey of Michigan ended in 1845, it was not the end of state-sponsored work. In February 1859, partly because of a large number of petitions for research on salt springs, the Michigan legislature initiated the Second Geological Survey and appointed as head geologist the University of Michigan's professor of geology, Alexander Winchell. The Second Survey was short-lived—it was defunded in 1863 when the Civil War put a huge burden on the state's finances—but Winchell nonetheless collected many paleontological specimens and made many detailed geological observations. He was recalled to the position of state geologist in 1869 when the survey was resumed.[74]

The Third Geological Survey continues to this day[75] to "facilitate basic and applied geological research to promote the best use of Michigan's geological resources for their social and economic benefits while protecting associated resource values and the environment."[76] The longevity of the state survey is a testament to Houghton's early commitment to improving both the economy and scientific understanding of his adopted Michigan.

More than a century and a quarter after Houghton's field notes were compiled, Rogers McVaugh of the University Herbarium echoed the importance of the reports and field notes from the geological survey for understanding the early holdings of the Herbarium. In an article published in the *Michigan Botanist* in 1970, he carefully detailed which specimens in the Herbarium could be traced back to Houghton's survey and included examples of the original labels from specimens collected by Houghton and his team. McVaugh used excellent detective work to identify when and who wrote the labels, analyzing the handwriting on them as well as other clues.[77]

While Houghton's presence and influence can be traced through collections in the university's museums and libraries, the only public memorial on campus to Houghton is the Professors' Monument, a cenotaph erected in 1846. Standing as a monument to four early professors who met untimely deaths, the Professors' Monument resides on the southeast side of the Hatcher Library.[78] The text honoring Houghton, translated from the Latin on the monument, reads:

In Memoriam
Douglas Houghton, M.D.
Professor of Chemistry, Mineralogy and Geology
in this university
And head of the geological survey
in this state
Learned in science, zealous in action
while fearlessly engaged in public work,
He perished by the overturning of a boat
in Lake Superior
drowned, never, alas, again to be seen
until the lake shall give up its dead
He died on October 13, 1845
at the age of 36
The Regents of the University
of Michigan
have had this monument erected.

Notes

1. The resolution is also published in *University of Michigan Proceedings of the Board of Regents, 1837–1863* (Ann Arbor: University of Michigan, 1915), 735. http://hdl.handle.net/2027/mdp.39015071494218?urlappend=%3Bseq=743

2. Of the hundreds of specimens collected by Houghton and his collaborators, only about 815 plant specimens, 80 birds, and 50 minerals remain in U-M's collections.

3. Alexander Winchell, "History of the Controversy about Dr. Houghton's Field Notes, November 11, 1858," Alexander Winchell papers, Box 19, Bentley Historical Library.

4. R. C. Allen and Helen M. Martin, "A Brief History of the Geological and Biological Survey of Michigan," *Michigan History Magazine* 6 (1922): 678.

5. University of Michigan. *Regents' Proceedings with Appendixes and Index, 1837–1864* (Ann Arbor: University of Michigan, 1915), November 1837, 17.

6. Ibid., 17.

7. Christopher Stefano, K. Erwin, and R. C. Ewing, "The University of Michigan Mineral Collection, Ann Arbor," *Rocks & Minerals* 88 (2013): 333.

8. Regents Proceedings, October 1839, 100.

9. University of Michigan Catalogue of the Officers and Students in the Department of Arts and Sciences 1843–1844 (Ann Arbor: Michigan Argus Office, 1843), 5–8, 12. http://hdl.handle.net/2027/mdp.39015071517406

10. Former university librarian Russell Bidlack did considerable research on Baron Lederer, much of which appears in "Four Early Donors of Books to the University of Michigan," *Michigan Alumnus Quarterly Review* 65, no. 14 (February 28, 1959): 110–22. Bidlack also discusses the gift of the first book that most likely was through a connection made by Douglass Houghton with Charles Borup. The collection had been advertised for sale in the January 1836 issue of the *American Journal of Science and Arts*, but no sale had been completed.

11. Regents Proceedings, November 1837, 19.

12. Regents Proceedings, January 1838, 33.

13. Regents Proceedings, December 1839, 105.

14. Regents Proceedings, December 1841, 205.

15. Report to the State Superintendent of Public Instruction for 1840. Cited in Regents Proceedings, January 1841, 147–48.

16. Jennifer Donovan, "Two Universities Collaborate to Preserve Historic Mineral Collection for the People of Michigan," *Michigan Tech News* (May 12, 2015). http://www.mtu.edu/news/stories/2015/may/two-universities-collaborate-preserve-historic-mineral-collection-for-people-michigan.html

17. Stefano, "Michigan mineral collection," 332.

18. R. Thomas Segall, "A Brief History of the Michigan Geological Survey," Michigan Geological Survey, Department of Natural Resources, accessed November 2014, p. 2. http://www.michigan.gov/documents/deq/GIMDL-GGGSDH_302337_7.pdf

19. Public and Local Acts of the Legislature of the State of Michigan, 1837, 14–15. http://hdl.handle.net/2027/mdp.39015071350584?urlappend=%3Bseq=20

20. The interactions between European-American explorers and indigenous peoples and the relationship between those interactions and greater continental politics is very complex; the scope and limitations of this essay do not allow us to give these complexities and the problematic way they have been treated by history the depth of discussion that they deserve. For an excellent treatment of this subject, see Michael Witgen, *Infinity of Nations: How the Native New World Shaped Early North America* (Philadelphia: University of Pennsylvania Press, 2012).

21. Public and Local Acts, 14–15

22. Helen Wallin, *Douglass Houghton Michigan's First State Geologist 1837–1845* (Lansing: Geological and Land Management Division, Michigan Department of Environmental Quality, 2004), 6.

23. Walter B. Hendrickson, "Nineteenth-century State Geological Surveys: Early Government Support of Science," *Isis* 52 (1961): 358.

24. Steve Lehto, *Michigan's Columbus: The Life of Douglass Houghton* (Royal Oak: Momentum Books, 2009), 7.

25. Edsel K. Rintala, *Douglass Houghton, Michigan's Pioneer Geologist* (Detroit: Wayne State University Press, 1954), 16.

26. The new species were named *Carex houghtoniana, Cyperus houghtonii, Malva houghtonii,* and *Synthyris houghtoniana*. The type specimens sent to Torrey (now at NY) as well as the isotypes (at U-M) can be viewed at the JSTOR Global Plants database, http://plants.jstor.org Anton Reznicek, email message to author, September 8, 2015.

27. Janice L. Rittenhouse and Edward G. Voss, "Douglass Houghton's Botanical Collections in Michigan, Wisconsin, and Minnesota on the Schoolcraft Expedition of 1832," *Michigan Botanist* 1 (1962): 61–70.

28. The source was previously thought to be Cass Lake, named for Lewis Cass (then governor of Michigan Territory), who led an expedition to find the Mississippi's source in 1820; Schoolcraft served as mineralogist and geologist on that expedition.

29. Rittenhouse and Voss, 61.

30. Witgen, *Infinity of Nations*, 3, 354.

31. Wallin, *Douglass Houghton*, 7.

32. Bela Hubbard, *Memorials of a Half-Century* (New York: Press of G. P. Putnam's Sons, 1887), 67.

33. Fire suppression, agriculture, and residential/industrial development have combined to nearly eradicate this ecosystem; prairie and savanna now cover less than 0.1 percent of their historical area in Michigan. Native Americans used fire as a land management tool from around 10,000 years ago until European settlement. See K. A. Chapman and Richard Brewer, "Prairie and Savanna in Southern Lower Michigan: History, Classification, Ecology," *Michigan Botanist* 47 (2008): 1, 28

34. Hubbard, *Memorials*, 85.

35. Segall, "Geological Survey," 1.

36. Rintala, *Douglass Houghton*, 34.

37. Hubbard, *Memorials*, 77.

38. Rintala, *Douglass Houghton*, 35.

39. Although there are few references to indigenous people in Houghton's field notes and almost none in his reports to the legislature, it is almost certain that they were instrumental in helping the surveyors locate mineral deposits and other valuable features. This may have been omitted from the notes and reports because the primary focus was on geological features and topography along with natural resources rather than anthropological information. A secondary motivation may have been a desire to frame the survey's work more as the triumph of European-American discovery and less as a demonstration of indigenous knowledge.

40. Hubbard, *Memorials*, 81.

41. A June 8, 1840, entry in Houghton's field notes states, "Being desirous to ascend the Tahquamenon River I encamped half a mile above its mouth and having procured a young Indian guide, with two of my voyageurs in a bark canoe I commenced the ascent at 6 o'clock p.m."

42. George N. Fuller, ed., *Geological Reports of Douglass Houghton* (Lansing: The Michigan Historical Commission, 1928), 23.

43. *Public and Local Acts of the Legislature of the State of Michigan, 1837/1838*, 120. http://babel.hathitrust.org/cgi/pt?id=mdp.39015071350592;view=1up;seq=126

44. Rogers McVaugh, "Botanical Results of the Michigan Geological Survey Under the Direction of Douglass Houghton, 1837–1840," *Michigan Botanist* 9 (1970): 213.

45. Regents Proceedings, April 1838, 44.

46. In this time period, nearly all physicians were also botanists—synthetic medicine had not been developed yet, so a good medical training required a comprehensive knowledge of plants.

47. Regents Proceedings, August 1839, 95.

48. Regents Proceedings, October 1839, 99.

49. Regents Proceedings, October 1839, 100.

50. Regents Proceedings, November 1839, 101.

51. Regents Proceedings, January 1841, 157.

52. Regents Proceedings, November 1839, 101.

53. Douglass Houghton to Jacob Houghton, January 17, 1841, photostat and transcript in Douglass Houghton papers, Bentley Historical Library.

54. Rintala, *Douglass Houghton*, 37–52.

55. J. K. Morton, "Observations on Houghton's Goldenrod (*Solidago houghtonii*)," *Michigan Botanist* 18 (1979): 31.

56. Douglass Houghton to William Woodbridge, July 22, 1840, photostat and transcript in Douglass Houghton papers, Bentley Historical Library.

57. Ibid.

58. Douglass Houghton Field Notes, 1837–1841, 241. Douglass Houghton papers, Bentley Historical Library.

59. *University of Michigan Encyclopedic Survey* (Ann Arbor: University of Michigan, 1957), 1476.

60. Douglass Houghton to Jacob Houghton, January 17, 1841, photostat and transcript in Douglass Houghton papers, Bentley Historical Library.

61. Rintala, *Douglass Houghton*, 72.

62. Douglass Houghton to Jacob Houghton, January 17, 1841, photostat and transcript in Douglass Houghton papers, Bentley Historical Library.

63. Regents Proceedings, April 1840, 127

64. Regents Proceedings, January 1841, 147.

65. Douglass Houghton to Jacob Houghton, November 14, 1841, photostat and transcript in Douglass Houghton papers, Bentley Historical Library.

66. Douglass Houghton to Jacob Houghton, January 17, 1841, photostat and transcript in Douglass Houghton papers, Bentley Historical Library.

67. Regents Proceedings, December 1841, 205.

68. Wallin, *Douglass Houghton*, 10.

69. Douglass Houghton to Jacob Houghton, January 17, 1841, photostat and transcript in Douglass Houghton papers, Bentley Historical Library.

70. Segall, *Geological Survey*, 1.

71. Segall, *Geological Survey*, 2.

72. Fuller, *Geological Reports*, 4.

73. J. Houghton Jr. and T. W. Bristol, *Reports of Wm.*

A. Burt and Bela Hubbard, esqs., on the geography, topography and geology of the U. S. surveys of the mineral region of the south shore of Lake Superior, for 1845 : accompanied by a list of workings and organized mining companies; a list of mineral locations . . . and a correct map of the mineral region . . . also a chart of Lake Superior, reduced from the British admiralty survey (Detroit: Printed by C. Wilcox, 1846), 21.

74. Segall, *Geological Survey*, 2.

75. Since 2011, the Michigan Geological Survey has been part of Western Michigan University's Department of Geosciences: http://wmich.edu/geologysurvey.

76. "Michigan Geological Survey: About." Western Michigan University, 2015. http://wmich.edu/geology-survey/about

77. McVaugh, *Michigan Botanist*, 213–43.

78. The other professors recognized are Joseph Whiting (died July 20, 1845), Charles Fox (died July 24, 1854), and Samuel Denton (died August 17, 1860).

MUSEUM BUILDING/S, 1841–1900

Mason Hall and the First University Museum

Kerstin Barndt

In 1841, four years after its move to Ann Arbor, the University of Michigan opened the doors to University Main Hall. Soon to be rechristened Mason Hall in honor of Michigan's first governor, Stevens Thomson Mason, the building was an unassuming four-story rectangle defying all architectural fashions of the time. It had been designed to accommodate various units and play several roles. In Mason Hall, religion, the recitation of classical literature, and exposure to natural history went hand in hand to educate the university's first cohorts.[1] Students lived here in dorm rooms and attended classes offered by the university's original College of Literature, Science, and the Arts. In one and the same building, they were able to attend religious services, check out books from the library, and study museum displays: Mason Hall became the first permanent home for the Cabinet of Natural History.[2]

In the following pages, I describe the ties that link the museum in Mason Hall to both internal and external projects and developments. Internally, Mason Hall furthers and reflects the university's emphasis on research, the process of disciplinary differentiation, and ultimately the shifting understanding of higher education itself. Externally, it connects the university to the state's foundational narratives, to its citizens' longing for knowledge and interest in practices of collecting, and to the nation as an imagined community. Michigan's two principal curricula—the classical, humanistic liberal arts and the evolving forms of natural scientific inquiry—were based on collections housed in Mason Hall. The museum served as a center where both strands, represented in the art collections on the one hand and collections of natural history and ethnology on the other, developed together and in dialogue with each other. Indeed, this dialog was personified in the influential figures of Alexander Winchell, a polymath natural scientist, and Henry Simmons Frieze, a trained classicist. As the driving forces behind the development of the museum in the second half of the nineteenth century, these two men consequently anchor this chapter's narrative, which follows the collections' growth from their origins in Mason Hall. As we will see, that growth eventually necessitated the construction of a new and independent University Museum building on State Street in 1881, and a designated art gallery in the University Library, which opened in 1883.

Writing about university collections in Berlin, Germany, Cornelia Weber has called for "an overarching historical and epistemological analysis of university collections and disciplines."[3] While scholars have concentrated on individual collections and museums, Weber deems a broader, transdisciplinary view essential to "extensively learn about the relevance of the collections for the development of different disciplines and—related to this—the success or failure of different types of collections."[4] Following Weber's lead, I stress how at Michigan collections of natural history and art history evolved together with the ethnographic collections during the first decades of the University Museum, mirroring each other's concerns and epistemological claims. At the heart of these claims we find an effort to develop new methods of inquiry guided by specimens and objects amassed in the museum, both in the humanities and in the sciences. At the same time, museum curators also embraced a vision of the University Museum as an institution in service to the state of Michigan, its citizens, its historical imaginary, and its future.

SCIENCE, THE ARTS, AND HISTORY

How should we imagine the museum in Mason Hall? The earliest photographs that have survived in the archive show the interior in the early 1870s as an open, multifloor gallery space adorned with wrought-iron balustrades (fig. 15). This new configuration of the galleries had been made possible by an initiative by Frieze while he served as interim university president from 1869 to 1871: the construction of a bridge building that closed the gap between Mason Hall and South College. With its signature neoclassicist dome at the center, this new building complex at the heart of campus was called University Hall, making room for a new chapel, lecture rooms, and a large auditorium, and integrating the museum with its north wing (formerly Mason Hall).

Floor-to-ceiling glass cases sat against the walls, showcasing collections ordered by disciplines: in the surviving photos, we can detect a number of rocks, as well as mounted birds. Standing on a pedestal in their midst, a cast of Apollo Belvedere anchors the displays in the classical canon of Western civilization.[5] In the nineteenth century museum, in other words, artworks took their place alongside specimens from nature, which were in turn flanked by historical objects—as if in dialogue to further a view of the collections as both aesthetically charged and scientifically useful.

To fully understand the prominence of the museum in Mason Hall, however, we have to take a step back in time and trace the development of collections and their dissemination across campus until this moment when the collections first entered the photographic archive.

From the late 1840s onward, a flurry of activities around collections and collecting made evident the centrality of the Cabinet of Natural History for the self-understanding of the young research university. Next to the showcase of minerals that Houghton was able to arrange before his untimely death in 1845, one could also find geological and zoological specimens. After Houghton's death, Silas H. Douglas and Regent Zina Pitcher aided Abram Sager in inventorying specimens from the Geological Survey. The three men had met as students at the Rensselaer Polytechnic Institute where Houghton had also studied. At Rensselaer, education reformer Amos Eaton had introduced the students to field work in botany, biology, and geology. While Sager labeled the zoological collection, Pitcher, himself an avid collector and botanist, sent herbaria to former U-M professor Asa Gray (by then at Harvard) for guidance in identifying the plants.[6] After Douglas had counted an astonishing 1,500 bird skins and applied arsenic to protect the specimens from invasive insects, the regents agreed to appoint a taxidermist who prepared a series of birds for the zoological cabinet.[7]

In Mason Hall, the University Museum provided a space where the classical liberal arts curriculum and new methods of scientific inquiry thrived side by side. Moreover, once the first cabinets were installed and minerals from the Lederer Collection and Houghton's geological survey work laid out, Mason Hall became Michigan's first *public* museum, competing only with a handful of eclectic (and private) dime museums in Detroit.[8]

From 1855 to the 1870s, Alexander Winchell and Henry Simmons Frieze anchored the University Museum in Mason Hall and personified the dialog between the natural sciences and the humanities (fig. 16). Winchell, originally hired as professor of physics and civil engineering, purchased the university's first meteorological instruments in 1854.[9] While Winchell was responsible for the natural history collections and the department of ethnography and relics, Frieze curated the art collections (see Choberka, this volume). A scholar of Virgil and Quintilian, Frieze taught Latin literature in cultural-historical context, laying the foundation for fields that would substantially grow in the coming decades: classical studies and archaeology as well as art history.[10] In addition, Frieze's aesthetic sensibilities and skills prominently included music. He regularly played the organ at St. Andrews Church, introduced music into the campus chapel services, and

Fig. 15. University Museum's cabinet displays, north wing of University Hall, photo print, c. 1875. Bentley Historical Library.

initiated the founding of the Choral Union and the University Musical Society.[11]

But more was arguably at stake in this collaboration than simply the relation between the arts and sciences, or between the future divisions of the university devoted to humanistic and scientific inquiry, respectively. What emerges, in addition, from the history of the nineteenth century collections is a shifting conception of the place occupied by religion in the realm of higher education. For example, when the Cabinet of Natural History first opened its doors to the public in University Hall, pieces of copper and other natural resources represented the fruits of field research, objects for future study, and the state's economic promise, to be sure. At the same time, however, the treasures amassed here also had religious implications. Around 1800, theologian William Paley had coined the notion of natural theology to argue that "divine wisdom could be discerned in nature."[12] In this vein, Louis Agassiz, the founder of Harvard's Museum of Comparative Zoology, proclaimed that natural history aimed for "the analysis of the thoughts of the Creator of the Universe, as manifested in the animal and vegetable kingdoms."[13] The objects collected during the various state surveys, then, would also have been considered representative of this divine order of the natural world—even on the nonsectarian campus of the University of Michigan where, in Mason Hall, the president and faculty led compulsory weekly Christian service in the chapel until President Angell made participation voluntary in 1871.[14] From the 1840s to the 1870s, the connections between Ann Arbor's Protestant denominations and faculty on campus were particularly strong, and tensions between different Protestant fractions influenced the ousting of President Tappan (see Blouin, this volume). Winchell was one of Tappan's most fervent critics on ethical and administrative grounds. In his studies and publications, however, Winchell constructed a bridge between

Agassiz's theological vision of natural history and Darwin's theory of evolution.

Advances in geological, biological, and humanistic scholarship reflected changing paradigms of religion, science, and ethics. With the discovery of geological formations and life forms, which greatly exceeded the temporal frame of biblical accounts, and with the publication of Darwin's *Origins of Species* in 1859, the history of life and of the universe moved to the forefront of scientific inquiry. While astronomy would soon be granted its own apparatus and space with the construction of the observatory, Winchell utilized the collections in Mason Hall to expound upon and expand Darwin's and other scientists' findings.

Meanwhile, the museum landscape at U-M continued to evolve outside of the walls of Mason Hall as well. Under President Henry Philip Tappan, the Ann Arbor campus expanded in the 1850s: new departments such as medicine, engineering, astronomy, and law began to offer classes.[15] The opening of the first chemical laboratory on the grounds of a state university and the inauguration of the Detroit Observatory set Ann Arbor's campus apart from other schools of higher education. At the center of these new disciplines stood a state-of-the-art scientific apparatus as well as deep collections for research and teaching. Taking the Cabinet of Natural History as a model, these collections were organized in museums of their own. Museum pioneers Sager, Douglas, and Pitcher played decisive roles in disseminating collections across campus. They successfully lobbied for a medical department and a chemical laboratory, and Sager and Douglas assumed leading positions in both institutions. In his new role, Sager founded the anatomical museum in 1851, which grew exponentially after the appointment of Corydon Ford in 1854 (see McIsaac and Girsch, this volume). At the same time, Douglas assembled the Museum of Materia Medica, a collection of pharmaceutical and chemical substances used for laboratory work and teaching in medicine, pharmacy, and chemistry.[16] With the spread of collections across departments, the University Museum began to disperse, and the term itself referred more flexibly to different collections housed in the Medical Building and in Mason Hall. In September 1863, the regents reported:

> The University Museum embraces the following valuable collections, illustrative of Science, Art and History . . . I. Natural History—Mineral Collection, Geological Collection, Zoological Collection, Herbarium; II. Museum of the Medical Department, The Anatomical Department, Materia Medica, full suites of gynecological instruments; III. Fine Arts and History, casts, engravings and photographs, medallions, copies of reliefs and busts.[17]

The University Museum had come into its own as a hub of scholarship, research, and university life.

CULTURES OF COLLECTING

The regents' overview testifies to the breadth and variety of the collections, but also to a certain disciplinary promiscuity characteristic of the mid- to late nineteenth century. Disciplinary affiliations were still in flux until the century's later decades, and the first natural scientists on campus began as generalists who over time would shape more narrowly defined disciplines. Frieze and Winchell exemplify the early, as yet weakly differentiated phase of this process. Stepping in after President Haven as interim president in 1869 and twice again while President Angell was absent, Frieze was a fervent defender of Michigan's state-funded, public education, which he saw as reaching from the common schools to graduate education, and building on the separation of church and state. Frieze also insisted that university education encompass *all* branches of knowledge—the liberal arts, including the study of world religions; the social and natural sciences; and professional schools. His account of the importance of liberal studies (as opposed to what he called "special" studies that lead to clearly identifiable professions) retains its relevance even today. On the occasion

Fig. 16. Henry S. Frieze and President James B. Angell with faculty, 1880. Frieze (left) and Angell (right) sit at the center of the first row. Joseph B. Steere and Mark Harrington stand in third row (second and fifth from left), photo print. Bentley Historical Library.

of the university's semicentennial in 1887, Frieze asserted that the College of Literature, Science, and the Arts in general and the department of philosophy in particular "is the highest department of educational work which is farthest removed from the appreciation and sympathy of men. Without it, however, we should cease to bring out and develop that class of minds which alone can keep us as a State on a level with the most enlightened states and countries."[18]

The museum displays in Mason Hall were a testament to this broad vision of university education. Here, Frieze worked side by side with Winchell who, as mentioned previously, was in charge of the collections of natural history and the department of "ethnology and relics." Divided by discipline, Frieze and Winchell shared scholarly breadth, an interest in natural, cultural, and political history, a love for music, as well as administrative ambitions. While Frieze pushed for the broadest possible vision of humanistic education and drew on classical literature and languages, material culture, and the arts, Winchell expanded the scientific curriculum through expeditions as well as museum and laboratory work.

Frieze oversaw the expansion of the art collection, which he had assembled on an extended trip to Europe. At Mason Hall, the Gallery of Arts found its home next to the Cabinet of Natural History. From 1857 on, the gallery was open daily for visitors, furthering the public function of the displays.[19] As we have seen, the term "University Museum" subsumed all collections instrumental in teaching and research. In the public eye, however, the term referred primarily to the collections housed in Mason Hall.

Under Frieze's direction, the Gallery of Art and History grew to encompass class gifts such as the cast of Laocoön; the Horace White Collection of gems and portrait medallions donated by Andrew D. White, professor of history and subsequently president of Cornell University; and Randolph Rogers's revered marble sculpture *Nydia, the Blind Flower Girl of Pompeii.*[20] Frieze's friendship with

Rogers had aided the purchase of the sculpture and other works by the artist.[21] Of these early collections, only *Nydia* has survived and remains on exhibit at UMMA to this day.

Working alongside Frieze, meanwhile, Winchell was deeply involved in the scientific collections, likewise overseeing a period of rapid growth.[22] In 1859, Winchell successfully petitioned the state legislature to reinstate the state's geological survey, which had lapsed since Houghton's death. His survey reports include detailed geological studies and descriptions of the state's mineral resources of economic interest: coal, gypsum, copper, salt, and iron.[23] During the Civil War, the survey was interrupted again but Winchell was able to continue his work in 1869. While Winchell's focus as geologist was primarily on mapping the state's natural resources, his team also surveyed fish, mollusks, and insects and introduced these specimens to the museum collection. Already in 1859, Winchell had begun taking students along on the survey, training them in fieldwork, collecting, and classification.[24] During years between surveys, he organized the first faculty-led student expeditions to the mining region of Lake Superior.[25] The Bentley Library holds a number of stereographs from the 1868 expedition showing not only the region's spectacular natural sights but also canoes and Native American guides, who accompanied the faculty-student team on their journey (fig. 17). The photos were most certainly used for public presentation of travel accounts such as those of Mark Harrington, published in *Michigan University Magazine*. Harrington was one of the students who participated in the expedition and the experience enticed him to work for the museum under Winchell, where he advanced to assistant curator from 1873 to 1876. In his travel account, Harrington calls on the reader to visit Mason Hall and its exhibits because "the best place to get an idea of its [Michigan's Upper Peninsula, K.B.] prospects and productions is in our Library and Museum."[26]

In 1866, the museum also hired J. T. Coleman as its first permanent "Taxidermist and Zoological Artist"—to great effect, as Winchell noted in his report to the regents. Coleman not only repaired and cleaned the plaster cast collection and the marble statue of Nydia, he also added "attractions" to the zoological gallery by mounting "a large number of specimens in their natural attitudes, and surrounded by their natural associations."[27] The first habitat displays in the history of the University Museum, these arrangements impressed Winchell as both educational and strikingly

Fig. 17. University expedition at Ford William, Hudson Bay, Canada, students and faculty in birch bark canoe with Mr. Clarke, the clerk, and Anishinaabe guides in the foreground, 1868, stereograph. Bentley Historical Library.

beautiful: "Groups of birds are disposed according to their native habits, over sea-side cliffs or mimic trees, or upon the moss-covered ground; and, while the zoological characters are perfectly preserved, to answer the purposes of scientific study, the disposition of the specimens produces a pleasing effect upon the eye."[28]

Considering the popularity of birding among amateur naturalists at the time and the early bird collecting successes by Sager and his colleagues, it is not surprising to find the first habitat displays centered on birds. Historian Mark V. Barrow notes that after the Civil War, an energetic "culture of collecting" developed that brought together "young and old, male and female, beginner and expert" passionate about natural life forms in an "extensive and heterogeneous community."[29] This culture, Barrow argues, accompanied a period of fundamental change in US history, a period of industrialization, education, population growth, and advances in technology from railroads to the telegraph, and from photography to new typesetting machines. Attention to nature, then, posed a counterpoint to the ethos of industrial growth, bolstering nature as a realm of leisure and aesthetic pleasure. Moreover, in the last decades of the nineteenth century, birders and other collectors sharpened their sense for destruction and loss. During this time, museums such as Ann Arbor's campus museum evolved as places that reflected these changes through exhibits of local fauna, including displaced or extinct species. Encountering a mounted wolverine in the University Museum, a visitor noted already in 1865: "The poor fellow is only a matter of tradition—being found now, seldom, if ever, this side of the Rocky Mountains. He is a fierce looking creature, and well becomes his name."[30]

ACADEMIC DISCIPLINES AND THE LIFE-CYCLE OF COLLECTIONS

Winchell and Frieze appear to have collaborated to great effect during the middle of the nineteenth century, but as disciplinary protocols developed and shifted so did the role of collections and museums, scientific apparatuses and laboratories. They all were key to a process of disciplinary professionalization and differentiation, which in turn changed how and what kind of on-campus collections did or did not flourish. As Mark Meadow notes with regard to scientific collections at the University of California, "changes in disciplinary practice, such as the general movement away from taxonomy in the natural sciences, and changes in disciplinary formations, seen in the fissioning and fusioning of departments across the university, can have . . . profound consequences for material collections."[31]

While some collections evolve alongside disciplinary change, others lose their historically specific use-value and become obsolete. At the University of Michigan, collections that obtained relative autonomy from departmental structures have proven to be more protected from dissolution than collections that remained within university departments. But the struggle to find the right administrative and disciplinary place for specific collections continues to this day.[32] It is also clear that independence alone does not necessarily secure adequate stewardship. Ever-tightening budgets, the elimination of curator positions, and the competition for campus space play important roles as well and have significantly impacted the care of some collections.

On the Michigan campus, the abrupt deaccession of the medical collections around 1950 shows the ultimate fate of material collections that had served their historical purpose in education (see McIsaac, this volume). The decomposition of the Randolph Rogers plaster cast collection in the basement of University Hall similarly provides an example of how collections that were once highly valued might become neglected during times of devaluation. Scholars interested in the history of science could have granted these discarded objects a second life as historically significant material witnesses of scientific practices. But in contrast to Harvard University, where David Pingree Wheatland began to assemble a collection of scientific instruments in the 1920s, the field of history of science was underdeveloped at Michigan at the time. To be sure, the university maintains an immensely rich historical archive of papers and photographs in the Bentley and University Libraries; but historical *objects*—of science or of history—are rare among these holdings.[33]

RELICS ON VIEW

Besides their central role in the university's internal history of evolving research disciplines, in teaching protocols, and in the general production of knowledge, as key components of a flagship state institution the collections also played a vital role vis-á-vis the public. While some university museums, especially those that, like U-M, served as a depository for state surveys, were also recognized as state museums later on, the University Museum in Ann Arbor never officially achieved this status. And yet, throughout the nineteenth century, newspaper reports presented the campus cabinets as Michigan's most treasured museum, noting that it attracted visitors from far and near. In 1861, a delegation from the Chicago Academy of Sciences convened in Ann Arbor to study the university's facilities. Hosted by President Tappan, the delegation praised the library, the chemical laboratory, and the university's medical, science, and art collections as "an honor to the State and evidence of the liberality, enterprise, and intelligence of its citizens."[34]

The delegation's report indicates that in their first formative decades, the state of Michigan and the university relied on museum collections and displays to project an image of the state as prosperous and progressive. While the collections provided the material basis for the development of academic disciplines, the displays also constructed an ideal image of the state's past and present as well as project its future. Through cultural and scientific displays, the university's exhibitions asserted a collective sense of exploration and new beginning, ideally combining, in museum scholar Sharon MacDonald words, "claims of the capacity to know and to govern."[35]

In this context, it is important to note that Ann Arbor's first museum collections accumulated not only discipline-specific specimens but also art works pertaining to Michigan history, ethnographical and archaeological artifacts, and so-called relics related to the Civil War, to Native American culture and history, and to slavery. Thus, the 1860s also saw the formation of a museum department devoted to "ethnology and relics." Objects in this collection ranged from an "Indian War dress" to "a portion of a pair of stocks employed in the punishment of slaves in Louisiana" to a "[p]iece of root from the apple tree under which General R. E. Lee surrendered to General U. S. Grant" and a "leaf of Magnolia Grandiflora from the tree growing by the tomb of Jackson."[36]

What do these objects – all of which must now be considered lost – have in common? According to the 1828 edition of *Webster's American Dictionary of the English Language*, a "relic" denotes "that which remains; that which is left after the loss or decay of the rest; as the relics of a town; the relics of antiquity; the relics of saints, real or pretended." The term also refers, the entry notes, to "the body of a deceased person; a corpse."[37] Moving from the sacred to the secular realm and translating one into the other, relics perform the work of mourning, memorialization, and nation building. Linked to the Civil War and slavery, many relics in the collections served to position the state university as an essential part of the union upholding Abraham Lincoln's Emancipation Proclamation.

While Civil War relics projected a patriotic vision of Michigan as a state that had won on the side of Lincoln's army and thus bolstered the state's foundational narrative of virtuous (and victorious) westward expansion, artifacts purchased from indigenous communities or extracted from burial grounds point to a simultaneous history of oppression. Thus, whereas the nationalizing role of the leaf from Jackson's tomb is hard to miss, the indigenous war dress occupies a far less obvious place in these collections—as do objects found in Native American burial grounds, including human remains.[38] Were these objects and remains also considered to contribute to the state's present and the future of the nation, or were they displayed as relics substantiating the trope of the "vanishing Indian" as propagated by Henry Wadsworth Longfellow's *Hiawatha* and others?[39] Most museum objects and displays suggest the latter. And yet, an Ojibwe canoe that Houghton brought to campus in the 1840s documents the skill and knowledge of Native Americans, which museum curators relied upon in their collecting efforts. It appears, then, that Native Americans were valued as contributing to *contemporary* Michigan culture and the educational mission of the museum. In the final analysis, however, the eclectic array of sculptures brought together in the 1890s gallery of art suggests (see Choberka, this volume) that Native Americans entered the historical consciousness of the university and the state on the premise of radical asymmetry. Shown as children in Rogers's sculpture *Indian Fisher Girl* (ca. 1866) or engaged in a dramatic equestrian fight in *The Last of his Tribe* (1881), Native American people were depicted as stereotypical "noble savages" on the verge of extinction, written out of the present and future of state formation.

Fig. 18. Randolph Rogers, Model of Cavalry Soldier, plaster, c. 1870. University of Michigan Museum of Art.

In the broader arc of U-M museum history, these early relics (now all lost) are linked to the display of Rogers's plaster casts for Detroit's Soldiers and Sailors Monument, which commemorated the loss of lives during the Civil War in the library's art gallery in the 1890s (fig. 18); both of these sets of objects connect, in turn, to the opening in 1910 of Alumni Memorial Hall and its multipurpose functions. That building served not only as art gallery and the location for the alumni association offices but also as a memorial to Michigan students and faculty who had died during the Mexican War, the Civil War, and the Spanish War. Tracing such an arc allows us to see that while collections may be lost, the cultural work they once performed can come to be resurrected in different forms that may seem more appropriate to the museum culture of another time. What we witness here, too, is a transition from full participation in the heroic and national memory culture that evolved after the Civil War to a more localized mode of commemoration that focuses on the university community in the twentieth century.[40]

In the process, as we have seen, once important collections such as the Rogers casts appear to lose their significance until they end up languishing in the damp basement of University Hall where they disintegrated. After Frieze's death in 1889, his successor as chair of classics, Martin Luther d'Ooge, inherited the care for the art collections but only until plans for a new art gallery in Alumni Hall took hold. In the new gallery, the expansive collection of original paintings bequeathed by Henry Clay Lewis in 1884 took center stage and U-M's first assistant professor of history of art, Herbert R. Cross, claimed curatorship. Cross "was not an admirer of Rogers's work, and was not the type to look after material which he had probably never seen and in which he felt no interest." Rather quickly then, the collection "passed from memory and . . . suffered."[41]

Collections in classical archaeology where revived under professor of Latin Francis W. Kelsey. He began purchasing historical artifacts in 1893 and directed excavations in Egypt and Tunisia in the 1920s (see Pedley, this volume). The many objects he brought to campus eventually led to the founding of a new unit independent of the art collections, the Museum of Classical Archaeology, later renamed Kelsey Museum of Archaeology.

GLOBAL COLLECTING AND THE RISE OF A NEW MUSEUM

While shifting disciplinary boundaries between classics, archaeology, and art history shaped the collections that Frieze had founded, the event that most transformed the university's original natural history and anthropological collections turned out to be Joseph Beal Steere's five-year global collecting tour, 1870–75 (see Barndt on Steere, this volume). Adding more than twenty-five thousand specimens to the university's collection and discovering bird species new to science, Steere, a U-M alumnus and student of Winchell, was awarded an honorary PhD degree and professorship upon his return to campus in 1876. After he opened the first boxes of specimens that his student had sent from Brazil, Winchell was convinced of the collection's value, as he stated,

> both in consequence of our almost complete destitution of South American specimens, and the striking character of Brazilian species and the unique features of that faunal region of the earth's surface. Armadillos, ant-eaters, monkeys, caymans, toucans, brilliant humming-birds are strange forms in a collection of North American animals; and when properly mounted, will add real interest to the museum.[42]

Winchell also pointed to geological findings by Steere that he thought could help solve a scientific dispute between Agassiz and his student Charles Frederic Hartt about the origin and evolution of the Amazonian formation. The dispute garnered great attention as it was linked to Agassiz's theistic anti-Darwinian stance, and Winchell was eager to contribute to its resolution.

During the first two years of Steere's collecting expedition, Winchell still found space in the newly reconfigured museum in Mason Hall (now part of University Hall) to exhibit some of Steere's exquisite finds. In subsequent years, however, it became dramatically clear that the museum had run out of space and manpower to adequately register, classify, and store the new material.

> If the whole of the collection, which Prof. Steere has brought home, remains permanently with us, it is clear that we must have more spacious quarters for our Museum . . . I think we have already reached the point where a new and commodious building for the storing of our collections . . . is a real and pressing need.[43]

At first President Angell and the regents contemplated a combined new home for the library and the museum, but in consideration of potential fire hazards Angell recommended that two independent buildings should be erected: a museum to exhibit the collections of natural history and ethnology and a library with a separate wing for the art gallery.

It took three years to convince the state legislature. When the appropriation for the museum finally passed in 1879, Rice A. Beal, the owner of Ann Arbor's weekly *Courier*, marked the memorable day by firing "25 guns" in celebration (even if, at $40,000, the amount was smaller than hoped for).[44] Beal, a cousin of Steere, had a stake in the collections that the naturalist had brought to campus as he had financially supported the expedition.[45] Steere, in return, had written over ninety travel columns from abroad for the *Courier* describing sights and people and recounting his findings. His columns had brought Steere local fame and many of his readers were eager to see the treasures firsthand in the new museum after having learned about them through Steere's reports,

Fig. 19. University Museums Building, architect William LeBaron Jenney, c. 1900, photo print. Bentley Historical Library.

thereby adding local pressure to the demand to exhibit Steere's collection. Moreover, the timing of Michigan's first museum building also coincided with a veritable national museum boom: the Metropolitan Museum in New York City had opened in 1872, Yale's Peabody Museum in 1876, the Peabody Museum of Archaeology and Ethnology at Harvard University opened the same year, and Agassiz's Harvard Museum of Comparative Zoology received a new wing. The year 1881 also marked the inauguration of the Smithsonian's first museum building for the US National Museum in Washington DC. During the same year, Detroit's newspaper magnate James E. Scripps departed on a grand tour of Europe. Not unlike Steere's columns, Scripps's newspaper reports about the art museums he visited along the way proved so popular with readers of the *Detroit Free Press* that initiatives for an art museum gained traction. A first temporary art exhibition was mounted in 1883, and five years later the Detroit Museum of the Arts welcomed its first visitors.

In Ann Arbor, the regents asked the university's very first chair of architecture and design, William LeBaron Jenney, to design the new University Museum. The space allotted to the museum was a prominent one: right next to University Hall on State Street. The building would become one of Jenney's most enduring contributions to campus architecture before he moved back to Chicago.[46] Jenney devised a compact building in Gothic-Italianate style with a tower at its center (fig. 19). At first, students applauded the museum as "an exception to the otherwise uninterrupted ugliness which prevails in our college building."[47] But due to dire times brought on by the Long Depression of the 1870s, the budget constraints on the construction were severe and many compromises had to be made.[48] If we look beyond the prevalent stories of the building's architectural shortcomings, however, we find a vibrant institution that continued to contribute to the burgeoning fields of biology, geology, paleontology, and anthropology. Through research, exhibitions, and public lectures about extinct and living species from Michigan and across the globe, the new museum investigated the laws of evolution from diverse disciplinary angles.

While Frieze's art collections would soon be exhibited in the galleries of the new library, the museum on State Street showcased artifacts, weapons, and wares from indigenous and prehistoric cultures. The cultural collections were furthermore expanded through the gift of the Chinese exhibit at the World's Industrial and Cotton Centennial Exposition and Frederick Stearn's gift of historical musical instruments.[49] In accordance with the museum's taxonomy for its natural history collections, the instruments were also presented in evolutionary progression from the ostensibly simple to the apparently more complex.

ADAM'S ANCESTORS

Although only the entrance of the new museum was adorned by a gothic frame, overall the façade echoed the entirely Gothic-inspired Oxford Museum of Natural History. In 1860, the latter had been the site of the famous Huxley-Wilberforce debate. Together with numerous other participants, biologist Thomas Henry Huxley and the Bishop of Winchester, Samuel Wilberforce, discussed the ethical consequences of Darwin's new theory of natural selection published in *On the Origins of Species* just a few months prior. Built twenty years into the debate about Darwin's theory, the museum in Ann Arbor paid homage to the towering figure with high reliefs depicting animal groups in three lunettes gracing the front of the tower: a rhinoceros fighting an alligator in the central vignette, a pterosaur in a smaller lunette underneath, and the encounter of a dodo and an extinct turtle, possibly a meiolania, in another (figs. 20 and 21).[50] The figures dramatized extinction, death, and survival in the animal kingdom and clearly alluded to the science of paleontology and theory of evolution.

The figures further suggested that this museum was one that took geological time and the history of life on earth as one of its main points of inquiry. With Steere's appointment as assistant professor of paleontology, the university for the first time had recognized paleontology as a field of study independent of geology. Steere had brought a number of fossils back from his global journey, which complemented the museum's collections of invertebrate fossils dating back to Houghton's and Winchell's survey work in Michigan. If Winchell had still resided on campus, this new professoriate would certainly have gone to him, as he had already described a number of new species of fossils (and would add over three hundred during the course of his life time). But Winchell had left Ann Arbor in 1873 to become the first chancellor of Syracuse University and then to teach at Vanderbilt University. Six years later, however, he returned to campus as professor of geology and paleontology and was able to shape the exhibits in the new University Museum. Steere, in turn, was named professor of zoology and took charge of the zoological collections.

Winchell had resigned from Syracuse due to insurmountable financial obstacles and Vanderbilt had dismissed him for his theory of human evolution. As an active Methodist, the naturalist strove to reconcile religion and science. Throughout his career, he held on to the idea of a divine design of life. This belief, though, did not prevent him from interrogating the scientific record of his day and from stating that human history preceded biblical chronology. Nevertheless, Winchell's studies on these so-called pre-Adamite human species had raised concerns at Vanderbilt. As David Livingstone notes, it is not quite clear which element of Winchell's theory was at the heart of the controversy: "Perhaps it was because his claim that the black races were of pre-Adamic stock seemed to Methodists to place them beyond the scheme of redemption—an embarrassing state of affairs to a denomination currently putting strenuous efforts into church outreach to the black community. Or was it on account of his insistence that Adam was descended from black ancestors and may himself have been of dark complexion?"[51] In the argument over one common human descent versus divergent, racially specific genealogies, Winchell sided with the supporters of one, monogenetic origin. Progressive as this seemed in retrospect, Winchell's pre-Adamite theory buoyed the ideology of race and white supremacy prevalent in the later decades of the nineteenth century. While of

Fig. 20. Front façade of University Museum depicting Pterosaur, photo print. Bentley Historical Library.

Fig. 21. Extinct species: dodo and ancient turtle, University Museum, photo print. Bentley Historical Library.

the same origin, the human species according to Winchell had developed distinct and hierarchically ordered racial types; and black and indigenous people were relegated to the bottom of the hierarchy. In this vein, he would assert that "just as far as the African diverges from the style of a white man, he approximates the lower animals."[52]

With his return to campus and with his fame as a proponent of his own kind of evolutionary theory, Winchell certainly pushed the interrogation of U-M's paleontological collections into the forefront of the new museum. Upon entering the museum, visitors would encounter the fossil collections on the first floor.

> We have the paleontological room on the first floor in the north end of the building with good lights and firm cases . . . These cases are all dust tight, and are models of beauty, combined with excellence that have called forth the encomiums of all and every person that has seen them. In the south room on the same floor we have the mineralogical room with twelve cases . . . where the different classes of minerals can be

Fig. 22. Taxidermy on view in University Museum, c. 1900, photogravure. Bentley Historical Library.

> examined. Beneath each is a cabinet tray for specimens for study, easily accessible, and they are the most practically arranged cases that have yet been devised for such purposes. In the main hall between these two rooms is ample space for the largest paleontological and mineralogical specimens.[53]

Next to impressive pieces of copper, the central space between the two larger rooms probably also featured the first vertebrate fossils such as the mastodon head that the regents had purchased at Winchell's request.[54] The geological and paleontological department also had a state of the art laboratory on the museum's fourth floor, filled with microscopes and photographic equipment, including for innovative "microscopical photography," pioneered by Winchell. Advances in visual technologies began to impact how faculty and students studied specimens. The museum was not only a place of three-dimensional objects but also one where new media and laboratory technologies were tested, refined, and made applicable for the classroom. In the museum, object-based research and knowledge found its broader public through exhibits, teaching, publications, and presentations. In 1898, the mineralogical collections moved to Tappan Hall to be closer to seminar rooms and the empty room was remodeled into a much needed lecture hall for classes and public events illuminated by glass lantern slides. Evolution was the main topic of the lecture series but with the arrival of Stearns's collection of musical instruments, public lectures also included music history.[55]

TAXIDERMY ART AND BIRD DISPLAYS

Under the curatorship of Steere, the second floor of the new museum displayed the zoological collections. One room was reserved for the exhibition of the expansive bird collection, the other showed mammal taxidermy, as well as amphibians and fish in alcohol.[56] In the center of the south room, mammal (and bird) groups mingled in striking oversized glass cases: in one, two snow owls framed an impressive walrus, in another a diverse group of antlered mammals greeted visitors with their bodily presence and startling gazes (fig. 22). South American and Pacific mammals, shells, and

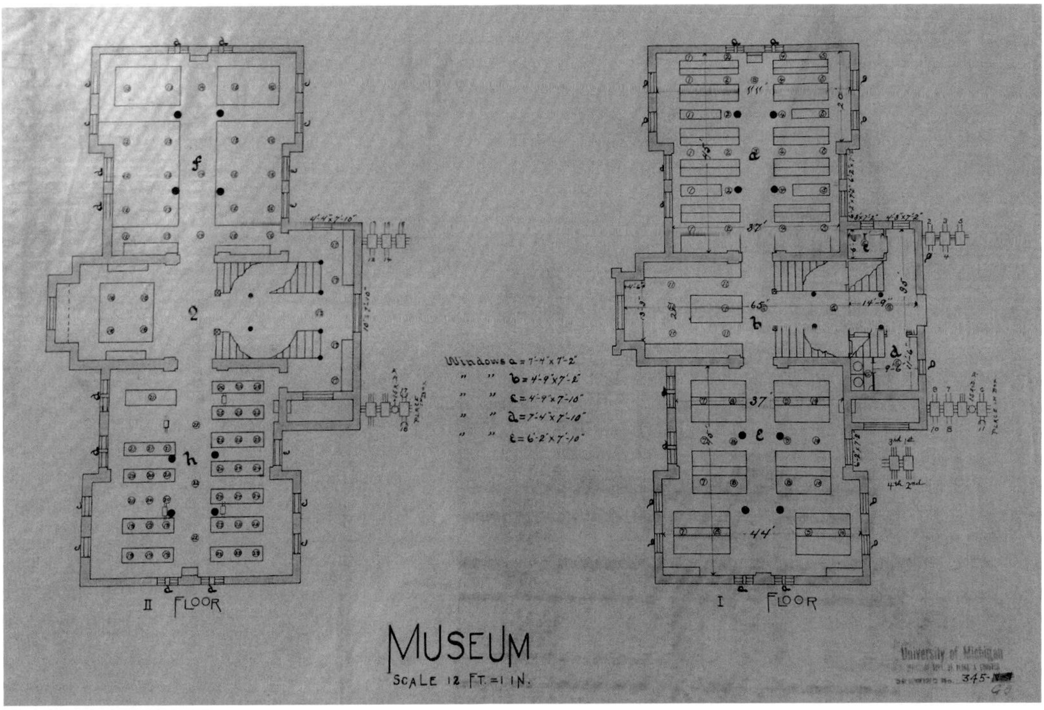

Fig. 23. Floor Plan of University Museum, architectural drawing by William LeBaron Jenney, 1879. Bentley Historical Library.

birds were also on exhibit from Steere's expeditions in the 1870s and again in 1883–84, among them the infamous Mindoro dwarf buffalo, the tamaraw (see Barndt on Steere, this volume). In the first decade, the museum's taxidermist, J. T. Coleman, had help his Detroit-based colleague James Hobson to prepare these displays and others; in 1895, the museum hired taxidermist and ornithologist Norman Wood who would remain with the museum for forty-eight years and become its first curator of birds. From the beginning of museum history at U-M, the bird collection remained the foundation of one of the core fields of zoological museum studies. By 1898, the collection had grown to 3,600 bird skins and 1,478 mounted birds.[57] The majority of this immense number of individual specimens was showcased in taxonomical order, but curators Norman Wood and Dean Worcester made sure to also exhibit groups in accordance with advances in mounting techniques and habitat display. Besides habitat and ecological principles, the displays of birds, their offspring, and eggs highlighted "protective coloration, albinism, parasitism, [and] individual variation."[58]

ETHNOGRAPHY AND ARCHAEOLOGY

The anthropology collections were the last to be moved from University Hall to the new museum building where they were installed a floor above the zoological exhibit, though galleries provided views to the floor below: "The third floor is a gallery story, having a large space open to the second floor (fig. 23)." The over one thousand objects of the Chinese Exhibit with its focus on Chinese cotton—from plant to the manufacturing of cotton and articles made out of cotton (as well as glass, silk, ivory, and porcelain) filled a prominent place on the floor's landing (fig. 24). The galleries in the adjacent rooms showed pottery, arms, carpenter tools, clothing, and art from indigenous cultures in the Americas, Taiwan (Formosa), the

Fig. 24. University Museum, view of habitat displays and artifacts from the Chinese Government Collection, c. 1900, photo print. Bentley Historical Library.

Philippines, and Alaska.[59] In his review of the exhibit, Harlan I. Smith, one of the first students to be trained in archaeology under Francis Kelsey, linked the museum displays to the evolving science of anthropology. He praised Steere's collections of ancient Peruvian vessels "of black, red and yellow earthen ware in innumerable shapes," and the pedagogical function of a synoptic series of casts from prehistoric American objects held by the Smithsonian Institution that had recently entered the University Museum: "By means of them the student can see the various stages in the evolution of the tools of man, from near the time when he had only his finger nails, and teeth down to the era when archaeology hands over her treasure to the historian."[60] Smith also noted that just around the time of his writing in 1893 new prehistoric villages were excavated in Kalamazoo while other indigenous mounds and remains were "fast disappearing as the result of the advance of our industries."

Archaeologist Wilbert Hinsdale would soon follow Smith's call to "salvage" the historical material traces of the region's indigenous culture and advance to the anthropology museum's first curator in Great Lakes archaeology (see Ford on Hinsdale, this volume). Meanwhile, objects pertaining to Native American culture from across the United States entered the museum. From the Harriman Alaska expedition, the museum secured and exhibited a ten-foot long totem pole taken from a deserted Tlingit home.[61] A review referred to the relic as one "from a declining custom," echoing the prevalent ideology according to which Native American cultures hover on the verge of extinction.

In the same year, the museum also boasted an exhibit of "new and some old curiosities, which are of general interest to the public. Among them are relics of slavery in the shape of iron knee-stiffeners . . . that made walking difficult and painful. In this case also are a Mexican axe and a hunting knife, and some Peruvian fractional currency . . . issued in Peru in 1873."[62]

The eclecticism of the assembled objects points to the main difference between this particular display and the other, more taxonomically ordered exhibits. In the university museum of the twentieth century where space to house and exhibit specimens from faculty expeditions became increasingly tight, curators questioned whether such "relics" should be collected at all.[63] The scientific paradigms and taxonomies that had developed throughout the nineteenth century and had shaped collecting disciplines from biology to paleontology, from archaeology to anthropology and art history, left objects of historical value such as those pertaining to slavery orphaned. As a relatively new independent field of study, modern history at the University of Michigan was taught as a text-based science with connections to English literature and political science rather than as a discipline that integrated collections of material culture. By the beginning of the twentieth century, the role of the museum with its capacity to participate in and refract questions of state and national identity had been supplanted by the ascendance of the natural sciences, and the natural model of evolution had crowded out history.

Not surprisingly, this development coincided with a generational shift in museum leadership after the death of Henry Simmons Frieze in 1889 and that of Alexander Winchell two years later. Both scholars had shaped the nineteenth century university collections to become cornerstones of both the academic and the public mission of the University of Michigan.

In 1893, Steere, an early student of Winchell's who had traveled the world and given the university collections in multiple disciplines a global scope, prepared the main museum exhibition for the World's Columbian Exhibition in Chicago. Steere owed to his teacher a vision of natural history as encompassing many disciplines, including ethnography. He also shared Winchell's aim to bridge Darwin's theory of evolution with natural theology. Partly due to his activism for prohibition in the name of religion, Steere was forced to resign from his curator and teaching position shortly after Winchell's death, opening the way for a new, younger, and decidedly secular generation of scholars to bring the University Museum into the twentieth century.

A survey of Michigan fauna held pride of place among the exhibitions that the University Museum had prepared for Chicago. This survey was a chance to fill in gaps in the collection regarding local mammals, birds, reptiles, fish, and amphibians. After the closing of the World's Fair, the complete set of mounted specimens returned to campus. Curators who followed Steere—notably Dean Worcester, Charles Adams, Jacob Reighard, and Alexander Grant Ruthven—consciously expanded upon this exhibit and defined the study of the biodiversity and ecology of Michigan's flora and fauna as one of the central assets of the University Museum. This decision did not prevent curators from also collecting nationally and internationally, but the political function the museum had served midcentury was now thoroughly replaced by an emphasis on biology. At the beginning of the twentieth century, the biological collections grew exponentially. Picking up the historical narrative of the university museum's disciplinary development here and extending it into the twentieth century, Carla Sinopoli shows in her essay below how this growth of the biological collections lead Ruthven to propose a radically new administrative structure for the university museum independent of the academic departments with which the collections were affiliated. This new status served Ruthven as a stepping stone for bringing to fruition his vision of a modern research museum building which opened in 1928 with state-of-the-art laboratories, storage facilities for tremendous research collections, and pedagogically enhanced public exhibitions.

Notes

1. In December 1841, the Proceedings of the Board of Regents note that the "Cabinet of Natural History and the Library have been transferred to two large rooms in the Main Building." Before this move, the collections were provisionally housed in Douglas Houghton's on-campus residence. Houghton and Abram Sager together prepared the first display cases for the Mason Hall museum. Besides the Lederer collection of European minerals, the cabinets showed rocks, fossils, minerals, and zoological specimens collected during Michigan's first geological survey. See Proceedings of the Board of Regents (PBR), 1837–1864, January 1839 (Report of Building Committee to the Board of Regents), and October 2, 1839, 100.

I thank UROP students Emma Kinery and Kathryn Zoller for their diligent research assistance and enthusiastic support of this project. My essay owes many insights to their findings.

2. In their plans for Mason Hall (University Hall), the Regents point to the chapel and the museum as two main functions of building.

3. Cornelia Weber, "From Anatomy to Zoology: Results on the History of University Collections Based on Trans-disciplinary Research," *The University Museums and Collections Journal* (UMACJ), vol. 3 (2010), 121.

4. Ibid.

5. Henry S. Frieze had purchased the casts and other "books, maps, engravings, photographs and terra cotta" in 1855–56 for the University of Michigan in Berlin and in Paris. See PBR, October 14, 1856, 652.

6. "Dr. Gray has consented to overhaul our collection of plants, arrange them according to their natural order and label them for the privilege of withdrawing from it one suite for his own benefit." Zina Pitcher to S. H. Douglas (typescript), September 28, 1846. University Museum records, Bentley Historical Library (BHL). Douglas had worked for Pitcher's office in Detroit and accompanied Houghton and Schoolcraft on survey expeditions. He served U-M in many functions, among others professor of chemistry, professor of Materia Medica, director of chemical laboratory, and dean of medicine. See Jerome C. Knowlton, "Michigan Portraits VII. Silas Hamilton Douglas," *Michigan Alumnus*, March 1, 1907, 251–53.

7. Silas Douglass to Zina Pitcher (typescript), September 1846. BHL.

8. "A Museum of Curiosities," *Detroit Free Press*, March 15, 1868, 1, and "The Story of Detroit's Museums," *Detroit Free Press*, May 2, 1890, 17. Old News Database, Ann Arbor Public Library.

9. PBR, 1837–1864, October 19, 1854, 575.

10. On Frieze's scholarly, pedagogical, and administrative accomplishments, see Francis Kelsey,"The Late Professor Frieze," *The Nation*, January 2, 1890.

11. Charles A. Sink, "The University Musical Society and the School of Music," *The University of Michigan Encyclopedic Survey*, vol. VI, 1126–28. Frieze presided over the University Musical Society from 1879 to 1881 and from 1883 to 1889; Winchell served in Frieze's absence from 1881 to 1883 and 1891 to hand the helm over to classicist and museum curator Francis Kelsey. Kelsey led UMS until 1927.

12. Paley, *Natural Theology*, 1802, quoted in Jon Roberts and James Turner, *The Sacred and the Secular University*, Princeton: Princeton University Press, 2000, 20.

13. Agassiz quoted in Roberts and Turner, *The Sacred and the Secular University*, 23.

14. Julie Reuben, *The Making of the Modern University*, Chicago: University of Chicago Press, 1996, 119.

15. Howard H. Peckham, *The Making of the University of Michigan*, Ann Arbor: University of Michigan, 1994 (1967), 40–43.

16. Pharmaceutical science developed from within the Chemical Laboratory under Douglas's leadership. The first classes in pharmacy were taught within the College of LSA in 1868, and the School of Pharmacy was established in 1876. The Museum of Materia Medica core set of around six hundred specimens arrived in 1856 from Paris. See PBR, 1837–1864, September 3, 1863, 1074–76.

17. PBR, September 3, 1863, 1070.

18. Henry S. Frieze, *The Relation of the State University to Religion. An Address delivered before the graduating classes, on June 26, 1887, at the semi-centennial of the University*, Ann Arbor: University of Michigan, 1888, 5; on the study of world religions, see 27; on the affirmation of faith, 37.

19. PBR, 1837–1864, December 2, 1857, 706.

20. Henry S. Frieze, *Catalogue of the Museum of Art and History in the University of Michigan*, Ann Arbor: University of Michigan, 1876, 3–4. The collections were displayed in Mason Hall, the Halls of the Alpha Nu Society and the Society of the Literary Adelphi.

21. Funds raised by the Ann Arbor Arts Association had helped to secure the purchase of the sculpture and furthered the relationship between the city and its campus museum. See Choberka, this volume. In 1881, Rogers donated a full set of his casts to the University of Michigan. See Millard F. Rogers, *Randolph Rogers. American Sculpture in Rome*, Amherst: The University of Massachusetts Press, 1971, 153–61.

22. Already in his 1866 report to the regents, Winchell noted the "rapid growth" of the collections from 47,677 to 66,897 specimens over the preceding three years. Adding to gifts by private collectors, students, and faculty, the geological survey and the expeditions to Lake Superior had yielded thousands of new specimens that were carefully recorded and systematized in catalogues. The growing collections required additional space and the move of the library into the new Law Building granted a brief reprieve. Alexander Winchell, "Statement of Operations in the Museum, in the Department of Geology, Zoology and Botany, and the Department of Ethnology and Relics for the Year ending 20th September 1866," Ann Arbor: University of Michigan, 1866, 8.

23. Alexander Winchell, *First Biennial Report of the Progress of the Geological Survey of Michigan, embracing observations on the geology, zoölogy and botany of the lower peninsula, made to the governor, December 31, 1860*, Lansing: Hosmer and Kerr, 1861; and Winchell, *Report of the Progress of the Geology Survey of the State Geological Survey of Michigan, presented to the State Geological Board, November 22, 1870*, Lansing: Hosmer and Kerr, 1871.

24. Winchell, Geological Survey of Michigan: 1859 Field Notes, Winchell papers, BHL, vol. 180.

25. PBR, 1864–70, September 24, 1867, Winchell reports to the president, 235.

26. Mark Harrington, "Our Lake Superior Trip IV," *The Michigan University Magazine devoted to College Literature and Education*, vol. 3 (May 1869), 321. Harrington's studies included zoology, botany, astronomy, and meteorology. In 1879, he was tapped to direct the university's observatory, a position he held until 1891.

27. Ibid., 9.

28. Ibid.

29. Mark V. Barrow, *A Passion for Birds. American Ornithology after Audubon*, Princeton: Princeton University Press, 1998, 10.

30. Rev. John Todd, "Strong Engine", *Michigan Argus*, December 15, 1865. Old News Database, Ann Arbor Public Library.

31. Mark A. Meadow, "Relocation and Revaluation in University Collections, or, *Rubbish Theory* Revisited," *The University Museums and Collections Journal*, vol. 3 (2010), xx.

32. The Museum of Zoology may serve as a counter example here. Bound to the vision of the new integrative vision of the Biological Sciences Building and the disciplines and Museum of Natural History it houses, the Museum of Zoology most recently moved back under the helm of the Department of Ecology and Evolutionary Biology, breaking the firewall of independence so forcefully claimed by one of its founding fathers, Alexander Ruthven in 1911.

33. The collection of historical microscopes held in the University Library's Special Collections may serve as an exception here. The microscopes, however, were never used in classrooms of the University of Michigan but came to the collection as a set.

34. "The Chicago Academy of Sciences at Ann Arbor," *Michigan Argus*, January 4, 1861. Old News Database, Ann Arbor Public Library.

35. Sharon J. Macdonald, "Museums, National, Postnational and Transcultural Identities," *Museum and Society*, vol. 1, no. 1 (2003), 1–16, here 3.

36. Winchell, "Statement of Operations in the Museum of the University of Michigan in the Department of Geology, Zoology, and Botany, and the Department of Archaeology and Relics for the Year Ending September 24th, 1868," PBR.

37. http://webstersdictionary1828.com

38. The first Native American human remains seemed to have entered the museum around 1870, brought in by no other than future museum curator Joseph Beal Steere just months before the Michigan alumnus departed on his first world collection expedition for the University Museum. Winchell Museum Report to the President, PBR, 1870–1876, September 1870, 87.

39. For an analysis of the figure of the vanishing Indi-

an see Philip J. Deloria, *Playing Indian*, New Haven: Yale University Press, 1994, 63–64, and Michael Witgen, *An Infinity of Nations,* Philadelphia: University of Pennsylvania Press, 2013, 10–13 and 317–20.

40. Kirk Savage, *Standing Soldiers, Kneeling Slaves. Race, War, and Monument in Nineteenth Century America*, Princeton: Princeton University Press, 1997. For his discussion of Randolph Rogers's Detroit monument see 86–87.

41. Orma F. Butler and Bruce M. Donaldson to John G. Winter, October 4, 1933. The internal U-M report provides a detailed account about the remains of the Rogers casts, Randolph Rogers visual materials series, BHL.

42. Winchell, "Report in the Museum of the University of Michigan in the Department of 'Geology, Zoölogy and Botany' and the Department of 'Archaeology and Relics' for the year ending September 19th 1871," PBR 1870–1876, September 25, 1871, 139–40.

43. Angell, President's Annual Report for the year that ended June 30, 1876. PBR, 1876–1880, October 17, 1876, 66.

44. *Michigan Argus*, May 2, 1879. Old News Database, Ann Arbor Public Library.

45. Beal also claimed to have 50 percent stake of ownership in the collections, which the university disputed. Beal prevailed and after he officially sold his part of the collection to U-M in 1879 (?), the collections were renamed Beal-Steere Collection.

46. For a discussion of Jenney's other Ann Arbor and campus-based projects, see Nancy Ruth Bartlett, *More than a Handsome Box. Education in Architecture at the University of Michigan, 1876–1986*, Ann Arbor: The University of Michigan College of Architecture and Urban Planning, 1995, 13–26.

47. Article in the *Chronicle* 1880, cited after Bartlett, 23.

48. The few historical accounts that mention the museum at all are all tainted by the architectural shortcomings. See for instance entry about the University Museum in Encyclopedic Survey (quoted in Bentley's general university timeline) and Wilfred Shaw, *The University of Michigan*, Ann Arbor: University of Michigan, 1920, 288.

49. Chinese Exhibit in 1885 and Stearns Collection in 1899.

50. I am grateful to Adam Routney, Jeff Wilson, and Janet Hinshaw whose expertise helped me to identify these animal depictions.

51. David N. Livingstone, *Adam's Ancestestors. Race, Religion, and the Politics of Human Origin*, Baltimore: Johns Hopkins University Press, 2008, 146.

52. Alexander Winchell in *Preadamites, or, a Demonstration of the Existence of Men before Adam*, first published in 1888, cited in Livingstone *Adam's Ancestors*, 190.

53. Andrew Climie, "The New Museum," *The Chronicle*, vol. XII, October 2, 1880, 3.

54. Lewis B. Kellum, "The Museum of Paleontology," *The University of Michigan. An Encyclopedic Survey*, Part VIII, 1490.

55. See *Michigan Alumnus*, vol. IV, 1898, 195. Jacob E. Reighard, professor of animal morphology, presented a series of public lectures on the "Evidence of Evolution" in 1904 and Ermine C. Case, professor of geology and paleontology, followed with "The Paleontological Evidence of Evolution" in 1907. Professor Albert Stanley lectured on Richard Wagner in 1904, see *Michigan Alumnus*, vol. X, 1904, 343.

56. By 1897 if not earlier, the alcohol collection had been moved to a "darkened room" with special fire protection. See *Calendar of the University of Michigan 1897–1998*, Ann Arbor: University of Michigan, 1889, 25–26.

57. *Calendar of the University of Michigan 1897–1998*, 25.

58. Ibid.

59. Ibid., 27.

60. Harlan I. Smith, "The Study of Man. Anthropology at the University of Michigan," *Detroit Free Press*, November 12, 1893, 14.

61. "Genuine Totem Pole for U. of M. Now on Exhibition in the University Museum," *Ann Arbor Argus-Democrat*, November 17, 1899. Accessed via Old News Database, Ann Arbor Public Library.

62. *Michigan Alumnus*, vol. VI (1899), 76–77.

63. "The museum at the University of Michigan was intended primarily for scientific collections, and the question which now confronts the authorities is, How far shall it receive objects which are rare or curious, but which have no special scientific value?" *Michigan Alumnus*, vol. X (1903–04), 147.

THE RISE AND THE FALL OF THE UNIVERSITY OF MICHIGAN'S MEDICAL MUSEUMS

Peter M. McIsaac

Today, few would argue that a massive collection of medical specimens should form the bedrock of a comprehensive medical education. Yet in 1850, when the Medical School at the University of Michigan (UMMS) was founded, things were very different. Anatomists were hired on the premise that they would build collections, and large amounts of money were regularly spent on procuring specimens, upgrading their appearance, and storing them in prominent locations. For decades, the university used its thousands of specimens to publicly promote its medical school. So why is it that by the end of the 1950s most of the medical collections at Michigan had so little perceived value that not only were they destroyed but apparently the accompanying archival records were as well? What is the story behind the importance—and then irrelevance—of the collections?

To answer this means dealing with a very incomplete material record—and bucking recent trends in medical practice and historiography whose narratives of "progress" have made the decades of collections-based medical teaching and research practically invisible. The U-M Medical School might indeed seem an unlikely place to tell such a story. As historians of medicine have shown, the university was one of a handful of institutions in the later nineteenth and early twentieth centuries that pioneered the key "modern" conceptual and institutional contours of North American experimental medical education and research.[1] As it pursued experimental research programs, so these narratives go, the medical school shifted to experimental laboratories that promoted hands-on scientific investigation and learning. However, this did not spell the demise of the collections that had been so laboriously built. Rather, from the 1890s to the 1930s, many of the UMMS's leading departments—including its laboratories—maintained and systematically expanded and differentiated substantial collections of human and animal specimens and objects (such as frozen human tissue sections and wax models of skin, glands, and embryos). Why did collections not only persist but continue to grow and specialize at precisely the time when the laboratory, and not the museum, seemed to be winning the day?

When examined relative to prevailing research and educational paradigms and physical location, it becomes possible to connect medical collections to those features of medical education and research at the University of Michigan that have informed its pursuit of innovation. This chapter focuses on two collections: the founding collections that later generations would name the Ford and Sager Anatomy Museum and the pathological collection that was consolidated and then expanded by one of the founding members of the International Association of Medical Museums (IAMM), Aldred Scott Warthin. After analyzing the changing contents and functions of U-M's first medical museum, the chapter sketches out the conditions under which the anatomical and pathological collections came into existence and then grew, becoming an integral element of anatomical research laboratories until the mid-twentieth century.

Thinking about the collections and their changing functions helps to illuminate the university's drive to establish and maintain a reputation as a leader in medical research and teaching. Acquiring important collections was also a viable means to ground the collector's professional medical career and anticipated legacy. Therefore, it was crucial to consider what went into the collections, where they were located, and how the collected specimens connected dissection room, medical laboratory, and classroom at respective points in time. This examination shows that specimens, as well as the practices involved in creating and using them, were surprisingly flexible and adaptable as new research paradigms arose. Their visual qualities, materiality, and relative stability allowed them to function in multiple ways (for example, as illustrations; as documentation of conditions; and, in the twentieth century, as material to be modeled, cut, and manipulated). As my analysis of specialized pathological collections in particular demonstrates, this flexibility goes some way in explaining the startlingly long life of medical collections. However, as I also show, the collections retained value only in relation to certain research questions—which shifted in relation to the changing shape of medical research as well as in relation to the careers, priorities, and resources of individual researchers at the University of Michigan.

COLLECTIONS AND THEIR PLACE IN EARLY MEDICAL EDUCATION

From nearly the moment it opened its doors in October 1850, UMMS sought to establish a substantial museum of medical specimens. Beginning in 1851, faculty meeting minutes recorded regular museum-related expenditures. Every few months, the faculty approved the purchase of new cases and jars for the expanding collections, which in turn soon required the creation of a dedicated space in the glass-domed attic of the medical school.[2] This placement reflected the centrality of the collections for teaching because the attic was reserved for the cornerstone of anatomical instruction: the dissection of cadavers. However, in the age predating anatomy acts and the willed donation of bodies to science, cadavers were often difficult to come by. Therefore, as many scholars have observed, extensive collections anchored a medical school's ability to teach one of its core subjects without having to resort to excessive body snatching (also known as "resurrectionism") and other dubious methods of obtaining cadavers.[3] At the same time, medical collections helped turn the lecture-heavy methods of nineteenth-century medical education to a more hands-on approach, as the materiality of most specimens could both document known cases and serve as examples of preservation techniques that students themselves could be taught to master. They had, in Erin Hunter McLeary's phrasing, both observational and experimental components.[4]

For all these reasons, medical collections represented a school's commitment to high-quality education. As late as 1910, medical education reformer Abraham Flexner wrote—in a report credited with articulating the principles of a modern, hands-on approach to medical education—that "the conclusive evidence of a lack of educational conscience or pride is the general absence of a decent museum."[5] In this light, it makes sense that for an ambitious medical school such as U-M the hiring of Corydon La Ford as university anatomist in 1854 was coupled with the building of collections. Soon after his arrival, Ford arranged for the donation of physician Edmond Carr's collections. Ford also worked to articulate principles according to which he would receive a regular salary as an anatomist by preparing and procuring additional specimens, in exchange for not engaging in private practice.[6]

Ford was not the only educator at UMMS with the expectation that collections-building was essential to the job and its remuneration. In the late 1850s, demonstrator of anatomy William Lewitt was hired on condition that one-fifth of his salary be used to provide new specimens; an 1867 report to the Board of Regents confirmed that by then, they amounted to over sixty individual specimens.[7] In that same report, authored by cofounder of the medical school Abram Sager, objects "donated" by a previous assistant in anatomy, Dr. Milne, were tracked alongside those made available by him and Ford.[8] While individual faculty often asserted ownership over specimens they personally prepared (until they were donated, typically in exchange for named recognition and sometimes monetary compensation), the commonplace linkage of explicit duties and salary to the creation of collections reveals the concerted institutional commitment to their establishment and expansion.[9]

The integral place of the collections in the operations and self-presentation of UMMS is evidenced by their century-long yearly discussion in the General Record of the University. The early records are especially revealing as a means to address students and other audiences with the aim of providing assurance (accurately or not) that the university was not engaged in body snatching and other unsavory practices.[10] Already in 1854–55, the Register touted the "important additions made to the Anatomical and Pathological Museum," citing both the dedicated labor of the appointed faculty and the external donations made by practicing physicians as sources for this milestone in the growth of the medical facilities as a whole.[11] Just four years later, in 1859, the collections had expanded in scope and number to contain hundreds of wet human and animal specimens, skeletons, skulls, and teeth, as well as plaster and papier mâché anatomical models; materia medica (pharmaceuticals) imported from Paris; and "Instruments used in Diseases of Females."[12]

In the ensuing decades, the collections increased in each of these areas, and also grew to include wax models imported from France and Germany, frozen sections, and microscopic slides. Specialized collections proliferated within the larger "umbrella" of the "Anatomy and Pathological Museum," though many of these were eventually split into named collections in their own right, as with the pathology collections under Aldred Scott Warthin (discussed below). Even though, in the 1920s, Warthin himself began to express doubts about the educational value of pathological specimens, UMMS continued to espouse a public commitment to growing its core collections, including those founded by Ford and Abrams, well into the 1930s. Only in 1960 did the legacy collections completely disappear from the General Record.

The General Records of the first decades are particularly illuminating with regard to the changing functions and status of medical collections within UMMS and claims to professional legitimacy and status.[13] What is striking about the makeup and state of the installation is not merely the logic of its proximity to the dissection areas but also its integral place in building the medical school's reputation. A case in point is the 1867 General Record, which, after enumerating the key types of specimens held in the museum, proclaimed:

> All the above collections are now arranged in Galleries, for the purpose of rendering them attractive, as well as accessible, both to students and visitors. The University thus affords a secure deposit for objects of value and curiosity, and it is therefore, to be hoped, that the Museum will receive accessions by donation from various sources.[14]

With a dual purpose of improving accessibility and visual appeal, well-ordered installations attested to the university's ability to value and preserve anatomical specimens. Such an approach addressed more than the potential students who were presumably the key audience of the General Records since practicing physicians were practically the only ones capable of acquiring and preparing medical specimens of this kind. The strategy of naming individual donors and their donations in publications and on labels attached to the displayed specimens not only rendered the museum as a site for the preservation and transmission of personal legacies but it also situated it as the nexus of what today might be called a professional network, a network whose participants conferred mutual recognition upon each other.

Further, because the display was designed to also appeal to visitors, the vision of a medical museum realized on a scale surpassing thousands of specimens, and built as a networked depository, signified more than just the ability to make high-quality educational materials available to students. In particular, it can be read as a mark of distinction vis-à-vis competing medical schools, most of which, in the nineteenth century, were so-called proprietary medical schools. Run on a for-profit basis, these schools tended to confer degrees on anyone who could pay; at the same time, they generally did not invest in trappings such as adequate dissection rooms, laboratories, and, as in the Flexner report cited earlier, museums.[15] As Lynn Morgan has shown with regard to embryological collecting, which continued as late as the 1930s and 1940s, "there was a time when a gift of exotic specimens—including human fetuses—would have been considered quite prestigious."[16] In addition to supplying the potential for unique research, the capacity of the U-M Medical Museum to amass recognized prestigious donations thus enabled it to display a kind of "vote of confidence" from practitioners from across the country. Extensive collections illustrating rare diseases promised to offer students and visiting professionals what they could otherwise typically get only by traveling to Europe in these years: exposure to the material manifestation of diseases in verifiable form. In particularly developed collections, diseases could even be shown as they appeared in various ways over time using standardized techniques.

At the same time, such strategies of linking institutional resources to professionals with no official university affiliation also confirm that university-based medical education at the time did not have a near monopoly on the production of new medical knowledge. As historians of medicine tend to agree, the paradigm of science-based, experimental research did not begin to be established in US medicine until the end of the nineteenth century. Until that point, in the model that prevailed (the so-called Paris Clinical School), physicians were to practice above all a keen sense of observation of patients and disease. As Erin McLeary has argued, this paradigm was highly conducive for the initial formation of medical museums whose aim was, in her words, "to correlate disease symptoms with specific pathological lesions uncovered after death" and whose examples could be culled from the very patients that practicing physicians regularly encountered.[17]

THE VITALITY OF MEDICAL SPECIMENS IN THE ERA OF EXPERIMENTAL MEDICINE

Yet the advent of a newer, research-based paradigm did not consign medical museums to the dustbin, particularly at the University of Michigan. Dating roughly from the 1870s and 1880s, when large numbers of US medical students and physicians began to return from working in German and Austrian medical labs, the new model placed scientific methods at the center of educational and research programs.[18] Instead of lectures and rote memorization, students learned by doing hands-on work in the labs and eventually in the clinical wards. Rather than primarily lecture, medical faculty focused on guiding students to explore basic research questions that they themselves increasingly pursued in the form of original laboratory or clinical research. This was by no means all that would go into the modern medical school, but for many historians today's medicine finds some of its core contours in the emergence of laboratory-based medical research and teaching.[19]

Despite the fact that UMMS has been widely recognized as one of the first US medical schools to implement this kind of science-based curriculum, there has been little investigation into the way these changes affected, and actually drew upon, the medical collections in the ensuing age of scientific medicine. The years that saw revolutionary change at UMMS—from roughly the 1880s until well into the 1920s—brought not just a continued acceleration of collecting activity but an increasing development of collections organized along the lines of medical subdisciplines.

In the early 1890s, this process of reorganization was just getting underway. By 1893, the collection of materia medica had become its own museum with a particular location, as photographs of that year indicate (fig. 25), but the anatomy collections contained a fairly wide range of specimens grouped together in single cases (figs. 26 and 27). It was only at this time, shortly before his death in 1894, that Corydon La Ford dedicated a collection of crania and teeth (still at the univer-

Fig. 25. The Museum of Materia Medica in the Old Medical Building, c. 1893, albumen print, photo by J. Jefferson Gibson. Bentley Historical Library.

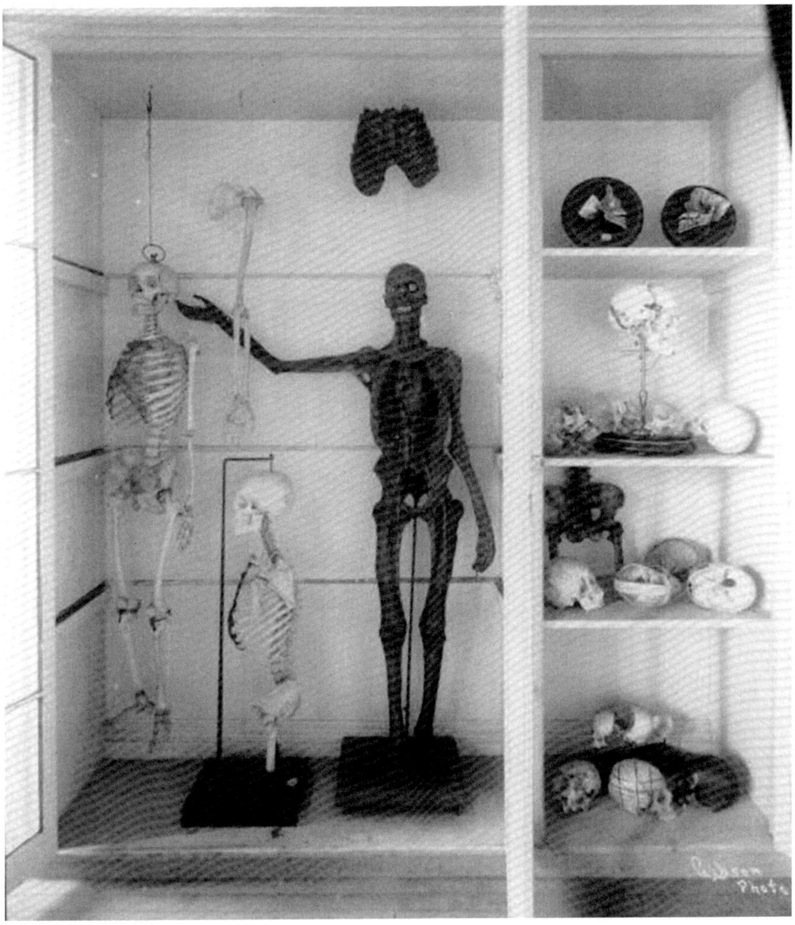

Fig. 26. A case in the osteology room, c. 1893, albumen print, photo by J. Jefferson Gibson. Bentley Historical Library.

Fig. 27. Anatomical models, c. 1893, albumen print, photo by J. Jefferson Gibson. Bentley Historical Library.

sity, albeit now in the Museum of Anthropological Archaeology) solely to the School of Dentistry.[20] By 1897, when a separate pathology museum (that eventually contained over 1,200 specimens) began to be advertised in the University Register, specimens related to abnormality and disease had been cleaved off from the once unified Museum of Anatomy and Pathology. And although embryologists George Streeter and G. Carl Huber eventually established a highly significant Embryology Research Collection (after 1900), in 1893 these specimens were spatially segregated but still contained within the Museum of Anatomy (fig. 28).[21] This collection, which was expanded by Bradley M. Patten until 1957 and by Alphonse R. Burdi from 1972 to 2004, illustrated that leading medical research paradigms could very well involve specialized and expansive collections as part of a laboratory-based research program.[22]

Specialized collections were so integral to laboratory settings from the 1890s to the 1930s that they often warranted little commentary, especially since medical collections still commanded scientific legitimacy. Far from becoming invisible in UMMS's public presentation, the university's legacy anatomy collections—then referred to as the

Fig. 28. Embryological models in the Museum of Anatomy, c. 1893, albumen print, photo by J. Jefferson Gibson. Bentley Historical Library.

Ford and Sager Collections—were a pillar of the facilities that the university proudly presented as cutting-edge research and teaching environments. Officially named "the Anatomical Laboratories, the Laboratory of Comparative Neurology, and the Laboratory of Human and Mammalian Embryology," there is little doubt as to the experimental underpinnings of these three units.[23] In the official descriptions, the labs are presented as extensive, covering multiple floors of the East Medical Building. In addition, they were outfitted with the necessary technical apparatus that enabled them to offer hundreds of medical students an experiment-based curriculum. As the U-M Bulletin entry for the general anatomy laboratories read:

> The work is facilitated by an osteological loan collection. . . . The several laboratories are [also] provided with projection facilities, x-ray exhibition cases, and sealed trays for the permanent exhibition of frozen sections. The teaching anatomical collection is stored in cases distributed throughout the dissecting rooms so that specimens can readily be used.[24]

Similar to the general anatomy labs, the histology, embryology, and neurology laboratories contained—alongside an active slide library and similar projection facilities—their own "extensive collection of embryologic models and models of the glands and other minute anatomic structures, made after the Born wax-plate reconstruction method, [which] are stored in cases conveniently placed in the laboratory."[25]

As these representative samplings show, the specimens were arranged so as to promote their ubiquitous and effective use. Moreover, the embryological collections remained valuable enough as to be listed alongside state-of-the-art research technology in the form of freezer exhibition equipment and X-ray apparatus, and for good reason. The invocation of Born models was a claim to scientific and pedagogical legitimacy, for as Nick Hopwood has shown, "it was plate modeling that in the decades around 1900 became the most important way to visualize complex structures in vertebrate embryos."[26] Particularly valuable for researchers at the time was not merely the models' ability to render minute structure visible but also their ability to be handled, worked on, even cut up—"as though they *were* bodies."[27] As Horace Davenport shows, the research of G. Carl Huber (1865–1934) and George Streeter (1873–1948) involved exactly this kind of work with wax plate embryological models at U-M.[28] The models' potential for experimental manipulation shows that museum specimens did not stand in simple opposition to laboratory-based research, particularly before World War II.

Collecting Pathologies: Aldred Scott Warthin

Another revealing instance of museums deeply informing medical research was the Pathology Collection at the University of Michigan, which rose and fell in line with the career and ambitions of Aldred Scott Warthin. A polymath researcher whose impact on medicine and the university went beyond his particular specialty, Warthin received a PhD in medicine from U-M before being appointed demonstrator in 1893, head of the pathology laboratory in 1895, and professor of pathology in 1903. Like many of his generation, Warthin's views of medicine, particularly medical research and teaching, were fundamentally shaped by postgraduate work he undertook in hospitals and laboratories in Vienna, Dresden, and Freiburg between 1893 and 1900.[29]

Today, Warthin is remembered as one of the first to discover the heritability of cancer in human beings (particularly certain forms of colorectal cancer), as the codeveloper of a stain to test for syphilis, and for pioneering work on the mononuclear phagocyte system (organs involved in the production of blood) alongside stints as vice president of the American College of Physicians and editor of the *Annals of Internal Medicine*. In his day, however, he was at least as well known for his public health work on the prevention of tuberculosis; his monograph on representations of death, based on his collections of rare medical texts and atlases (now part of the libraries' special collections); and his books, one on the effects of mustard gas, and *The Creed of a Biologist*, which advocates eugenic breeding. Nearly forgotten is his creation of a pathology collection that, over the course of its three decades of existence, contained over 1,200 specimens, and his leading role in founding the International Association of Medical Museums (IAMM), the predecessor organization for today's International Academy of Pathology.[30]

Warthin created the Pathology Collection almost immediately after he took over the Pathology Laboratory. Even before he became president of IAMM in 1910, Warthin used his experience at the university to plead for the broad necessity of expansive medical collections. As shown by a summary of a 1908 talk that he gave, advocating medical museums:

> The speaker felt that the simplest way in which to illustrate the necessities of the case [for medical museums and exchange among them] was to speak of his personal needs at the medical school at Ann Arbor, MI. This town, owing to its situation, far inland and somewhat isolated, furnishes a unique clinical material, rich in pernicious anemia, leuchaemia etc, but almost lacking, for instance, in typhoid. . . . He believed, too, that a uniform system of museum classification might be elaborated, which would help greatly towards that standardization of medical education which was the ideal of some of us, in which the practice followed in the German schools might be attained, the student passing in his "*Wanderjahre*" from one university to another during the completion of his medical course. In short, he felt that the organization of this Association filled an imperative want.[31]

For Warthin, medical collections were an indispensable resource for research and teaching. Museums' possibilities for preservation enabled both the study of those cases available locally and the procurement of relevant specimens from any number of sources otherwise made inaccessible in the expanse of American geography. Yet more than underpinning research, the museum appeared to be key for emulating the German system of scientific medicine so admired by the pioneers of modern US medical education: by learning museum methods, a student could obtain a kind of "lingua franca" of the body and disease that permitted seamless movement and exchange among institutions.

Yet Warthin was not only a thinker with respect to the museum's place in modern medical research and education. As his numerous contributions to IAMM and professional meetings indicate, Warthin was a highly skilled museum practitioner. He regularly presented findings on preservation techniques and museum equipment ("Demonstration of a special museum jar," 1908; "On methods of preservation of specimens in which natural colors have been retained," 1910, 1913; "On the molasses plate method used at the University of Michigan," 1913) as well as on topics such as the utility of museum specimens for teaching reactions in organic chemistry ("On the demonstration of micro-chemical reactions in museum specimens," 1913/15).[32] In professional settings, Warthin developed exhibitions culled from his university collections to present innovative research on tuberculosis (*Exhibit of the University of Michigan at the Pathological Museum at the Congress of Tuberculosis*, 1908) and, as late as 1929, techniques and results related to the staining of spirochetes (which represented a breakthrough in testing for syphilis and other forms of disease, and are used to this day).[33] Even though neither pathological specimens nor more than sparse documentation remain in the University of Michigan's archives, these examples testify to some of the ways that museum work was both the basis for the mode of inquiry and a means to demonstrate results.

In line with trends taking place in Germany and the United States, Warthin envisioned using university medical specimens for the concerted education of the general public. Responding to a debate on public medical museums initiated through IAMM, Warthin noted that:

> at Ann Arbor the Medical School has been planning to convert the old Museum into a State Museum of Hygiene that would be especially arranged for public use. He remarked on the interest that was shown by the public in the little anatomical museums seen at all German fairs.[34]

Although he criticized the fairground anatomical shows (panoptika) for arousing "morbid interests," he strongly believed that techniques could be found for presenting similar anatomical material in ways that would direct "this interest into useful channels" and that would attract cooperation from "the Associations for the Prevention of Tuberculosis."[35]

In 1913, Warthin used his presidential address at the XVII International Congress of Medicine in London to argue that university medical schools should regard public medical museums as their third-highest priority, following museums for medical teaching and museums for research.[36] Although I have located no record confirming that a public museum of hygiene was realized in Ann Arbor, Warthin's convictions eventually resulted in visitors coming to his offices to view material on occupational diseases that he hoped would form the basis of a state museum dedicated to industrial hygiene.[37] As he expressed in his presidential address, Warthin believed that improving the public's understanding of human anatomy was key to ameliorating many of the problems faced by modern society.[38]

For investigators such as Warthin, the idea that medical collections offered the key to shared production and exchange of knowledge among professionals—and between professionals and the lay public—was rooted in conceptions of the diseased body itself as a kind of pathology collection. This view emerged early for Warthin in connection with the insights he gained during his first round of postgraduate work in Vienna. In a letter addressed simply to "Dear Friend," he described the Viennese clinical environment:

> In this maze of courts and passages . . . can be seen every known disease. . . . It is simply wonderful. There is no other place where one can study disease as one can here, and it is studied in the only true way—from the bed to the post-mortem table. . . . Here one gets true objective knowledge. There can be no more theorizing or mild speculation. Disease becomes a sure thing, an entity which has a natural history. . . . And such a chance for study! These Viennese are veritable pathological museums; each body usually shows a dozen different conditions. . . . And it is due to the life of these people . . . consequently the bodies pay the penalty of an abnormal life.[39]

For Warthin, the thrilling success of Viennese methods depended on bodies that were not just heavily diseased but were so marked by—in Warthin's view—deplorable ways of life that cause and effect were preserved side by side and in their unfolding development. Relying as it did on deep-seated convictions about the relationship of order and visuality, the metaphor of the body as pathological museum practically entailed its study through the corresponding techniques developed to enable the preservation and display of medical specimens.

The neatness of this correspondence betrays its basis in a worldview and prevailing epistemology, which was subject to shifts in prevailing scientific questions and method. As important and persistent as museum techniques were for Warthin, he himself began to experience and articulate the limits of the ability of museum approaches to address ongoing challenges facing medical teaching and research. As correspondence from the 1920s shows, Warthin betrayed increasing skepticism toward medical museums for purposes that included the teaching of pathology. His experience had taught him that involving the specimens in teaching "greatly lowers their museum value, and the mortality of specimens so used is very high,"[40] as he wrote to McGill's dean of medicine, Charles F. Martin, in 1923. He was even more frank in his 1924 letter to Harvard's S. B. Wolbach, noting that the considerable expense of maintaining collections was all the more disappointing in that he had "seen no medical museum anywhere in the world in which, from the standpoint of teaching, more than 50% of the preparations had not become simply sentimental junk."[41] Museum staff and budgets, he wrote, would be better used for turning up interesting fresh pathological samples or lantern slides.[42] Slides and photographs, it is worth noting, had long been part of Warthin's approach to teaching, with articles appearing on the relative advantages of photographic enlargement for teaching as early as 1913.[43] As important as museum methods had been for Warthin's very vision of medicine, empirical outcomes had apparently caused him to temper his commitment to them. Temper, but not abandon. In his letters to Martin in the 1920s, Warthin stressed that he continued to believe "quite strongly in the central museum as a place of preservation for type specimens and for the historical preservation of the records of observation and research."[44]

That UMMS has maintained neither collections of the kind that Warthin continued to find most useful nor those of anatomy and embryology certainly has something to do with issues of cost and utility—problems that irritated even Warthin. Yet, as he noted in his missive to Wolbach, medical museums had been amassed at UMMS only in particular departments: anatomy and pathology.[45] In an environment in which "the University of Michigan makes no especial assignment of appropriation of separate budgets for the medical museums," they thrived only as long as individual faculty invested time and resources in their maintenance.[46]

Given the collections' ties to individual labs and personal careers and research programs, it is hardly surprising that their cycles of value and obsolescence were intimately related to faculty lifecycles. More surprising is that although faculty members such as Warthin—whose careers depended on the creation of medical collections—are remembered today, that they ever worked with collections has been nearly forgotten. The self-evidence with which specimens once figured as key components of experimental laboratory environments and teaching regimes has become a thing of the past, leaving behind only traces that defy full reconstruction. In an age primed to regard "laboratory" in certain narrow terms, medical museums have become for all practical purposes invisible—in spite of the vital role they played in the emergence of modern medicine at medical schools such as the University of Michigan's.

Notes

1. Kenneth M. Ludmerer, *Learning to Heal. The Development of American Medical Education,* Baltimore: Johns Hopkins University Press, 1985, 29–38; Thomas Neville Bonner, *Becoming a Physician. Medical Education in Britain, France, Germany, and the United States, 1750–1945,* New York: Oxford University Press, 1995, 253–55, 264–69; Howard Markel, "'An Example Worthy of Imitation': University of Michigan Medical School, 1850–2000," *JAMA* 287.7 (February 2000): 915; Sabine Hildebrandt, "Lessons to be Learned from the History of Anatomical Teaching in the United States: The Example of the University of Michigan," *Anatomical Sciences Education* 3.4 (July/August 2010): 203–4.

2. Medical School of the University of Michigan, 1850–2010, records; vol. 1, 1850–1875. Bentley Historical Library.

3. See Hildebrandt, "Lessons to be Learned," 203–4; Michael Sappol, *A Traffic in Dead Bodies. Anatomy and*

Embodied Social Identity in Nineteenth-Century America, Princeton: Princeton University Press, 2001, 4–6; Markel, "An Example," 916. Also Horace W. Davenport, *Not Just Any Medical School: The Science, Practice, and Teaching of Medicine at the University of Michigan, 1850–1941*, Ann Arbor: University of Michigan Press, 1999, 9–11.

4. Erin Hunter McLeary, "Science in a Bottle. The Medical Museum in North America, 1860–1940," PhD diss., University of Pennsylvania, 2001, 30–35.

5. Abraham Flexner, *Medical Education in the United States and Canada*, New York: Carnegie Foundation, 1910, 82. Also quoted in McLeary, "Science in a Bottle," 13.

6. *Proceedings of the Board of Regents of the University of Michigan 1837–1864*, Ann Arbor: University of Michigan, 1863, 590, 600, 1133. See also Elizabeth Martha Farrand, *History of the University of Michigan*, Ann Arbor: University of Michigan, 1885, 134. Hildebrandt, "Lessons to be Learned," 203–4. Later in his career, Ford began to articulate principles according to which medical researchers would understand their primary allegiances to their medical school, in exchange for which they would be paid. In the nineteenth century, these notions helped to break the influence of an older system that saw instructors paid directly by students in their classes, which tended to keep standards low. In the twentieth century, these notions exerted pressure on medical school professors who aimed to maintain private practices alongside their institutional duties. See Ludmerer, *Learning to Heal*, 206–13. Also Markel, "An Example," 916–17, 919.

7. *Proceedings of the Board of Regents of the University of Michigan 1864–1870*, Ann Arbor: University of Michigan, 1867, 206–7. I am grateful to Kerstin Barndt for drawing my attention to this document. Upon Lewitt's hiring in 1858, his salary was $250 per year, of which $50 was to be committed to building the museum. See *Proceedings of the Board of Regents of the University of Michigan 1858–1863*, Ann Arbor: University of Michigan, 1867, 36–37.

8. *Proceedings of the Board of Regents of the University of Michigan 1864–1870*, Ann Arbor: University of Michigan, 1867, 206.

9. According to Farrand, Abram and Ford were paid $1,000 and $3,215.83 for their respective medical collections. See Farrand, *History of the University of Michigan*, 204.

10. Donald F. Huelke discusses claims in the University Register that UMMS had an adequate supply of legal bodies in the context of strategies designed to quell local concern about body snatching, for instance the sourcing of bodies far from Ann Arbor. According to Erin McLeary, such claims in the University Register were also important in attracting prospective students, for whom competition was fierce. These arguments suggest that the University Register was also designed for communication with the public outside the immediate university community. Donald F. Huelke, "The History of the Department of Anatomy the University of Michigan, Part I, 1850–1894." *U-M Medical Bulletin* 27 (1961): 17–24. McLeary, "Science in a Bottle," 38–40.

11. *State University of Michigan. Catalogue of the Officers and Students of the University of Michigan: 1854–55*, Ann Arbor: E. B. Pond, 1855, 37–38.

12. *State University of Michigan. Catalogue of the Officers and Students of the University of Michigan: 1859*, Ann Arbor: University of Michigan, 1859, 56–57.

13. In using the General Records as historical documents, I want to read them in rhetorically critical fashion. This means being sensitive not merely to the possibility of overstatement, as might certainly happen in documents of self-presentation and advertisement, but also understanding that one of their core aims ultimately remains linked to the communication of the legitimacy and professional status of the faculty and its work. Seen this way, the General Records offer a view not only into the changing placement, installation, and organization of the respective collections over time but also their (functional, symbolic) ties to the prevailing paradigms in medical education and research. I seek to track not merely the reorganization, branching, and migration of specimens as collections are built, maintained, and later discarded at Michigan but also align the changing function of collections with emerging specialties and the changing nature of academic medicine at Michigan and the United States more broadly from 1850 to roughly 1950.

14. *University of Michigan. Catalogue of the Officers and Students*, Ann Arbor: University of Michigan, 1867, 88.

15. See Ludmerer, *Learning to Heal*, 11–15; Hildebrandt, "Lessons to be Learned," 203.

16. Lynn Marie Morgan, *Icons of Life. A Cultural History of Human Embryos*, Berkeley: University of California Press, 2009, 58. Morgan's direct example, the fetus collection in the biology department at Mt. Holyoke College, represented an outgrowth of larger collecting efforts initiated by Johns Hopkins Medical School in the early twentieth century. See Morgan, *Icons of Life*, 3.

17. McLeary, "Science in a Bottle," 13.

18. See Ludmerer, *Learning to Heal*, 29–38. Bonner, *Becoming a Physician*, 231–54.

19. As Ludmerer's seminal *Learning to Heal* argues, the emergence of the modern research university, the reorganization of primary and secondary schools, the growth of philanthropic funding, and the advent of clinical training in teaching hospitals are all additional crucial factors shaping modern medical research and education.

20. This collection, today known as the Ford-Mitchell Collection, would eventually, in 1928, be transferred to the University Museum and later to the Museum of Anthropological Archaeology. See *University of Michigan. Bulletin of General Information 1935/36*, Ann Arbor: University Registrar, 1935, 15–16. As the dental and anthropological collections are receiving their own treatments elsewhere, I do not discuss this collection here.

21. See the website of the National Museum of Health

and Medicine. http://www.medicalmuseum.mil/index.cfm?p=collections.hdac.collections.burdi

22. This collection could not be studied in depth as the objects and the related records were transferred to Washington, DC, in 2004.

23. *University of Michigan. Bulletin of General Information 1933/34 pt. 11–18*, Ann Arbor: University of Michigan, 1933, 42.

24. *University of Michigan. Bulletin of General Information 1933/34 pt. 1–10*, 57.

25. Ibid., 57–58.

26. Nick Hopwood, "'Giving Body' to Embryos. Modeling, Mechanism, and the Microtome in Late Nineteenth-Century Anatomy," *Isis* 90 (1999): 491. As Hopwood points out, other institutions such as the Carnegie Institute in Washington—led by one-time Michigan embryologist Franklin Mall—and the University of Göttingen also maintained extensive collections of these models as late as the 1960s. See Hopwood, "'Giving Body' to Embryos," 492.

27. Original emphasis. Hopwood, "'Giving Body' to Embryos," 491.

28. Davenport, *Not Just Any Medical School*, 96–97, 105–6.

29. See Henry T. Lynch, "Aldred Scott Warthin, M.D., Ph.D. (1866–1931)," *CA-A. Cancer Journal for Clinicians* 35.6 (1985): 345.

30. "Early Years," Website of the International Academy of Pathology. http://www.iapcentral.org/home/history/early-years. Accounts of the change in the organization's name and mission coincide chronologically with the decline of collections at the University of Michigan, with the 1950s serving as the time in which the word "museum" gained a significantly negative ring. "IAMM Transitions to IAP," website of the International Academy of Pathology. http://www.iapcentral.org/home/history/tranisitioning-to-the-iap. Accessed September 23, 2015.

31. Aldred Scott Warthin, "Summary from Discussion II," *Bulletin of the International Association of Medical Museums* 2 (1908): 11–12.

32. Aldred Scott Warthin, "Demonstration of a Special Museum Jar," *Bulletin of the International Association of Medical Museums* 3 (1909): 6, 18; Aldred Scott Warthin, "On Methods of Preservation of Specimens in Which Natural Colors Have Been Retained," *Bulletin of the International Association of Medical Museums* 3 (1911): 22–23; Aldred Scott Warthin, "On the Molasses Plate Method Used at the University of Michigan," *Bulletin of the International Association of Medical Museums* 4 (1913): 62; Aldred Scott Warthin, "On the Demonstration of Micro-Chemical Reactions in Museum Specimens," *Bulletin of the International Association of Medical Museums* 4 (1913): 56–58; Aldred Scott Warthin, "On Methods of Preservation of Specimens in Which Natural Colors Have Been Retained," *Bulletin of the International Association of Medical Museums* 4 (1913): 63–64.

33. Aldred Scott Warthin and A. C. Chronister, "A More Rapid and Improved Method of Demonstrating Spirochetes in Tissues (Warthin and Starry's Cover-glass Method)," *American Journal of Syphilis* 4 (1920): 97–103.

34. Aldred Scott Warthin, "Response to the Paper: 'On the Value of Public Medical Museums, and on the Possibility of Establishing Such a Museum in Each of the Larger Cities,'" *Bulletin of the International Association of Medical Museums* 3 (1911): 60.

35. Warthin, "Response," 60.

36. Aldred Scott Warthin, "The Ideals and Functions of the International Association of Medical Museums. Presidential Address. XVII International Congress of Medicine, London, August 1913," *Bulletin of the International Association of Medical Museums* 5 (1915): 19–22. In Warthin's eyes, state-funded universities had special obligations and opportunities in meeting the challenge of lay education.

37. Aldred Scott Warthin, "Symposium on the Value of Exhibits Illustrating Occupational Diseases," *Bulletin of the International Association of Medical Museums* 4 (August 1913): 75–77.

38. Warthin, "Ideals," 21–22. Warthin cited the example of the exhibit on the anatomy of the human body at the International Hygiene Exhibition in Dresden of 1911 in support of his thinking, writing, "No wonder that crowds [in Dresden] constantly awaited their turn to see this wonderful exhibit. I believe that few things would do more for the race than such an exhibit placed in every large city."

39. Aldred Scott Warthin, Vienna, to "Dear Friend," August 2, 1893, personal correspondence, 1893–1894. A. S. Warthin papers, Bentley Historical Library.

40. Aldred Scott Warthin, Ann Arbor, to Dr. Charles F. Martin, September 7, 1923. Box 1, Medical Museums. Warthin papers, Bentley Historical Library.

41. Aldred Scott Warthin, Ann Arbor, to Dr. S. B. Wolbach, March 10, 1924. Warthin papers, Bentley Historical Library.

42. Ibid.

43. Aldred Scott Warthin, "On the Teaching Value of Enlargements of Microphotographs," *Bulletin of the International Association of Medical Museums* 3 (1911): 61.

44. Warthin to Martin, September 7, 1923, Warthin papers, Bentley Historical Library.

45. Warthin to Wolbach, March 10, 1924, Warthin papers, Bentley Historical Library.

46. Ibid.

ART AND UNIVERSITY EDUCATION IN TRANSFORMATION

Disciplinary Histories and the University of Michigan Museum of Art

David Choberka

Since Professor Henry S. Frieze first acquired a collection of art objects in 1855, the history of the art collection of the University of Michigan has intertwined with the university's disciplinary history. Initially assembled by Frieze as a teaching collection for the university's courses in classical languages and civilizations—at that time the core of the liberal arts curriculum—the art collection was reorganized and subdivided over the following century and a half in ways that reflect both the decentering of that classical education from the core of the curriculum and the ongoing history of diversification, specialization, and professionalization in the academic disciplines. From its original role as a material repository of the aesthetic and cultural traditions that lay at the center of the classically oriented liberal arts education, the art collection drifted through the late nineteenth century as its original pedagogical purpose gave way to extensive disciplinary transformations. In the twentieth century, the collection and the art museums found new direction, at first in tandem with the formation of the various disciplines that conventionally study works of art and art historical traditions, only to be transformed again in the era of museum education and academic interdisciplinarity into a multifaceted repository of visual culture with broad relevance to the great variety of disciplines that engage with visual culture today (fig. 29).

1855–61: A CLASSICAL TEACHING COLLECTION

Although, as discussed elsewhere in this volume, the University of Michigan was founded in Detroit in 1817 and relocated to its current campus in

Fig. 29. The audience enjoys "A Sense of Steel: Dances for di Suvero," an original U-M Department of Dance performance, in front of the expanded University of Michigan Museum of Art in 2011. University of Michigan Museum of Art.

Ann Arbor in 1837, in many respects the 1850s were its formative years. The Constitution of Michigan of 1850 redefined the office of regent as an elected position and established that the regents would select a president to lead the university. In 1852, Henry P. Tappan became the university's first president—and introduced a bold new vision that set the institution on the path that defined the next decades. Three years later, the regents granted Henry Frieze, a professor of Latin, leave to travel to Europe for a year, provided that he use a portion of his salary to hire a substitute instructor for his classes and then use the remaining portion to purchase "specimens of ancient art illustrative of the classics, thus laying the foundation for a classical museum in the University."[1] Frieze returned the following year with a collection of engravings, photographs, and, most spectacularly, plaster and terracotta copies of important works from museums in Naples, the Vatican, and the Louvre.[2]

These acquisitions were part of a flurry of collecting that began at the university in the mid-1850s. In addition to starting the collection that over the next nearly one hundred years would be called the Fine Arts Collection, the Art and Antiquities Collection, and the Art and History Collection, the university was establishing collections of anatomical, chemical, geological, and natural history specimens. A Cabinet of Natural History and library were already growing in the North College Building (now Mason Hall), and in late 1856, when the regents welcomed the arrival of Frieze's purchases, they also appropriated funds to furnish a Gallery of Fine Arts for the new collection in the South College Building, which stood where Haven Hall stands today.[3] In his annual report to the regents at the end of 1857, President Tappan explained that the Gallery of Fine Arts "is daily open to visitors and attracts much attention" and that "Professors of Ancient Languages, History, and Antiquities find even in this very limited collection, very valuable aids in illustrating the subjects of instruction assigned to them." He remarked also on "the hope that it will prove the beginning of a movement destined to enrich the University with many treasures of Art."[4]

When Tappan had presented his vision for the university's mission, structure, and curriculum in his first annual report to the Board of Regents in 1853, he included the development of all manner of teaching collections as an indispensable part of the plan (see Blouin, this volume). In his thorough and lengthy proposal to model the university on what was then known as the "Prussian model," he contrasted a staid, parochial, and dogmatic system of education that predominated in Europe with a reformation-influenced Prussian system that both educated enlightened and free individuals and provided them with practical knowledge for the world's emerging professions, sciences, and trades.[5] Tappan's interest in the practical and useful side of education matched perfectly with, and was likely a related phenomenon to, the late nineteenth-century vogue in learning from museum objects. Governed by what museum historian Steven Conn calls an "object-based epistemology," collections and museums were established across the country to provide the public with tangible and visual access to the objects of study of the modern world's rapidly specializing academic disciplines and the vast systems of classification by which that material was being rationally ordered.[6] Tappan made the connection himself as he called for the university to expand libraries, laboratories, and collections of all kinds "so that every form of knowledge may be reached and illustrated."[7]

As the teaching collection for the study of classical languages, cultures, and history, the objects that Frieze acquired in 1855 can be said to have had a special status in Tappan's vision of also educating free individuals. Tappan described it as especially important for a university in the United States—with its fledgling democracy, diverse immigrant population, and unprecedented economic successes—to promote ideals and values beyond the merely utilitarian, commercial ones that had driven the nation's early prosperity. The university, he proclaimed, should attend to "the spiritual cultivation of the human being" so that, on the one hand, the nation's growth as an economic power might be guided by higher principles, knowledge, and taste and, on the other, that it might itself become also a great producer of culture for the world. Foreshadowing the formation of the art collection and its importance to the university's core academic mission, Tappan presented ancient Rome as an example of a civilization whose economic and political power had faded with time but whose spiritual achievements remained for all to behold in the form of arts and culture.[8]

In 1858, the university's growing library and the Art and Antiquities Collection moved to the North Building (or Mason Hall) where they were united with the Cabinet of Natural History, creating a comprehensive university museum that displayed the objects of study for many of its disciplines (see essays by Barndt and Sinopoli in this section).[9] Concurrent with the installation, Professor Frieze produced the first catalog of the Art and Antiquities Collection, none of which remains in the present-day collection of the University of Michigan Museum of Art.[10]

The university's collecting activities and establishment of a public museum placed it (along with a handful of other leaders in higher education such as Harvard and Yale) at the forefront of a movement that would transform the US cultural landscape in the last four decades of the century. Indeed, the university's nascent museum preceded America's golden age of museums, which began around 1870. Historian Victor J. Danilov lists the University of Michigan, along with Dartmouth and Bowdoin College, as the first institutions in the nation to establish art collections.[11]

With their goal of providing students and the public with access to the great works of the classical and European traditions, these early university collections featured copies. However, even as art collections and prestigious museums began to populate the US cultural landscape, the definition of what constituted a work of art worthy of collection and display was changing. Teaching collections of replicas became increasingly obsolete as art museums pursued original works of fine art.

The sculpture *Nydia, the Blind Flower Girl of Pompeii* by the Ann Arbor-raised artist Randolph Rogers, acquired in 1861 and the longest standing object in the university's art collection, is a transitional figure between the two modes of art collecting. Its neoclassical marble rendering fit perfectly with the copies that predominated in the collection, but it was an expensive original work of art. Its acquisition was made possible by the first partnership between the art collection and the citizenry of Ann Arbor when, in the fall of 1859, members of the Rogers Art Association raised the funds to acquire *Nydia* from one of the era's most celebrated American artists.[12] Professor of Greek language and literature J. R. Boise remarked at the time that the acquisition of *Nydia* marked a significant point of departure in the university's foray into collecting fine art. Though it was popular with the public and a source of increasing notoriety for the university's art gallery, *Nydia* had an ambivalent connection to the collection as a teaching object. While relevant to a study of the classical aesthetic and perhaps to the history of Pompeii, the source material that inspired the piece was a work of contemporary fiction. This introduction of an object of modern English literature into a classical teaching collection foreshadowed significant changes to come.

1861–83: DISCIPLINARY TRANSFORMATION

Even as Frieze was assembling the first art collection, the disciplinary structure of the university was becoming increasingly differentiated, specialized, and professionalized, and the classical curriculum that Frieze's collection was selected to serve was losing its primacy as the core of a liberal arts education. One of the most important innovations of the university system instituted under the Tappan administration in 1852 was the establishment of a scientific course of study in what was then known as the Department of Literature, Science, and the Arts. The curriculum was divided into just two tracks: the classical and the scientific, each with several specialized courses of study. From its inception, the establishment of a scientific course of study within a liberal arts curriculum was considered a major innovation and point of distinction from the curricula at prestigious peer institutions such as Harvard. A marked difference between the two courses of study was that students in the scientific course were not required to study ancient languages as part of their general education.[13] With the study of Latin and Greek still the centerpiece of general education for the classical course and for the diversifying humanities that proliferated within it, these disciplines retained their guiding position within the liberal arts curriculum. However, the establishment of the scientific curriculum in 1852 signified the beginning of a fundamental shift in the trivium—the interrelated disciplines of rhetoric, grammar, and logic that formed the basis of a traditional liberal arts education—away from its foundation of Greek and Latin.

Through the latter half of the century, many new departments and programs were established. New disciplines began to take shape in the increasingly anachronistically titled classical course of study, as history, literature, and politics, became their own professionalizing humanities or social sciences. From a single course of study in 1861, the students of 1883 could choose from nineteen specialized course instructions, including—from the increasingly differentiating humanities and newly forming social science disciplines—the Classics, Modern Languages and Literature, English and Rhetoric, History, Philosophy, the Science and the Art of Teaching, and Political Economy.[14] The most important developments in the decentering of classical studies were the formation of the study of modern English as language and literature and its establishment as the basis for the general education curriculum in rhetoric and critical thinking. From the moment that Frieze first assembled the art collection, Latin and Greek were beginning slowly to creep from the center of the trivium to the constellation of specialties arrayed around the newly ascendant core of a trivium based on modern English.[15]

The diversification and professionalization of the humanities was mirrored through the same decades by the proliferation of programs in the scientific course and then by the establishment of professional schools.[16] The formation of these specializing and professionalizing disciplines was a local expression of national trends that can also be seen in the number of academic and professional organizations that were founded in these decades.[17]

Concurrent with this proliferation of academic disciplines at the university (and throughout the cultural and institutional landscape of the United States) was that of collections. In the university's North Building, geological, mineralogical, natural history, anatomical, and anthropological collections quickly filled the limited display space. Proceedings of the Board of Regents and annual reports by the university president throughout the latter 1850s, 1860s, and 1870s indicate that building these collections was a high priority and that space constraints as well as the lack of state fiscal appropriations for acquisitions were ongoing concerns.[18]

These physical and fiscal constraints constituted the most serious limitations to the growth of the art collection in the 1860s and 1870s, but the nature of art collecting in the United States was also changing and with it the sorts of status that the university could achieve with its own collection. From the beginning, the university's presidents and the regents discussed the prestige that the growing collections bestowed on the institution and expressed interest in competing with other institutions to maintain this status.[19] Most pointedly, at the September 1865 meeting of the Board of Regents, President Haven proclaimed that the university had the largest museum collections in the nation and outlined the steps necessary to maintain that lead: improve museum spaces, acquire funding for acquisitions, and diversify and more clearly define the purviews of the curatorial positions that oversee the collections.[20] Part of the period's trend toward professionalization, this rationalization of the curating of the university's collections led the following year to Frieze's official appointment as "Curator of so much of the Museum as relates to the fine arts," a formalization of the position that he had already filled since he first formed the collection.[21]

The designation of "fine arts" rather than something more akin to "antiquities" or "history" points to the changing nature of art collecting in this period, which witnessed massive growth in public institutions across the nation propelled by a wave of philanthropic capital generated by the industrializing north and the westward expansion of the railroads. In what scholars have variously identified as an era of philanthropy or an era of professionalization stretching through roughly the four decades that followed the US Civil War, these two interconnected trends shaped and propelled the development of the nation's entire public institutional landscape. Many of the landmark art collecting institutions east of the Mississippi were founded in these years with philanthropic backing: the Metropolitan Museum of Art (1870), Philadelphia Museum of Art (1876), Art Institute of Chicago (1882), and Detroit Mueum of Arts (1883). These institutions' collections of fine original works of art marked a change in the role of art collections as vehicles of public education and prestige.[22]

Frieze himself observed that broader sentiments about the purposes of an art collection had left his teaching collection behind. He lamented in his "Introductory Notice and List of Contributors" to the second edition of his catalog of the collection (1876) that a general disinterest in collecting copies of art for teaching purposes prevailed among the "friends of art" and that both donations and monetary support for acquisitions had been scant in the twenty years since the collection was first formed.[23] The 1876 catalog's half-page summary of the new additions to the collection since 1855 lists a donation of plaster replicas of the 1700-piece Horace White collection of gems and portrait medallions, which had been acquired in 1857, and the Rogers work *Nydia,* followed by only the occasional random gift of photographs and engravings. Three exceptions are the more substantial gifts from alumni: a full-size replica of *Laocoön*, given by the class of 1859, being the most noteworthy; followed by a copy of Murillo's *Madonna of the Immaculate Conception* from the class of 1866, which remains as the second longest standing piece in the collection; and a replica of Houdon's statue of Benjamin Franklin from the class of 1870, which was displayed outdoors. The use of the art collection as a venue for alumni to contribute to the university and to memorialize their time as students points to the transformation of the collection from a teaching tool to a vehicle for other interests.

The same trend can be seen among the university's leadership. In 1859, the regents invited the president and full professors to place their portraits in the fine arts gallery.[24] Two years later, they expanded the celebratory and memorializing scope of the gallery with invitations to important figures in state government to install their portraits.[25] While these portraits were not included in any of the catalogs that Frieze or any of his successors produced, they indicate the introduction of new interests into the university's collecting and displaying of art that diverged from the pedagogical and curatorial purposes that had initially guided the collection.

1883–1910: PROFESSIONALIZATION AND PHILANTHROPY IN THE GILDED AGE

The years 1883 and 1884 brought some significant changes. In 1883, a new library building was built, the third floor of which included galleries for the sculpture and the paintings in the art collection. In 1884, the first major bequest of art was made to the university by Henry C. Lewis of Coldwater, Michigan, though it would not be physically transferred until 1895. The early 1880s also saw a further step in the differentiation, professionalization, and bureaucratization of the university's museums. In 1882, the regents more clearly defined the authority of the president over the relationships between the museums themselves and between the museums and the university. At the same meeting they further defined the curatorial positions overseeing the collections, including a curator of the Museum of Fine Art and History, a position that Frieze would occupy until his death in 1889, followed by professor of

Latin Martin Luther D'ooge.[26] While as late as 1879 Frieze's express curatorial and pedagogical intentions focused on building the teaching collection of inexpensive copies, to the extent that the collection had a guiding ideology through the last two decades of the nineteenth century, it was the combined interest in prestige and representing the nested histories of the university, state, nation, and what might be called the West that prevailed. Intentions to serve university and public education remained, but pedagogically and curatorially the collection lacked the guidance that the new discipline of art history would begin to provide after 1910.

Philanthropy, a sincere amateur interest in public education through art, and pure serendipity seem to have been the driving forces that brought the Henry C. Lewis bequest to the university. Of the over four hundred paintings and seventy sculptures, which included numerous copies of important European pieces, ninety-five works remain in the collection.[27] Lewis was a wealthy banker who had never received any formal education and whose interest in collecting art was sparked by his travels in Europe. His establishment of a public art gallery in Coldwater in 1869, the contents of which he subsequently bequeathed to the university, is a singular and relatively early example of the intersection between new wealth and philanthropy in the Gilded Age.[28]

The arrival of the Lewis collection in 1895 filled the space on the third floor of the library building. In the sculpture gallery, the additions of the Lewis collection joined an also significantly sized collection of works that had been donated by Randolph Rogers in 1885. Comprising the contents of his studio in Rome, the Rogers donation contained mostly maquettes and full-size plaster casts for many of his notable public artworks, several of which remain in the collection.[29] A combination of copies of classical works and neoclassical renderings of great figures of US and Michigan history, Civil War and Emancipation imagery, and Native Americans, the display that came together in the sculpture gallery in the Old Library in the 1890s appears eclectic. But it was a materialization of the intersection between the collection's original classical educational intentions and the dual and tightly interrelated cultural purposes of the period: (1) educating the democratic public and (2) presenting to that public a history of the nation, the states, and its institutions that places them at the forefront of Western cultural development.

Figure 30 shows us that sculpture gallery. It is centered on *Nydia*, which is ringed by an array of great classical and neoclassical works. The plaster model for a monumental portrait of General Lewis Cass by American sculptor Daniel French looms over *Nydia*'s shoulder. Cass was a territorial governor in Michigan and a leading figure in the founding of the state. Behind Cass, we can see the full-size *Laocoön* that had been donated by the class of 1861. Another photo shows the painting gallery where the massive collection of paintings from Lewis was housed (fig. 31). The walls are filled with paintings to the ceiling and metal railings house even more paintings that are leaning in rows on the floor. Among them stands a group of life-sized Civil War soldiers.

While visual culture had been important for dressing the story of America in the trappings of classical civilization and high European culture since at least the eighteenth century, the visual strategy took on renewed significance in the aftermath of the Civil War and the neoclassical bent of the age of philanthropy. No longer representing only the material inheritance of the classical world, the collection had changed into what in retrospect appears to be an eclectic visual representation of the various threads that composed the story that the university and similar institutions tended to tell about themselves in this era—inheritors of the civilizing spirit of the classical world, establishers of a new Athens in the wilderness of the new world, central institutions of the new order being built out of the conflict of the Civil War. A narrative of the nation's manifest destiny in the time of westward expansion and the removal of the Native American population appears in the elegiac representations of Native

Fig. 30. The sculpture gallery on the third floor of the old Library Building, 1898, photo print. Bentley Historical Library

Fig. 31. Plaster models from the studio of Randolph Rogers share space with paintings from the Lewis bequest and the university's growing collection of portraits in the Old Library Building, 1898, photo print. Bentley Historical Library.

Americans, while the presence of the memorial imagery of the Civil War hallowed the virtues of the victorious and ascendant United States. It was not just in the art collection that such preoccupations made their appearance. The proceedings of the Board of Regents through the 1860s reveals a general interest in collecting artifacts of Native American cultures and of the US Civil War, including the trappings of southern slavery.[30]

By the beginning of the twentieth century, the art collection had departed significantly from the pedagogical intention for which it had first been formed, and the university found itself with a mixture of a historical installation, a classical teaching collection, and the beginnings of a fine art collection. The Lewis bequest, especially the paintings, established the university as the caretaker of a significant collection of European and American art and put the university ahead of rival collecting institutions Harvard and Princeton in moving into the new world of collecting fine original works of art.[31] European and American art of primarily the sixteenth through nineteenth centuries constituted two of the primary areas of growth for the collection through the first half of the twentieth century. Though the first pieces were not acquired as part of the university's art collection until 1930, interest in the other early major growth area, Asian art, was first shown at the university in 1885, when materials that had been in China's exhibit at the New Orleans Exposition, 1884–85, were displayed in the university's galleries.[32]

Public interest in the art collection was significant. In the September 1908 proceedings of the Board of Regents, the education committee reported that the gallery had been visited by 13,341 visitors the previous year.[33] That same year the cornerstone was laid for Alumni Memorial Hall, a building that would provide the university with devoted space for a museum of art. The growing importance of the collection to the local community was shown in 1909 when the Ann Arbor Art Association was formed, a partnership between faculty and citizens that supported acquisitions and exhibitions until the late 1960s.[34]

1910–46: ART HISTORY AND MUSEUM INSTITUTIONALIZATION

In 1910, Alumni Memorial Hall was completed, and what was then starting to be called the university's Art and Archaeology Collection was transferred from the Old Library to larger, dedicated galleries in the new building. The building itself, which also housed the Alumni Association, was a monument to the Gilded Age's cultural and institutional trends. A testament to the growing prestige of the university, the beaux-arts construction was built with the support of fifteen hundred alumni and the major philanthropic backing of four men successful in business in Michigan: Ezra Rust, Dexter M. Ferry, Simon T. Murphy, and Arthur Hill (only the first of whom was an alumnus).[35] The building was dedicated as a memorial to alumni who had served in the Civil War and the Spanish-American War; a memorial plaque still hangs in the building. Signaling the university's intent to build a fine art collection, an acquisitions committee was constituted, consisting of the university president, a regent, the curator of the art gallery, and a member of the Alumni Memorial Hall Building committee.[36]

When the Department of Fine Arts was established the following year (renamed History of Art in 1959), the position of curator of the art gallery was transferred from the faculty in Greek and Latin to the new program, which was also furnished with offices and classrooms on the main floor of Alumni Memorial Hall.[37] The transfer brought new art historical curatorial expertise that would guide the collection's growth through much of the twentieth century. Building on the strengths of the Lewis bequest, significant acquisitions in American and European art were made over the next three and a half decades. The collection also became increasingly diversified during this period, with major acquisitions of Asian art and the beginnings of a collection of African art, the museum's modest collection of Latin American pieces, and the first forays into modernism.

The years 1910 to 1946 were important ones for the rationalization of the university's collections and museums, and for the university as a whole. In 1928, the Ruthven Museums Building was completed, a large structure that housed separate Museums of Anthropology, Paleontology, and Zoology and a Herbarium, along with a Department of Visual Education (that later evolved into the Exhibit Museum, now Museum of Natural History; see Sinopoli, this section). Most important for the Art and Archaeology Collection, in 1928 the Museum of Classical Archaeology was established in exhibition space in Newberry Hall. The following year, the archaeological objects (essentially objects from the ancient and classical Mediterranean basin) were removed from Alumni Memorial Hall—a division in the curatorial purviews of the two institutions that continues to this day. The two museums were temporarily reunited between 1941 and 1946, and then formally established as two separate institutions: the University of Michigan Museum of Art and the Museum of Archaeology (renamed the Kelsey Museum of Archaeology in 1953).[38]

The rise of art history as a self-standing discipline guided the growth of the art collection through this era of institutionalization. Certainly the study of art history was not a new development. Faculty in various area studies and in history often included some engagement with art in their class curricula. In his *Encyclopedic History of the University of Michigan*, Wilfred B. Shaw cites a 1912 survey of art in American colleges that showed that eighty-three courses in the history of art were being taught in other departments, all of which we could characterize as language/area studies or history.[39] Even as far back as the mid-1850s, the university had Alvah Bradish, professor of the theory and practice of art, on its faculty for a few years. Nonetheless, the formation of art history as a discipline within the new Department of Fine Arts constituted an important step in the formalization and professionalization of the university's art museum.

Herbert Richard Cross curated the collection until 1922, followed by Bruce M. Donaldson. According to Shaw, Cross emphasized Renaissance and later European art and advocated specifically for the classical division of the collection to be left to the classical studies departments.[40] Such an attitude may have helped pave the way for the establishment of the Kelsey Museum of Archaeology. Cross's tenure as curator also helped to focus the art museum's early collecting goals, though acquisitions would remain sparse until the 1930s. This would change under the curatorship of Donaldson. He had a background in the study of decorative and applied arts and maintained Cross's attention to European painting while adding contemporary American art to the museum's collecting interests. From 1930 on, a rapid increase in acquisitions through museum purchases and major gifts expanded the already significant and sizeable collection of European and American art up to the nineteenth century and launched the beginnings of what would rapidly become one of the finest collections of Asian art at a university museum. Those modest beginnings turned into a bonanza when Margaret Watson Parker bequeathed her exquisite collection of 679 Asian and Western works of art to the university in 1936 (see McNamara, this volume).[41]

Acquisitions through these years also point to the diversification of the collection beyond these primary areas, a feature that becomes increasingly pronounced at the end of the twentieth century. In the 1930s, the federal government began giving the museum an extensive collection of New Deal art from the Federal Arts Project and Works Project Administration. In recent years, these have proven a tremendous resource for classes in history and American culture, and interdisciplinary and comparative classes about such topics as the representations of women and work. The 1930s also saw the modest beginnings of collections of art from Africa and experimental modernist works from Europe and Mexico.

Moving into the middle of the twentieth century, the university's art collection had been transformed by ongoing professionalization and institutionalization; most importantly, with the curatorial guidance from the discipline of art history, the museum had embarked on the path toward creating an excellent university art museum.

1946–70: THE UNIVERSITY OF MICHIGAN MUSEUM OF ART

The institutionalization of the art collection was completed in 1946 when the collection housed in Alumni Memorial Hall was officially named the University of Michigan Museum of Art (UMMA) and put under the authority of a professional museum director. The first two directors, Jean Paul Slusser (1946–56) and Charles H. Sawyer (1957–72), infused the collecting and curatorial sensibilities of the art history faculty, who still served as

Fig. 32. Between 1957 and 1974, the apse of Alumni Memorial Hall was outfitted with a Unistrut system that provided significantly more display space. The functional modernism of the Unistrut system was at odds with the Gilded Age Beaux-Artes architecture of the building and symbolized the forward-looking intention of the Director Sawyer, photo print. Bentley Historical Library.

the museum's curators, with a distinctly modernist sensibility and a focus on professionalizing museum practice and increasing the orientation of the museum toward university education. Increasing access to spaces in Alumni Memorial Hall allowed these directors to embark on a more ambitious program of exhibitions and to increase the area devoted to displaying the permanent collection. In 1946, the Alumni Association moved from the building, and in 1949 the Department of Fine Arts relocated to Tappan Hall.[42]

Jean Paul Slusser was a professor of drawing and painting in the College of Architecture and Design and had been an active member of the Ann Arbor Association before becoming director of UMMA in 1946.[43] Slusser brought to the museum his interest in the modernist movements that were still reshaping European and American art at that time and a commitment to UMMA's role as a teaching resource for the university. On the eve of his becoming director he observed: "A University museum of art, is primarily a device for the furtherance of teaching. Its function is to conserve and display the various art works belonging to the institution, and to supplement them with showings of other material pertinent to the work pursued in the courses dealing with the theory, practice and history of the visual arts."[44] Slusser envisioned the pedagogical uses of the collection as restricted primarily to art history and not as a central teaching resource for liberal arts education as in the early Frieze era, but his intent points to an abiding theme in UMMA's development as a university museum since its founding in 1946 that continues to grow and expand.

In 1947, the regents approved a small annual appropriation for acquisitions that Slusser put to great use in building an excellent collection of European and American modernist works on paper, including a surprisingly good collection of German Expressionist works, graphic arts of the Parisian belle époque, incredible works by Goya, and a fine selection of Japanese woodblock prints.[45] Gifts from Slusser himself supplemented the collection's growth in these areas. UMMA's Asian art collection grew significantly through Slusser's tenure as well, as the Parker bequest arrived between 1948 and 1955 and the museum received its first gifts of Asian art from the Plumer collection, portions of which also went to the Museum of Anthropology. In 1953, UMMA established its first dedicated gallery of Asian art to house its growing collection, which today stands as one of the finest university collections in the nation.[46] The growth of UMMA's collection of Asian art during these years coincided with the growth and disciplinary division of the academic programs at the university studying that part of the world, part of a larger trend of increased attention to Asia in the early years of the Cold War.[47]

Charles H. Sawyer's directorship began in 1957 with a renewed mandate from the regents to build the collections and to relate them more broadly to university learning, especially the growing area studies programs (fig. 32). Sawyer, who was both an art historian and an experienced museum professional, attended to building the collection of Western art to parallel the impressive Asian collection, focusing especially on building a collection of European works on paper highly instructive for the classical art history curriculum. His tenure also saw significant growth in the museum's collection of Indian and Persian art.[48]

The importance of Sawyer's tenure to the educational mission of the museum extended significantly beyond these areas of collection growth. In 1959, he began forming a Museum Practice Program,[49] itself a new academic discipline focused on training future museum professionals to understand and improve upon the work that museums do. Under Sawyer, the program focused largely on art museums and curatorial practice. Generally called museum studies today, such programs have played a central role in the professionalization of museum staff over the last half-century and contributed significantly to the growth of museum education as an increasingly important focus of museums.[50]

1970–2009: DIVERSITY, INTERDISCIPLINARITY, AND A CHANGING EDUCATIONAL MISSION

In the last decades of the twentieth century, UMMA continued to build on its strengths in European, American, and Asian art and made significant strides toward establishing extensive new collections in photography and African art. In the same years, the professionalization of museum work and the influence of the discipline of museum studies took shape in the form of a professionalizing staff and reorientation of museum practice toward university and public education.

By the 1970s, annual acquisitions by UMMA significantly enhanced existing areas and added important new collections in photography and African art, a trend that continues into the year 2017. From the early 1970s, UMMA built its photography collection, beginning with nineteenth-century works, and over time building a collection of approximately three thousand photographs, including many contemporary American works. The latter are frequently accessed by university classes exploring American culture, society, and all manner of representations of ethnicity, gender, age, class, and place.

The acquisition of African art had been an abiding interest of the museum for several decades and, like photography, was increasingly regarded as a priority from the 1970s on. The interest in collecting African art coincided with the growing attention to it within art history and broader interests in academia and was part of the broader context of the 1970s, when social movements, such as the Black Power movement, and popular culture, such as the television miniseries *Roots*, aroused increased attention to African cultural heritage. The Department of Afroamerican and African Studies, for instance, was established at the university in 1970. The importance that the institution and its benefactors placed on growing this area of collection was clear when the Friends of UMMA acquired the LaMore Collection of African and Oceanic Art for the museum in 1972. Evan Maurer, who became director in 1981, was himself a scholar of African art, and his relationship with collector Helmut Stern was pivotal in building the excellent collection that UMMA continues to grow today.[51] Other noteworthy areas of growth in the collection through these years were the establishment in the 1990s and 2000s of a truly world-class collection of Korean art, acquired with the support of Sang-Yong Nam. This is yet another example of the way collection growth mirrored disciplinary formation at the university. In 1990, the first Korean language courses were offered, and in 1995 a Korean studies program was established. In the 2000s, the Nam Center for Korean Studies was founded with the same philanthropic backing that helped build UMMA's collection.[52]

The 1970s through 2000s also brought a huge growth in areas of the collection that have proven most useful to the interdisciplinary, comparative, and intersectional academic programs that developed through the same years, namely contemporary art that engages with the cultural and social landscape that students grow up in today.[53] The humanities in general were transformed through these years by increasing attention to a much wider array of media and forms of expression, including a "visual turn," and by attention to the social and cultural construction of identity that gender and postcolonial studies fostered. By the early 2000s, the faculty across the humanities and beyond were trained and prepared to engage with the art collection as a core teaching resource, perhaps more so than they had been even when Frieze's collection stood momentarily at the core of the trivium.

The capacity of the university's art museum to meet that interest relied to a great extent on the professionalization of its staff and the ascendancy of museum education as a guiding discipline. Under Director Evan Maurer (1981–88), UMMA ended its reliance on history of art faculty to provide the museum with its curators and instead established full-time curatorships over the collection as professional positions within the museum (an unusual model for U-M museums, where faculty-curators have typically split positions). Interestingly, this change reflected and still reflects the tendency within the discipline of art history to regard museum work, because of its engagement with the public, as a different kind of labor, and perhaps lesser kind, than the pursuit of pure academic research. While the change initially insulated the museum from the discipline that had guided it through most of the century, it set in place the professional and institutional structure that has allowed the museum to engage more fully with the campus and the community as a self-standing cultural institution with vast relevance to all kinds of learning.

The museum's commitment to its expanding educational mission was first signaled in 1975 when the docent program was formed under the directorship of Brett Waller. In the following decade, Director Maurer advanced the growing emphasis on museum education when he created the position of curator for education. Throughout the eighties and nineties, however, the primary audience for these concerted efforts to make the art collection educationally relevant was the public and schoolchildren. The most important step in the commitment to the museum's university educational mission came in the 2000s under Director James Steward (1998–2009), who, in close collaboration with his staff and with the university administration, envisioned a greatly expanded institution with the physical space and the orientation toward and commitment to education to make UMMA an important center for learning across the university.[54]

2009–17: AN ART COLLECTION FOR UNIVERSITY LEARNING

With the art collection growing to greater than twenty thousand objects, the greatest single acquisition of Steward's directorship can truly be said to be the new University of Michigan Museum of Art itself. The new building that opened in 2009 added the 53,000-square-foot Maxine and Stuart Frankel and the Frankel Family Wing to the renovated 41,000 square feet of Alumni Memorial Hall and provided ample space for permanent European, American, Asian, African, and contemporary collection galleries, and temporary exhibition spaces that have been used in numerous collaborations with the university's other collection institutions, such as the Museum of Anthropology, the Kelsey Museum of Archaeology, and the Clements Historical Library.[55]

The museum's commitment to embedding the collection in university instruction extends beyond these physical spaces. Starting under the directorship of James Steward, UMMA's education staff began a concerted effort to increase the

Fig. 33. A university history class views art in UMMA's paper study room. University of Michigan Museum of Art.

museum's importance to research and learning on campus (fig. 33). This aim was far from isolated to the University of Michigan. Since the 2000s, numerous philanthropic cultural institutions, most notably the Andrew W. Mellon Foundation and the National Endowment for the Humanities, have diligently worked to support university art museums in engaging with the humanities and other disciplines on the interdisciplinary and comparative terms that have opened up the vast trove of visual culture held by UMMA to all forms of inquiry. Under the directorship of Joseph Rosa (2010–2016), UMMA has received significant grants from both organizations and used the resources to increase the number of undergraduates visiting the collection with classes per year to over seven thousand in 2015, an increase in three years of over 400 percent. From medical students engaging with art to explore the topic of empathy to English composition students exploring representations of adolescence and coming of age through Japanese woodcuts, African objects for initiation rites, and contemporary art about America's media landscape; from ROTC students learning about the history of the representation of warfare to psychology students using contemporary photography to discuss their assumptions about gender and identity, the encyclopedic collection that UMMA has accumulated by intention and by happenstance over the previous century and a half has become a virtually limitless resource of visual culture for a diverse array of disciplines far beyond those that guided its assembly.

Notes

1. University of Michigan Board of Regents, *Proceedings of the Board of Regents, 1837–1864*. Ann Arbor: University of Michigan, 1915. http://hdl.handle.net/2027/mdp.39015010367012, March 29, 1855, 604.

2. *Board of Regents, 1837–1864*, October 14, 1856, 649.

3. *Board of Regents, 1837–1864*, 649. See also Wilfred Byron Shaw, *The University of Michigan, an Encyclopedic Survey*. Ann Arbor: University of Michigan Press, 1942–1958, 1481; and *UMMA: A Beacon for Art*. Ann Arbor: University of Michigan Museum of Art, 2009, 21.

4. *Board of Regents, 1837–1864*, December 3, 1857, 706.

5. Henry Philip Tappan, *Report to the Board of Regents of the University of Michigan*, Ann Arbor: Cole & Gardiner, 1853, 3–27. For a scholarly analysis of Tappan's conception of the Prussian model and how it informed the early structure and curriculum of the university, see James Turner and Paul Bernard, "The Prussian Road to University? German Models and the University of Michigan, 1837–1895," in *Intellectual History and Academic Culture at the University of Michigan*, ed. Margaret A. Lourie. Ann Arbor: Horace H. Rackham School of Graduate Studies, University of Michigan, 2002, 9–38.

6. Steven Conn, *Museums and American Intellectual Life, 1876–1926*. Chicago: University of Chicago Press, 1998, 4–9.

7. Tappan, *Report to the Board of Regents*, 17.

8. Ibid., 20–21.

9. *Board of Regents, 1837–1864*, June 24, 1858, 745–46 and December 22, 1858, 813. See also Shaw, *The University of Michigan*, 1481.

10. *Descriptive Catalogue of the Museum of Art and Antiquities in the University of Michigan*. Ann Arbor: E. B. Pond, 1857. There are some uncatalogued pieces in UMMA's storage that may have been part of this early display.

11. Victor J. Danilov, *University and College Museums, Galleries, and Related Facilities: A Descriptive Directory*. Westport, CT: Greenwood Press, 1996, 4.

12. Shaw, *The University of Michigan*, 1482.

13. Ibid., 425, 631.

14. *University of Michigan. General Register*. Ann Arbor: University of Michigan, 1861, http://hdl.handle.net/2027/miun.aaf1187.1861.001, 39–41. *General Register*, 1883, http://hdl.handle.net/2027/miun.aaf1187.1883.001, 38–64.

15. See *General Register*, 1852, http://hdl.handle.net/2027/miun.aaf1187.1852.001, 24 and 31, and Shaw, 545–69.

16. Department of Engineering (1858); Department of Law (1859); Homeopathic Medical College (1875); College of Dental Surgery (1875); School of Mines, including the first courses in architecture (1875); School of Pharmacy (1876); Naval Architecture and Marine Engineering (1881); School of Nursing (1891); and a newly consolidated Department of Engineering (1895). See "Departmental History," Bentley Historical Library, accessed December 7, 2015, http://bentley.umich.edu/legacy-support/umtimeline/departmental.php

17. The American Medical Association (1847), American Institute of Architects (1857), American Philological Association (later Society for Classical Studies) (1869), Modern Languages Association (1883), American Historical Association (1884), American Association of Museums (later American Alliance of Museums) (1906), to name just a few.

18. On the question of space, see *Board of Regents*, December 1860 (President Annual Message, Section III—Library and Museum), June 1862, September 1862, September 1864, and September 1866. See also Beacon, 27, 30. On appropriations, see *Board of Regents*, June 1863, September 1866, September 1867, August 1869, and September 1870.

19. *Board of Regents*, June 1858 and September 1864 (President's Address).

20. *Board of Regents, 1864–1869*, September 26, 1865, 113–14.

21. *Board of Regents, 1864–1869*, June 28, 1866, 155.

22. On the transformation of art museums and art collecting, see Conn, *Museums and American Intellectual Life*, 192–232.

23. Henry S. Frieze. *Catalogue of the Museum of Art & History in the University of Michigan*. Ann Arbor: University of Michigan, 1876, 1–2. See also his complaints to the Board of Regents about the lack of funds and interest in copies as acting president in 1870 and 1879 in *Board of Regents*, March 1870 and *President's Report, 1879*, 32.

24. *Board of Regents*, March 1859. See also Shaw, 1482.

25. *Board of Regents*, December 1861. See also Shaw, 1482.

26. *Board of Regents*, October 1882. See also Shaw, 1481, 1483. The designation Curator of the Museum of Fine Arts and History in this particular instance is evidence of the fluidity of the labels applied to the collection through these years, which was referred to in various combinations of art, history, antiquities, and archaeology, until the formal institutionalization of the museums in the twentieth century.

27. Shaw, *The University of Michigan*, 1482; *UMMA: A Beacon for Art*, 34–36.

28. For a thorough study of Henry C. Lewis, see Carole Simon Drachler, "The Gallery and Art Collection of Henry Clay Lewis," PhD diss., University of Michigan, 1980.

29. *President's Report to the Board of Regents*, October 1892; *UMMA: A Beacon for Art*, 29–30.

30. September 1862, September 1864, September 1867, September 1869.

31. Princeton's painting collection developed from the 1920s, and the Fogg Museum of Harvard was founded first in 1895, primarily with a collection of copies. On Princeton, see "History," Princeton Art Museum, ac-

cessed December 7, 2015 (http://artmuseum.princeton.edu/about/history). On Harvard, see Danilov, 10–11.

32. *Catalogue of the Chinese Collection of Exhibits for the New Orleans Exposition. 1884–5*. Shanghai: Statistical Dept. of the Inspectorate general, 1887. See also Beacon, 52. This collection is still held by the University of Michigan Museum of Anthropological Archaeology.

33. *Board of Regents*, September 1908, 345.

34. *UMMA: A Beacon for Art*, 36.

35. "Alumni Memorial Hall," Millennium Project, University of Michigan, 2011, accessed December 7, 2015, http://umhistory.dc.umich.edu/mort/original/1925/17 Alumni Memorial Hall/index.html; Beacon, 38–39.

36. *Board of Regents*, March 1910.

37. Shaw, *The University of Michigan*, 577, 1481, 1483.

38. *Board of Regents*, November 1928; Shaw, 1483; *UMMA: A Beacon for Art*, 42.

39. Shaw, *The University of Michigan*, 576–77.

40. Ibid., 578.

41. See section on Parker. See also *UMMA: A Beacon for Art*, 42–47.

42. *UMMA: A Beacon for Art*, 41, 48. Some space was still allotted for university lectures until a major renovation was undertaken in 1966–67, after which UMMA became the sole occupant. See Shaw, 1485; *UMMA: A Beacon for Art*, 55.

43. Shaw, *The University of Michigan*, 1484–85.

44. Jean Paul Slusser, "Museum of Art is Created," *Michigan Alumnus* 52:25 (August 24, 1946), 497–98.

45. Shaw, *The University of Michigan*, 1486; *UMMA: A Beacon for Art*, 43, 56.

46. *UMMA: A Beacon for Art*, 42, 48, 52–54; Shaw, 1486.

47. In 1936, the Oriental Languages Program, which had focused on languages of the Near East since the nineteenth century, added Chinese and Japanese to its curriculum. In 1947, the university's Center for Japanese Studies was established, and in 1948 the Oriental Languages and Literature department was split into separate Near Eastern and Far Eastern Language and Literature departments. In 1961, the Center for Chinese Studies was established.

48. *UMMA: A Beacon for Art*, 50, 56.

49. An earlier "Museum Science" curriculum had been offered by the University Museums in the Ruthven building since the early 1930s, and was formalized by the Graduate School in 1938 (Guthe papers, Bentley Historical Library). This nondegree granting course sequence seems to have largely dissolved after Guthe's 1944 departure, and Sawyer was likely unaware of its history when he created the art history-oriented Museum Practice Program.

50. *UMMA: A Beacon for Art*, 51.

51. Ibid., 63, 65, 68.

52. "History of the Nam Center for Museum Studies," Nam Center for Korean Studies, accessed January 25, 2016, http://www.ii.umich.edu/ncks/aboutus/history

53. The program in American Culture, for instance, turned its attention toward exploring the diversity of cultures in America from the 1970s on. In 1973, the Women's Studies program was established; in 1988, the Program in Intergroup Relations followed and in 1995, the Institute for Research on Women Gender.

54. *UMMA: A Beacon for Art*, 71.

55. Ibid., 84–85.

MUSEUM BUILDING/S, 1900–2017

Carla M. Sinopoli

In this chapter, I continue the story of the University of Michigan research and exhibiting museums begun by Kerstin Barndt above. Her contribution traced the story of the university museums over their first century as they developed from singular to plural in a changing state university during the first decades of a new state and a young nation. In that period the university museum was central to the university's conception of itself—part of Henry Tappan's essential "apparatus of knowledge" (see Blouin, this volume). Through nineteenth-century geological and biological surveys, museum curators also played a critical role in state making: mapping, collecting, cataloging, and curating the resources of the state.

By the first decade of the twentieth century, when Barndt's chapter ends, the University Museum Building—though only in its third decade and containing only a portion of the university's collections—was already overfull. And disciplinary authority had begun to shift from museums to academic departments, part of larger academic trajectories in the development and separation of academic disciplines in Europe and the United States.[1] As discussed in this chapter, the museums themselves were becoming similarly subdivided by discipline and material. With some exceptions (i.e., Harley Harris Bartlett, Walter N. Koelz; see catalog entries, this volume), the wide-ranging naturalists who characterized nineteenth-century museum collecting and display were being replaced by twentieth-century zoological, botanical, paleontological, and archaeological specialists.

In this chapter, I trace the evolution of the university museums from roughly 1900 to the present. Paralleling the narrowing of focus discussed above, I do not attempt to explore the full scope of the museums, which have been ably examined in other chapters in this volume. Instead, I primarily focus on those museums that after 1928 came to be grouped together under the official title "the University Museums." This grouping consistently included all the biology museums (Zoology, Paleontology, and the Herbarium) as well as the Anthropology museum.[2] It occasionally included the Museum of Classical Archaeology. The first four of these constituted the by-then standard occupants of North American Natural History museums, all based on scientific collecting and classification of objects and specimens[3] (while the positioning of the cultural discipline of anthropology in "natural" science museums would become problematic within the discipline over the first half of the twentieth century, these debates had not reached Ann Arbor in the early 1900s).[4] In contrast, the university's Classical Archaeology Museum and collections were considered more closely aligned with the arts and humanities than with the sciences. However, the accumulation of large research collections derived from fieldwork shared fundamental similarities with work of the natural history museums, likely contributing to the Museum of Classical Archaeology's occasional realignment. As in Barndt's chapter, I focus this discussion on two senses of building: museum buildings and the building of disciplinary (museum) structures, beginning my story at the start of the twentieth century and ending it by looking toward the present and the continued relevance of the University of Michigan museums.

THE 1928 MUSEUMS BUILDING

The construction and form of the university's third museum building owe their existence to the remarkable efforts of a single individual, whose name the building eventually came to bear. Alexander Grant Ruthven came to the University of Michigan in 1903 as a doctoral student in herpetology (see Miller, this volume). After earning his PhD in 1906, he was hired as curator of zoology in the University Museum. He soon became frustrated by the poor overcrowded conditions of the 1881 Museum Building, with collections wedged into every nook and cranny (including an unused elevator shaft), and unable to accommodate the rapidly growing natural history collections that he and other scientists were bringing in. He was equally frustrated by the decentralized structure of the museum's organization. Responsibility for the collections fell under multiple faculty in diverse academic departments, each with varying levels of commitment to the collections and with little interest in the museum as a coherent whole. And although not documented, it seems likely that this modern scientific biologist was also unhappy with the disparate displays that occupied valuable space in the building, including an entire floor of the museum devoted to displays of the Chinese Government Collection from the 1884–85 World's Fair and the Stearns Musical Instrument Collection—both far removed from Ruthven's zoological interests.

Ruthven's assessments of the museum's poor condition and lack of institutional stature were accurate. Certainly, the infrastructure had become woefully inadequate to the growing scientific collections and the needs of modern scientific research. And the enthusiasm that had heralded the opening of the Museum Building only a few decades earlier was gone. While numerous visitors still came to the museum each year,[5] the centrality that collections had held in the intellectual life of the university was weakening[6] due to the lack of coordination within the museum, the growth of potentially competing academic departments, and the diminishing importance of museum collections and object-based learning in the increasingly specialized American academy.[7]

We can never know the answer, but it is at least reasonable to argue that if not for Ruthven's intervention and fierce commitment to museum-centered, collection-based, scientific research, the university's natural science museum(s) would not have survived. In December 1912, Ruthven, then chief curator of the University Museum, wrote to the regents, asking them to rename the University Museum as the Museum of Zoology and to appoint him its director. His request was successful. Among other consequences, Ruthven's strategic move both positioned him as the university's most powerful advocate for the museum and formally separated Zoology Department faculty from responsibility for the collections, defining museum-based researchers and collections as distinct from, and—in Ruthven's eyes—far superior to, their peer departments.[8]

Even before he took up the position of director, Ruthven had begun lobbying for a new museum building.[9] From 1911 until funds for the building were finally approved in 1925, he wrote dozens of letters making his case for the new museum to individual regents, university presidents Hutchins (1909–20) and Burton (1920–25), and influential citizens and avocational scientists (several of whom he appointed as honorary curators in the museum). An unattributed article published in the *Michigan Alumnus* in 1923, likely written by Ruthven,[10] made a more public argument, deploring the shameful state of the existing museum and its negative consequences for the university's reputation among peer institutions.

Over the years that Ruthven advocated for a new museum, his vision for it greatly expanded. Not surprisingly, he initially prioritized the research needs of the zoological collections. Indeed, shortly after becoming director of his renamed museum—which also had responsibility for anthropological collections—Ruthven wrote to a number of potential donors of anthropological objects, noting in one letter that "this Museum is now a Museum of Zoology, and consequently no attempt is being made to obtain anthropological material."[11] His zoological passions were also evident in an early, undated letter to the Board of Regents that Ruthven wrote shortly after he was named director (probably in 1913), noting his

Fig. 34. Geology collections on display in the Natural Science Building, where they were moved in 1915, photo print (undated). Bentley Historical Library.

desire to build a "great research museum comparable to the Museum of the Comparative Zoology at Harvard."[12]

Whatever his personal loyalties, Ruthven did not entirely ignore other collections. In the same undated letter referenced above, he also noted the need to properly store the Geological Survey collections of the Department of Geology, a need that was in part addressed when the collections were given a separate space as the "Museum of Geology" in the Geology Department in 1915 (fig. 34).[13] By 1923, after the establishment of the Herbarium (1921) and Museum of Anthropology (1922), Ruthven come to envision a *museums* building that would house the several natural science museums, "the collections in zoology, anthropology, geology, and botany," and that would be largely dedicated to research, with a limited number of exhibits that could be "maintained without much effort or expense."[14]

Nor did Ruthven's drive for a new building lead him to neglect research. During the period of his directorship, the staff of the Museum of Zoology was extraordinarily productive, conducting numerous scientific expeditions in Michigan and around the globe and adding to the rapidly growing collections. Field research and publication were supported by donations from the museum's honorary curators: affluent and influential amateur scientists whom Ruthven cultivated both as fellow researchers and as advocates for his ambitious goals. Ruthven also reinvigorated the university's longstanding status as a research arm and repository for the State of Michigan, and he was named chief naturalist of the new biological division of the state survey that he helped create in 1905 (see Miller, this volume).

After several unsuccessful attempts, funding for the new Museums Building was finally approved by the state legislature in May 1925 ($900,000 for the building and $310,000 for land) and Ruthven turned his attention to designing it. He did not believe there was a good model for museum buildings anywhere in the United States.[15] Given complete authority by the regents to design the building and oversee its construction, Ruthven had a very clear vision for the museum, which he described two years before the building was approved in an April 1923 letter to President Burton. The building should have multiple wings and be of three to four stories, with "laboratories, ranges, aquariums rooms, rooms for experimental purposes, lecture rooms, seminar rooms, and properly designed studios for photography, painting and drafting." Exhibit spaces would be segregated from research spaces and kept to a minimum, and laboratories should be designed to secure the maximum of north light.[16] Appended to the letter was an estimate for the size and distribution of space in his imagined building, with 68,950 square feet designated for zoology; 11,100 square feet each for paleontology and anthropology; 5,500 square feet for botany; 32,200 for exhibits; and additional space for classrooms, photography labs, and so on.[17]

As plans for the building moved forward and construction began, Ruthven's pencil-drawn sketches became the template for the layout of the research spaces. Even after the hiring of Albert Kahn as the building's architect, Ruthven was involved in every decision about its design and construction (fig. 35). Extensive correspondence, now in the Bentley Historical Library, documents his role in such details as the disposal of excess soil from the foundation's excavations, the installation of a telephone switchboard and electric lighting, the crafting of ornamental doors and sculptures, and the design of collection cabinets and lab facilities.

Ruthven's vision for the museum building was largely achieved. When it opened on June 14, 1928, the Museums Building was a four-story, V-shaped structure, with tall windows and distinct wings dedicated to research and exhibition (figs. 36 and 37). Below the name "University Muse-

Fig. 35. Construction of the new Museums Building, architect Albert Kahn, 1927 or 1928, photo print. Bentley Historical Library.

Fig. 36. The Museums Building, 1928–1930, photo print. Bentley Historical Library.

Fig. 37. Although not installed until 1940, Carleton Angell's puma sculptures quickly became distinctive and memorable elements flanking the entrance to the Museums Building. Carleton Angell's studio, c. 1939. Museum of Paleontology.

ums" on the building's façade was a quote from the famous nineteenth-century naturalist Louis Agassiz, founder of Harvard's Museum of Comparative Zoology: "Go to Nature; take the facts into your own hands; look and see for yourself." Thomas Barbour, then director of Harvard's museum, delivered a lecture at the building's opening, celebrating "the usefulness of the university museum," with precious research collections and dedication to the training of "a few selected [graduate] students who wish to study some special field alongside one whose life is devoted to its cultivation."[18]

When Ruthven envisioned his building, his goal was to design a structure that would last for seventy-five to one hundred years. However, like the building it replaced, this one also quickly filled, as curatorial research expeditions rapidly added to the collections. Within a decade of its opening, pleas for additional collection space began.[19] By 1947, substantial parts of the Herbarium collection had been moved to the nearby Museums Annex Building, and by 1960 the Herbarium was completely gone from the building (see catalog entry, this volume). Nonetheless, Ruthven was prescient in his estimate of the building's duration as a museum facility: by 2018 all natural history collections will have been moved out of the building to a new off-campus facility and the exhibitions will relocate to a new museum space, ending a run of ninety years.

WORKING OF THE UNIVERSITY MUSEUMS

By the time the Museums Building opened in 1928, the trajectories of the natural science museums (zoology, paleontology, herbarium, and anthropology) had largely separated from those of the humanities-focused collections (classical archaeology and art). But not entirely. They would come together under different administrative umbrellas in a variety of ways in subsequent decades, as for example in the inclusion of the newly created Museum of Classical Archaeology (in Newberry Hall) with its natural history counterparts in the University Museums division from 1928 to 1940.

The catalog entries in section 3 of this volume trace the histories of individual museums. Here I explore their changing organizational structures and interrelations, both within the Muse-

Fig. 38. Hall of Evolution in the Museums Building, c. 1930. Museum of Natural History.

ums Building and across the university. These changes—which echo larger trends in the roles of university museums and scientific collections nationwide—allow a consideration of the legacies of the complex structures that Ruthven helped to establish and of the challenges the university has faced in conceptualizing the role and place of the museums and their sizeable collections within its changing institutional framework. I focus this discussion largely on the museums that occupied the Museums Building, though I expand the lens as appropriate.

1928–45: THE FIRST DECADES OF THE UNIVERSITY MUSEUMS

At its opening, the Museums Building housed the staffs and collections of four museums: the Herbarium and the Museums of Zoology, Anthropology, and Paleontology. The first three were administratively autonomous and reported to the university president; the Museum of Paleontology was part of the Department of Geology.[20] The minerals of the Lederer Collection and the 1837 Geological Survey never moved to the Museums Building and remained physically and administratively within the Department of Geology, where they became increasingly marginal as the intellectual focus of the discipline changed.[21] An administrative, publication, and maintenance staff served all four museums. As Ruthven had envisioned, the museums were research units—dedicated to what he viewed as the core of museum work: "the increase of knowledge" and training of scientists.[22] A Department of Visual Education was created to serve the museums by making exhibitions to diffuse the knowledge that the scientists generated and to teach biological principles (fig. 38) to undergraduates and, to a lesser extent, K–12 schools and the general public.

The director of University Museums coordinated decision-making about their shared building, staff, and exhibits among the four museums resident in the Museums Building and advocated for new facilities for the Museum of Classical Archaeology in Newberry Hall. Ruthven served in this role from 1928 to 1935 and was succeeded by the Museum of Anthropology director Carl E. Guthe when his presidential duties made the dual roles impossible. By that time, the Department of Vertebrate Genetics was also part of the University Museums structure, though not physically in the Museums Building.

The Museum of Classical Archaeology left the University Museums group in December 1940 when a regental bylaw change formally recognized the University Museums as "an administrative unit of the University composed of associated organized museums, together with a museum Service Unit."[23] A Museums Advisory Committee was formally constituted at this time, consisting of the directors of the five associated units.[24] The committee was in operation from 1940 through 1945, a period that spanned US engagement in World War II. During the war, activities of all the museums slowed dramatically—constrained by the loss of personnel to war service and restrictions on travel, fuel, and materials, which constrained both research expeditions and museum visitors.

Carl Guthe left the university in 1944 to direct the New York State Museum. During the nine years he stood at the head of the University Museums, he was an influential and unifying figure. Like Ruthven, he fiercely advocated for the research and graduate training focus of the museums.[25] However, his own interests soon began to expand from his anthropological scholarship (where he made seminal contributions to the university and the discipline as a whole) to include a commitment to the more public face of museums and the training of museum practitioners. Under Guthe's leadership, the Graduate School approved the creation of a seven-course curriculum in Museum Science in 1938, with coursework on museum theory, collections, and visual education, among other topics.[26] This began the university's long, albeit discontinuous, involvement in museum studies teaching, which continues in today's Museum Studies Program.

1945–56: MUSEUM REALIGNMENTS

After Guthe's departure, the position of director of the University Museums was not filled and the university administration soon became concerned about the lack of coordination among the science museums and the lack of clarity on their relations with academic departments and the larger university community. In late 1945, a higher-level Museums Advisory Board was created to address these concerns. Members included the museum directors, the dean of the Graduate School, the director of the University Library, and the dean of the College of Literature, Science and the Arts (LSA). In addition, an operating committee consisting of museum directors was constituted and charged with responsibility for the operation of the Museums Building and exhibitions.

This convening of the advisory board marked an important turning point in the history of the university's research museums, and a key step in an ongoing conversation about the appropriate institutional place of museums within the university structure—a conversation that had been largely muted during Ruthven's charismatic leadership. In its first year (1946–47), the board—chaired by LSA dean Hayward Keniston—instituted a number of changes. It created a hierarchy of museum appointments (director, curator, associate curator, etc.) with equivalence in expectations and ranks to academic departments. It also confirmed the museums' role in graduate training (often a sore point in departments). In reviewing museum appointments, the board highlighted the low salaries—and consequent low morale—of museum staff compared to their faculty counterparts. While celebrating the importance, dedication, and scholarly contributions of the curators, the board also defined the museums as being much like the library—a "service department for other units"—by virtue of their responsibilities to collect, classify, and care for biological and anthropological materials. The advisory board further lamented the lack of attention to exhibits, highlighting them as the weakest part of the museums' activities, and authorized the creation of a new curator of exhibits.[27]

Among the most fraught issues that contributed to the creation of the advisory board concerned the teaching responsibilities of museum curators. Curators had long trained individual doctoral students within the museums. Over time, many curators had taken on increasing teaching responsibilities in departments and, in some museums, split appointments in both department and museum had begun to be common. In April 1948, the university bylaws were amended to formalize these relations, permanently situating all instruction in the "appropriate teaching department."[28] Rather than resolving tensions, however, new questions emerged. If curators were expected to also hold faculty lines, which unit should be the primary institutional home of the curator-professor?[29] Where should tenure rest? Would it be possible to continue to have specialist curators who were only in a museum, where they addressed specific research priorities and did not have teaching responsibilities? And, perhaps most significantly, should the museums continue to report directly to the president or should they now be part of the undergraduate college where the departments were based? These questions would take another eight years to resolve.

Initially, it was the dean of LSA who did not want the museums to come into his college—arguing that the college was dedicated to undergraduate teaching and its executive committee was not "competent to make intelligent decisions in matters affecting research policies."[30] The museum directors, in contrast, desired the benefits of more standard faculty appointments and at the end of 1949 submitted a proposal to join the College of LSA.[31] They were determined to retain the administrative autonomy and research missions of their museums but wanted additional resources and structural equivalence to faculty in salary, permanence of appointment, health insurance, and other "rights and privileges" that curators still lacked.[32] No action was taken on this proposal and the Museums Advisory Board was disbanded after 1948.

In addition to its work on the structure of the research museums, the most consequential, and publicly visible, work of the 1945 to 1948 advisory board was in the creation of the curator or "prefect" of exhibits and the hiring, in 1949, of Irving G. Reimann to fill that role (fig. 39). Reimann was an experienced museum professional and he quickly began to expand the exhibits staff and update, enlarge, and restructure the quality and public visibility of the public displays in the Museums Building (see Museum of Natural History entry, this volume). He also quickly began to complain about the inadequacy of the Museums Building for exhibits and to advocate for a new museum building dedicated solely to exhibits and fully separate from the research museums.[33] While Ruthven's vision had been that curators from research museums would be only minimally involved in exhibitions as reviewers of scientific content, this was in fact never the case, and many research museum curators were active participants in Reimann's expanded exhibition program, as indeed they had since the building opened in 1928. This was especially the case for the Museum of Paleontology, as many of the large fossils on display were important research specimens.[34] Museum of Anthropology curators were also active in the development of short- and long-term exhibitions, including the building of the Native American dioramas, which were to become so controversial by the end of the twentieth century (see below).[35]

The Museums Advisory Board was reactivated in the spring of 1953, chaired by the new LSA dean Charles Odegaard (1953–58). In this iteration, the board did not have representation from the library but did include the dean of the School of Natural Resources.[36] In February1954, both the board and member museums were expanded, briefly bringing together the art and science museums under a single umbrella. The Kelsey Museum of Archaeology, Museum of Art, and Institute of Human Biology were added to the University

Fig. 39. Irving G. Reimann in the diorama studio, c. 1967. Bentley Historical Library.

Museums division and provision was made for "such other units as may from time to time be included by the Board of Regents."[37]

The work of this second iteration of the Museums Advisory Board was, like that of its predecessor, consequential. In 1956, after extensive reviews and discussions,[38] the four museums in the Museums Building and the Kelsey Museum were placed under the College of Literature, Science and the Arts; the Institute of Human Biology was merged into the Department of Zoology; and a new LSA museum, the "Exhibit Museum," was created, with Reimann as its first director.[39] For all but the Kelsey Museum, which has its own building and combined research and exhibition focus, this realignment formalized the distinctive museum structure that persists until today: with a public museum (now called the "Museum of Natural History") that lacks curators and collections, and four "research museums" that lack a visible public face on the U-M campus and that, despite their rich histories, extraordinary collections, and high international research profiles, are largely unknown to the wider university community.

The many shifts in the structure of the University Museums in the middle decades of the twentieth century (which are by no means all detailed here) are less interesting for their ever-changing committees, individuals, and museums represented than for revealing that, by this time, the museums had come to be understood as problems that required administrative attention. A number of factors likely contributed to this, including the ongoing space and resource needs of the museums; tensions between the museums and their peer departments, which had increasingly supplanted them as the core academic centers of both teaching and research; and the loss of focus on objects and collections in much scientific research. In biology, research interests and funding sources had increasingly shifted to genetic and cellular biology, creating "the precarious place of specimen collections in the life sciences" and the loss of many collections.[40] Disciplinary transformations had also led to much of anthropology abandoning material culture and collection-based research by the mid-twentieth century,[41] though collections remain essential to archaeology.[42] While both Dean Keniston and his successor Dean Charles E. Odegaard took pains to explicitly affirm the value and importance of the collections and of the museums that held them, their roles and status no longer made the same sense to the larger university structure or held the same centrality that they had in 1817, 1837, or even in 1928, when the Museums Building opened.

1956–2006: CONSOLIDATION, COMPRESSION, AND REALIGNMENT

In many respects, business continued as usual in the museums over the five decades following the 1956 administrative realignment. The occupants of the Museums Building continued to share decision-making about the building, renaming the Operating Committee the University Museums Building Committee. They established an advisory committee for the new Exhibit Museum, consisting of the five museum directors, in order to "administratively reorganize already existing relationships" and "guarantee advice to the Exhibit Museum" on its staffing and exhibit content.[43] Neither of these committees was long-lived, though records are unclear on when they were dissolved. At the same time, while the Exhibit Museum continued to have strong relations with the research museums, its new autonomy allowed it to expand beyond their research foci to add exhibitions on geology, astronomy, and human physiology.

In other respects, things changed significantly. At one of its first meetings, the Museums Building Committee discussed whether annual reports of individual museums would still appear in the President's Report and determined that it "is probable there will be no University Museums Report after this year."[44] Former president Ruthven's museums no longer had a direct relation with the university president, nor could they as easily harness a single voice to communicate their common interests to higher administrative levels within the college or university. The individual museums now separately reported to the LSA dean and, it was hoped, their strongest connections would be to the departments in which the faculty-curators held half-time positions and whose research and teaching requirements determined faculty-curator tenure and promotions. The research museums and their faculty-curators continued their commitment to graduate student teaching as they took on increasing roles in the education of undergraduates.

Over succeeding decades, individual museums experienced different trajectories: some growing dramatically, others contracting. These are described in the individual catalog entries and will not be recounted here.[45] Each of the research museums (with their peer departments) appointed a succession of nationally and internationally respected scholars—attracted by the rich collections—who made significant contributions to their respective disciplines and student training. Each museum continued to conduct research in Michigan, while also working internationally and engaging in scientific collaborations with scholars from around the globe. Collections continued to grow, with the direction of growth driven in large part by individual curators' research interests rather than a drive for encyclopedic coverage. And the direction and stature of the museums and their relations with their core departments were inextricably tied to larger disciplinary trends, affecting the value—or lack of value—attributed to collection-based research.

The role, status, and proper administrative home of the museums were not fully resolved with the 1956 move to the university's liberal arts college. Already in 1958, Dean Odegaard and Museum of Art director Charles Sawyer corresponded about the possibilities of forging a new relation among the three museums holding cultural collections: the Museum of Art, the Kelsey Museum of Archaeology, and the Museum of Anthropology. Sawyer argued for the value of maintaining these as three separate institutions rather than a single "Museum of Cultural History." His letter, which reveals only one side of the conversation, hints at Dean Odegaard's interests: "We hope these brief comments may be helpful in stimulating the kinds of discussion you have in mind in plotting the future roles of the cultural museums within the University."[46] We do not know what Odegaard's hopes were; the museums, with their distinct disciplinary foci, remained separate, though they regularly interacted through shared collections,[47] collaboration in exhibitions (held at the Museum of Art[48] and Kelsey Museum), and variously intersecting research interests. On a few occasions, the museums themselves attempted bottom-up "reunification" efforts, such as in the early 1990s, when discussions took place over creating a shared museum infrastructure focused on the study of evolution.

As the stature of academic disciplines and departments as the premier academic centers of teaching and research continued to strengthen over the late twentieth century, university officials, deans, and department chairs repeatedly sought to understand the whys and rationales of the research museums and the roles and responsibility of their curators. Museums as institutions, and especially nonexhibiting museums, were often largely unfamiliar to the individuals taking up these roles, who typically came from academic departments and often had little familiarity with the research, care, or ethical obligations of museum collections. As leaders came and went, the museums regularly responded to questions and concerns about their institutions, collections, and value. This repeated need to explain their purpose likely was intensified by the specialized nature and largely internal focus of the research museums, but was far from unique to the University of Michigan.

The museums in the renamed (in 1968)[49] Ruthven Museums Building continued to interact over building needs and the space conflicts that inevitably arose as collections outgrew the 1928 building. Herbarium collections, as noted above, left the building in the 1960s, ceding their fourth floor space to the Museum of Anthropology. The Museum of Zoology received funding from the university and the National Science Foundation to add a third wing to the building in the mid-1960s (see catalog entry, this volume). Over time, individual museums also received off-campus space for their collections, though all maintained a core presence on the central campus until 2001, when the Herbarium was moved to a "temporary" facility in south Ann Arbor. This facility will soon become the home of all the original Museums Building inhabitants, uniting them again in a single modern research and collection facility (see below).

Despite frequent reviews and discussions of the problems of museum organization, the structure established in 1956 persisted unchanged until 2011 (see below). On occasion, other issues brought one or another museum to the attention of the larger university community and diverse publics. The most visible examples of this lie in controversies about the collections and displays associated with Native Americans in the Museum of Anthropology and Museum of Natural History, respectively. As discussed in Caple and Williams's contribution (this volume), the collection of Native Michigan materials goes back to the very founding of the university and state, and to the work of the Geological Survey during the period of political expansion and accelerated taking of Native lands. The collecting of materials from archaeological sites, including of human burials, intensified with the creation of the Museum of Anthropology in the early 1920s, with its focus on documenting the prehistory of Michigan's indigenous people (see Hinsdale biography entry, this volume). This research, like much early museum collecting at U-M and elsewhere, was grounded in the desire to document and preserve "disappearing" resources, whether archaeological sites and Native material culture and peoples or, as Ruthven argued, Michigan's threatened botanical and zoological resources,[50] and in larger disciplinary developments within twentieth-century American anthropology. The (recently renamed) Museum of Anthropological Archaeology holds the largest archaeological collections from the state of Michigan, including materials from more than 2,000 archaeological sites, and the remains of more than 1,300 ancestral Native Americans.

Nearly two decades before the passage of the Native American Graves Protection and Repatriation Act (NAGPRA) in 1990, Native students and community members had protested problematic museum exhibitions and the university's holding of Native American human remains.[51] With the passage of NAGPRA, and particularly following conflicts over repatriation that intensified in the late 1990s, complaints about both museums grew. The reasons and histories of these conflicts merit scholarly study in their own right[52] and are not reviewed here. In 2011, after several years of dialogue and review, the Museum of Natural History (the renamed Exhibit Museum) removed the then sixty-year old Native American dioramas, agreeing that their portrayals of Native peoples were unhistoricized and problematic. The dioramas were put in storage where they are today visited by classes interested in the complex history of museum display and portrayals of indigenous peoples. Tensions over repatriation with the Museum of Anthropology intensified throughout the 1990s and early 2000s.[53] They ultimately led to the university administration becoming more involved in the management and decision-making over repatriation through the establishment of a committee in the Office of the Vice President for Research[54] and the hiring of dedicated NAGPRA staff in that office and the museum in 2009.

Overall, there was both loss and gain in the five decades after the 1956 reorganization of the natural science, exhibits, and anthropology/archaeology museums. Under Reimann's leadership, the Exhibit Museum thrived, experiencing dramatically expanding audiences at its new exhibitions and vibrant programming (such as the Friday evening film series), then declined after his retirement. For the research museums, the stage on which they operated (as museums) within the university had contracted considerably, even as the influential profiles of their scientists extended far beyond the university's walls. While disciplinary changes may have devalued collections for parts of this period, by the late twentieth century, objects and specimens had once again taken center stage as disciplinary centers shifted again, in ways discussed in the next section. Throughout all these changes, the museums remained centers of intellectual innovation, whose curator/faculty played important roles in disciplinary transformations, conducted innovative research, and trained generations of leading scholars. As the museums' scholarship continued and evolved, new collections continued to be added. Today, the collections have grown to more than twenty million objects and specimens. The Museum of Zoology remains by far the largest, with some thirteen million specimens; approximately 1.75 million specimens are curated in the Herbarium; three million each in the Museum of Paleontology and Museum of Anthropological Archaeology; and ca. 110,000 objects in the Kelsey Museum of Archaeology.

2006–2017: NEW STRUCTURES, INVESTMENTS, AND OPPORTUNITIES

While the starting date of 2006 for this final phase of the museums' continuing story is arbitrary (fifty years after the LSA museums reorganization), the ending date is not. The year 2017 is not merely the university's bicentennial. By the close of the year, virtually all the collections of the research museums will have moved out of the overcrowded Ruthven Museums Building. Excluding a subset that will remain in laboratories on central campus for research and teaching, the bulk of the collections of the Herbarium, Museum of Zoology, Museum of Paleontology, and Museum of Anthropological Archaeology will be housed in a modern, secure, climate-controlled facility—the Museums Research Center (MRC)—located about six miles south of the University of Michigan central campus. For the first time since 1960, the collections of the four research museums will again be housed in a single structure, albeit admittedly at a distance from the heart of the University.

In this section, I briefly review some of the developments in the museums over the last decade and look to their future, concluding with an optimistic vision for the continued value of the museums and their collections to the University of Michigan. All the changes I discuss here began well before 2006. For example, the digitization of the collections and their dissemination via the web was initiated in the 1990s in all the museums. The Museum of Zoology was particularly innovative in this, launching its impressive Animal Diversity Website in 1995 (see Museum of Zoology catalog entry, this volume). Here, I mainly look beyond the beginnings of these developments to their present and the future. However, before doing so, one additional organizational change should be noted, which further connected two of the museums to their closest department. In 2011, the Museum of Zoology and Herbarium were placed under the Department of Ecology and Evolutionary Biology. Their directors were made associate chairs and now report to the department chair.

Digital Infrastructure

As noted, as digitization technologies developed, the university museums quickly recognized the opportunities that digitization could provide—both for the management and tracking of collections and for their dissemination to larger scholarly and public audiences. Given the decentralized organization of the museums, these efforts were initially undertaken independently, with each museum and, in some cases, separate divisions within museums, developing or adopting their own databases, supported by variously skilled collections staff and, financially, from a mix of external grants and reallocation of internal resources. The result was a proliferation of siloed database systems that were difficult to support and could not communicate with each other. In addition, as digital imaging projects expanded in each of the museums, so too did concerns about the need to secure the preservation of the thousands of images being created. In 2009, in an unusual coming together of the museums, the College of LSA created a multi-museum committee to specifically address the challenges and opportunities of digitization and the resulting needs for digital infrastructure. In addition to representatives from the museums, the committee included digital infrastructure and imaging specialists from the University Library, LSA, and the Department of History of Art.

Fig. 40. Research space in the new Research Museums Facility before collection move, 2016. Photo by Carla Sinopoli.

The work of the committee resulted in the 2012 appointment of an IT specialist dedicated to enhancing the security of digital museum records and improving existing infrastructure. The committee also recommended the movement of the museums data to an LSA College-supported shared database system to support both collection management and collection research. An initial attempt to adopt a single system to serve all the LSA museums was not successful, and today the museums are migrating their many databases to two more specialized systems. The biological museums are using a system developed specifically for biological collections, while the two archaeology museums are using a software more appropriate to the needs of their collections, and which has also been recently adopted by the Museum of Art.[55] As these systems develop, the University Library will provide public access and search capacities to all the databases so that individuals will be able to search across collections and virtually reunite historic collections—such as those of the 1837 Geological Survey—that have, over time, been physically and administratively dispersed among multiple museums.[56]

All the university museums are currently working to digitize their entire collections. In addition to communicating collection information within the university, these efforts make the collections accessible worldwide through standard data-sharing networks developed for zoological and botanical collections, and via websites and virtual exhibitions. Given the enormous scale of the collections, these projects will require years to complete. Already large portions of the collections are seen by millions of viewers per year, visited by researchers, educators, students, and interested individuals around the world.

Physical Infrastructure

The last decade has seen impressive investment in museum building projects at the university. In 2009, with the support of generous donors, U-M opened new wings of the Kelsey Museum and Museum of Art, the first dedicated museum facilities constructed on the University of Michigan campus since 1928. As discussed in the respective catalog entries, these additions added significantly to exhibition space, improved collection storage, and allowed the museums to significantly expand their engagements with larger university, regional, and global communities. A new Biological Sciences Building is scheduled to open in 2018 on central campus just north of the Ruthven Museums Building. It will contain the offices and research laboratories of the curators of the three biology museums and will be the home for a new Museum of Natural History. The museum will retain some popular exhibitions from the current museum (as well as Carleton Angell's pumas that have guarded the entry to the Museums Building since 1940; Figure 37 above), but in a reconceptualized setting and alongside many new displays. Ruthven's Museums Building will close its doors (at least as a museums building) in 2019.

By that time, the collections of the four research museums originally in the Ruthven Building will be housed in their new facility south of campus. The design of the Museums Research Center and relocation of the collections are massive undertakings and the university has invested considerable resources in bringing them about—more than twenty times the $900,000 it took to construct Ruthven's Museums Building. Like all major changes, the new facility brings with it opportunities and anxieties: opportunities to better care for and more fully document the collections in vastly improved storage and research spaces (fig. 40). The anxieties include worries about distance and how or if the many students who work with the collections will be able to access them in their new location, and whether the collections, and the staff who care for and study them, will become even less visible to the larger university community once they are situated far from central campus. This future is still to be written, and no doubt will be revisited in 2037 when the university museums celebrate the bicentennial of the creation of the Cabinet of Natural History.

CONCLUSION: THE CONTINUING PROMISE OF THE UNIVERSITY MUSEUMS

As this essay and volume elaborate, the history of the University of Michigan museums is long and complex. For the natural science museums in particular, the story has been one of both contraction and expansion. From being essential to the founding vision of the university and a key part of Henry Tappan's apparatus of learning, the research museums have been encompassed by the growth of academic disciplines that were only beginning to take shape when the university was founded. At the same time, the collections and research profiles of the now several museums have grown remarkably from their beginnings as the Cabinet of Natural History in the mid-nineteenth century and now count among the largest and most significant university museum(s) collections in the world.

There is today considerable, and indeed growing, promise in the contributions that the museums and their collections and exhibits can make to students, scholars, and diverse publics. Here, I highlight several current trends that I suggest are "re-centering" the museum collections at the heart of the university's mission. The first is apparent in this volume itself: the increasing interest in the study of the history of museums and collections in their larger cultural and political contexts. University museums have received comparatively little attention in the sizeable and rapidly growing scholarly literature on museum history. However, such museums, and perhaps particularly the University of Michigan museums, are important topics for historical research, related as they are to the larger trajectories of the American academy and state-building enterprises, as well as to the complex and changing disciplinary trajectories that they have shaped and been shaped by. The impressive success of the latest iteration of the University of Michigan's interdisciplinary Museum Studies Program points to the growing academic interest in museums among undergraduate and graduate students and faculty from dozens of academic disciplines.

The collections of the University of Michigan museums have mostly been generated by the field research of curators, affiliated researchers, and students, and have been key to knowledge production in anthropology, classical archaeology, zoology, botany, and paleontology. They will continue to serve these roles into the future, both through research that generates new collections and through repeated returns to old collections with new questions and analytical methods. Indeed, the research value of the collections will only continue to increase, as development and site destruction threaten archaeological sites, and climate change and associated habitat loss increase species displacement and extinction.

Moving beyond the disciplinary lenses of individual researchers, the collections today speak to scholars and students in many additional areas beyond those of the museums' core disciplines. The recent "material turn"[57] in the humanities and humanistic social sciences has reignited interest in material culture and encouraged scholars to take things—including the artifacts and specimens in museum cabinets and exhibitions—seriously as topics of study. Since the late 1980s, expanding research, policy interests, and findings in the study of biodiversity have fostered a return to museum collections within biology, and the potentials of new methodologies for the analysis of "big data" allow for new crosscutting of interdisciplinary research agendas in the biological sciences. To a significant extent, historical museum collections from systematic research expeditions, such as the Michigan geological and biological surveys, constitute the record of historical biodiversity and are increasingly critical for its study. News reports of previously unidentified species being found in the storerooms of museums appear frequently (often a side benefit of inventorying and digitizing projects),[58] though they are matched by lamentations about the declining resources nationwide for research on and care of natural history collections.[59]

Collection-based research and exposure is also expanding the museums' roles in undergraduate education, part of a larger (re)turn to object-based and experiential learning in the academy. Every year, thousands of undergraduates engage with the museums in exhibition spaces, museum teaching and store rooms, and research laboratories. The Museum of Art has especially excelled in this area, with major federal grants to support engaged learning (see Choberka and catalog entry, this volume). However, it is far from alone, and all the museums actively share their collections with hundreds of students each year, through teaching with collections, hosting class visits, and facilitating student research and collection projects.

In a reversal from the more common focus of museums studies literature on exhibitions and public engagement, this chapter has largely focused on research museums and collections. However, I would be remiss not to note the essential role of exhibitions in the continuing promise of the university museums. The exhibition spaces in the Museum of Art, Kelsey Museum, and, soon, Museum of Natural History (not to mention the impressive efforts of the Matthaei Botanical Gardens and Arboretum, Sindecuse Museum of Dentistry, and other university galleries and display venues) have created new opportunities for public engagement, interdisciplinary collaborations, and the creation of innovative and challenging exhibitions. As the original museum in Mason Hall did when it opened in 1840, today's university museums continue to serve as places where town and gown meet, new knowledge is created, and Tappan's vision of an expansive university can continue to be realized.

Notes

1. Chris Gosden and Frances Larson, *Knowing Things: Exploring Collections at the Pitt Rivers Museums 1884–1945* (Oxford: Oxford University Press, 2007), 150–51; Samuel J. M. Alberti, "Molluscs, Mummies and Moon Rock: The Manchester Museum and Manchester Science," *Manchester Region History Review* 18 (2007): 108–32; Susan L. Star and James R. Griesemer, "Institutional Ecology, 'Translations' and Boundary Objects: Amateurs and Professionals in Berkeley's Museum of Vertebrate Zoology, 1907–39," *Social Studies of Science* 19 (1989): 387–420.

2. In the United States, early anthropology was understood as belonging in "natural" history museums, a product of colonial and racial views of colonized indigenous communities and early anthropology's "scientific" classificatory epistemologies. There is a vast literature on this topic; relevant works include Tony Bennett, *The Birth of the Museum: History, Theory, Politics* (London: Routledge, 1995), 180; George W. Stocking, ed., *Objects and Others: Essays on Museums and Material Culture* (Madison: University of Wisconsin Press, 1985); Ruth Philips, *Museum Pieces: Toward the Indigenization of Canadian Museums* (Montreal: McGill-Queens Press, 2012).

3. Robert E. Kohler, "Finders, Keepers: Collecting Sciences and Collecting Practices." *History of Science* 45 (2007): 428–54.

4. Steven Conn, *Do Museums Still Need Objects?* (Philadelphia: University of Pennsylvania Press, 2010), 32.

5. Ruthven letter to the regents, ca. 1911 (Museums records, Bentley Historical Library).

6. "Our Overcrowded and Inadequate Museum: Some Reasons Why the University Must Have a New One," *Michigan Alumnus*, March 8, 1923, no. 320 (no author attributed).

7. Steven Conn, *Museums and American Intellectual life, 1876–1926* (Chicago: University of Chicago Press, 1998), 20, 26.

8. See Miller, this volume. Relations between the Zoology Museum and the Department of Zoology were competitive and marked by conflicts and resentments over control of resources and training of graduate students. Ruthven again prevailed and was appointed as chair of the Zoology Department *and* director of the museum in 1927.

9. S. Smith, University secretary letter to Ruthven, January 24, 2011 (Museums records, Bentley Historical Library).

10. "Our Overcrowded and Inadequate Museum."

11. Ruthven letter to Mr. Baird Broomhill, November 2, 1914 (Museums Papers, Bentley Historical Library). This is one of several such letters Ruthven wrote.

12. Ruthven memo to the Board of Regents (Museum Papers, Bentley Historical Library). The undated document has a handwritten note saying "ca. 1911." The memo must date to 1913 or after, as Ruthven describes himself as director of the Museum of Zoology.

13. Ibid.; though it is worth noting that this move also had the effect of moving the collections *out* of his Zoology Museum; see also Christopher J. Stefano, Kathy Erwin, and Rodney Ewing, "The University of Michigan Mineral Collection, Ann Arbor," *Rocks and Minerals* 88.4 (2013): 328–38. DOI: 10.1080/00357529.2013.796754

14. Ruthven letter to President Burton, April 1923 (Museums records, Bentley Historical Library).

15. Ruthven letters to President Burton, September 22, 1920, and April 23, 1923 (Museums records, Bentley Historical Library).

16. Ruthven letter to President Burton, April 1923, (Museums records, Bentley Historical Library).

17. Totaling more than 129,000 square feet, these numbers are impressively close to the 125,000 square feet in the finished building ("Distribution of Space in the University Museums Building and Museums Annex," January 1954 [Museums records, Bentley Historical Library]).

18. Address by Thomas Barbour, published in *The University Museums Building of the University of Michigan* (Ann Arbor: University of Michigan, January 1, 1929), 21, 23.

19. Carl E. Guthe, "Memorandum on the Quarters of the Museum of Anthropology, April 20, 1939" (Museums records, Bentley Historical Library).

20. The Museum of Paleontology was formally separated from the Department of Geology in 1941.

21. C. J. Stefano et al., 328.

22. Ruthven, 1926, "Toward the Development of a Museum Policy," in *Report of the Director of the Museum of Zoology to the Board of Regents, 1925–1926* (Ann Arbor: University of Michigan, 1926), 42.

23. *Report of the University Museum to the Board of Regents, 1940–1941* (Ann Arbor: University of Michigan, 1941), 1.

24. Ibid., 1.

25. Carle E. Guthe, "The Museums at Michigan," *Michigan Alumnus* (1937): 469–77.

26. Letter to Guthe from Graduate School, May 18, 1938 (Museums records, Bentley Historical Library); this was a relatively early program for US universities; see Carol Malt, "Museology and Museum Studies Programs in the United States: Part One," *The International Journal of Museum Management and Curatorship* 6 (1987): 165–72, 167.

27. *Report of the University Museums to the Board of Regents, 1946–1947* (Ann Arbor: University of Michigan, 1947), 1–5.

28. *Bylaws of the Board of Regents,* as amended to April 1, 1948 (typescript on file, Museums records, Bentley Historical Library), 2–3.

29. Unlike many institutions, shared positions at the University of Michigan are typically 50–50 splits, so "home departments" for shared positions must be negotiated.

30. Letter to Provost Adams from Dean Hayward Kenison, January 3, 1950 (Museums records, Bentley Historical Library).

31. "A statement from the directors of the museums concerned in the proposal to place the museums in the literary college," December 30, 1949 (Museums records, Bentley Historical Library).

32. Ibid.

33. Irving G. Reimann, letter to the Advisory Committee for the Exhibits Museum, March 30, 1956 (Museums records, Bentley Historical Library).

34. Letter to Dean Odegaard from L. B. Kellum and E. N. Goddard, January 12, 1956 (typescript on file, Museums records, Bentley Historical Library), 2.

35. Raymond Silverman and Carla M. Sinopoli, "Besieged! Contemporary Political, Cultural, and Economic Challenges to Museums in the Academy as Seen from Ann Arbor," *University Museums and Collections Journal* 4 (2011): 19–31.

36. "Second June Meeting 1953," *Proceedings of the Board of Regents, 1951–1954*, 981.

37. "February Meeting 1954," *PBR, 1954–1957*, 1248.

38. Summarized in a letter from Dean Odegaard to Vice President M. L. Niehauss, January 31, 1956 (Museums records, Bentley Historical Library).

39. The Museum of Art did not move to LSA and remained under the central administration.

40. Mary E. Sunderland, "Modernizing Natural History: Berkeley's Museum of Vertebrate Zoology in Transition," *Journal of the History of Biology* 46 (2013): 369–400; Steven Conn, *Do Museums Still Need Objects?*, 50–51.

41. Steve Conn, *Do Museums Still Need Objects?*, 30–35.

42. The high status of University of Michigan anthropological archaeology in the 1960s and 1970s, coupled with dynamic leadership, allowed the Museum of Anthropology to thrive and grow (see catalog entry, this volume).

43. Minutes of the first meeting of the Advisory Committee of the Exhibits Museum, March 29, 1956 (Museum records, Bentley Historical Library).

44. Minutes of Museums Operating Committee, February 1956 meeting (Museum records, Bentley Historical Library).

45. Unfortunately, in the absence of the earlier compiled annual report of the university museums, records on individual museums' trajectories are much harder to find for this period.

46. Letter from Charles E. Sawyer to Dean Charles Odegaard, February 5, 1958 (Museum records, Bentley Historical Library).

47. Such as the Plumer collections of Asian ceramics, parts of which are in the Museum of Art and parts in the Museum of Anthropology.

48. For example, of the Museum of Anthropology Koelz Collection Tibetan thangka paintings, which were exhibited at the Museum of Art in 1980; Carolyn Copeland, *Tankas from the Koelz Collection* (Ann Arbor: Center for South and Southeast Asian Studies, 1980).

49. December meeting 1968, *Proceedings of Board of Regents*, 1585.

50. Ruthven, Report of the Director of the Museum of Zoology, 1922–1923 (Ann Arbor: University of Michigan), 9.

51. Documents on file, Museum of Anthropological Archaeology; also, Silverman and Sinopoli, 2011.

52. See Amy Lonetree, *Decolonizing Museums: Representing Native American in National and Tribal Museums* (Chapel Hill: University of North Carolina Press, 2012), 160–65, for a perspective on these events.

53. *Michigan Daily*, March 18, 2005. https://www.michigandaily.com/content/u-return-burial-remains-tribe

54. *The University Record*, October 19, 2009. http://ur.umich.edu/0910/Oct19_09/07.php

55. The UMMP, UMMZ, and UH are using Specify, a software developed at the University of Kansas and supported by the National Science Foundation, while the cultural museums have adopted CollectiveAccess.

56. Indeed, the three biology museums recently received a grant from the National Science Foundation to digitally reunite the collections of three important historic U-M expeditions: the Michigan Geological Survey, the Beal Steere 1870–1875 expedition, and the Asian collections of Walter Koelz and Rup Chand (see entries, this volume).

57. See, for example, Peter N. Miller, *Cultural Histories of the Material World* (Ann Arbor: University of Michigan Press, 2013) and Dan Hicks and Mary C. Beaudry, *The Oxford History of Material Culture Studies* (Oxford: Oxford University Press, 2010).

58. E.g., Richard Conniff, "Meet the New Species," *Smithsonian Magazine* (August 2010), access at http://www.smithsonianmag.com/40th-anniversary/meet-the-new-species-748819/?no-ist; Sydney K. Brannoch and Gavin J. Svenson, "A New Genus and Species (*Cornucollis gen.* n. *masoalensis* sp. n.) of Praying Mantis from Northern Madagascar," *Zookeys* 55 (2016): 65–81; Erik Olsen, "Museum Specimens Find New Life Online," *New York Times*, October 19, 2015, http://www.nytimes.com/2015/10/20/science/putting-museums-samples-of-life-on-the-internet.html

59. Richard Conniff, "Our Natural History, Endangered," *New York Times,* April 1, 2016, http://www.nytimes.com/2016/04/03/opinion/ournatural-history-endangered.html?_r=0

PART II

Collectors, Archivists, and Curators

Biographical Essays

CORYDON LA FORD (1813–94) AND THE MUSEUM OF ANATOMY

Martha Girsch

The 1817 document that created the University of Michigan (the "Catholepistemiad") in Detroit envisioned a collective of various professorships in the humanities and sciences, including one specifically designated in Medical Sciences.[1] Within a few decades of the university's 1837 establishment in Ann Arbor, the Medical School opened.[2] The first of the university's professional schools, it housed five distinct professorships, narrowing the broad scope of "medical sciences" into more specialized designations, such as anatomy and surgery. Among the first physicians appointed to the Medical School faculty was Dr. Moses Gunn, an anatomist and surgeon from upstate New York. An 1846 graduate of Geneva Medical College, Dr. Gunn possessed a clear and demonstrated passion for the study and teaching of human anatomy; he arrived in Ann Arbor and began giving anatomy lectures at the university within just two weeks of his college graduation.[3]

Fig. 41. Corydon La Ford, portrait by R. D. Palmer, 1860s. Bentley Historical Library.

After a few years, Dr. Gunn suggested that the professorship of anatomy and surgery be split into two separate academic disciplines to better reflect their growth and specialization. Upon the regents' approval, Dr. Gunn nominated his college friend Corydon La Ford, a fellow anatomist, for the position of professor of surgery (fig. 41). Officially notified of his appointment in the spring of 1854, Dr. Ford would emerge as a formidable intellect, a devoted pedagogue, a gifted anatomist, and an asset to the Medical School's nascent anatomical collections—key to the acquisition, curation, and preservation of some of the university's earliest collections and libraries.

Corydon La Ford was born into a Puritan farming family in upstate New York in 1813. An early bout with poliomyelitis (infantile paralysis) left young Corydon with a paralyzed leg, rendering him incapable of performing necessary physical tasks required on the farm. His physical limitations did not hinder his growing mind, however; Corydon was an excellent student and as a teenager became a teacher at his local school. At the age of twenty-one, he left home to study medicine under a number of prominent New York physicians, and before age thirty he was a student at Geneva Medical College.[4] It was during his years at the college that Corydon struck up the friendship with Geneva alumnus Dr. Gunn, which would eventually lead to his position at Michigan.

While still a medical student in New York, Corydon demonstrated a burgeoning interest in anatomical collections and curation: he took a post as librarian and curator of the college's small medical museum to pay his room and board. This early exposure shaped Corydon's approach to the acquisition and preservation of medical specimens. Furthermore, the freedom to explore and pore over the collections and books in the library was wonderful for the ambitious physician. He even spent a summer employed by the college, making the anatomical preparations for the museum's specimens.[5] Thus, with professional relationships forged and a wealth of hands-on anatomical knowledge amassed over two years of study at Geneva Medical College, Corydon received his doctorate in 1842.

Through the end of that decade and into the beginning of the next, Dr. Ford served in various faculty positions and universities and hospitals throughout the East Coast. Following his spring 1854 appointment to the University of Michigan Medical School faculty, Dr. Ford began lecturing as professor of anatomy in October of that same year. He quickly became one of the most beloved and revered lecturers on the medical campus. His students remembered him as a serious man whose physical presence was preceded by the soft tapping sounds of his distinctive gold-capped cane in the hallway outside the amphitheater where lectures were held. Ford was devoted to ensuring the growth and development of the curricula in the Medical School; upon learning that the university could not or would not fund the acquisition of certain medical models and illustrations for use in the classroom, he offered to pay for these pedagogical materials out of his own pocket and be reimbursed at a later, unspecified date.[6] He also encouraged the regents to fund a school of homeopathy within the larger University Medical School so that his students might expand their education beyond allopathic medicine. In this essay, I focus on his contribution to the founding of the Anatomical Museum at the Medical School.

Before delving into the history of this museum and Dr. Ford's role in it, it is important to consider our modern experiences of museums and exhibitions. In this digital age, it can be difficult to conceive of collecting, cataloging, and learning from objects without the aid of electronics. Students and scholars who visit the university's museums and libraries today encounter digital and electronic technology that function as cataloging and research tools to augment or round out their experience. Digital image repositories and videos that demonstrate dissections or preparations are integral to today's learning process. But in the nineteenth and early twentieth centuries, students relied on hands-on interaction with objects, models, and specimens. Looking backward, then, we may begin to understand the importance of Dr. Ford's collection, preparation, and exhibition of anatomical specimens for his students and fellow researchers.

The existing medical pedagogy and the teaching practices in Dr. Ford's era greatly benefited from his vision. Students at that time had two approaches to learning: listening (to lectures) and hands-on learning—seeing, touching, dissecting, and reassembling. Dr. Ford's students gathered twice weekly for lectures in a high-ceilinged amphitheater, hoping to get a good view of the professor as he completed dissections to accompany his presentations. The medical students' visceral experience with handling and examining objects, however, often involved exhuming corpses from nearby graves to study their internal workings. This macabre practice was mostly swept under the rug and deemed part of the "necessary inhumanity" of the study of anatomy.[7] While the practice of grave robbing would persist for years after the establishment of Dr. Ford's anatomical museum, the collection gave medical students an alternative route to get their eyes and hands on anatomical specimens.

An analysis of one of the only known remaining photographs of the Anatomical Museum supports the argument that Dr. Ford's vision was of a purely educational collection. The photo, taken by Ann Arbor photographer J. Jefferson Gibson circa 1893, depicts a wall with built-in shallow, glass-doored shelves that house a variety of anatomical specimens (fig. 42). Present are skeletons, some appearing to be complete (including one that appears to be a small child) and others disassembled; heart models; cross sections of brains and other organs; and specimens preserved and

Fig. 42. Display case in the Anatomical Museum, c. 1893, albumen print, photo by J. Jefferson Gibson. Bentley Historical Library.

floating in murky test tubes. Conspicuously absent is anything we would associate with a public museum or gallery today: there are no labels, no signage, no exhibition descriptions. The placement of specimens behind closed doors discouraged visitors from casual observation or handling. Furthermore, the specimens are stacked almost haphazardly, with some objects even obscuring others; this was not a gallery where items were specifically curated to provide an exhibition narrative. Instead, it was a collection of items intended for hands-on study, where a medical student might search for an anatomical specimen from a certain body part, locate it behind the glass door, and remove it for personal study. Students and faculty in the Medical School were thus the main beneficiaries of the Anatomical Museum. They had the perfect combination of knowledge of the material, interest in its function, and educational motivation for its observation—just as Corydon Ford had intended when he opened the museum.

Dr. Ford continued to serve the Anatomical Museum as curator and the Medical School as professor from 1869 to the end of the nineteenth century. He had planned to retire in 1894 at the age of eighty-one as professor emeritus, but he suffered a stroke in April of that year while walking home from his last lecture before retirement. During the university's memorial service, students and colleagues lauded Dr. Ford as an extraordinarily gifted anatomist and pedagogue who had perfected the art of teaching in the over one hundred courses he had offered:

> Anatomy was the "pons asinorum" of medical students until Ford made it as fascinating as poetry and fiction. . . . All that he said reached the ear and all that he did reached the eye at the same time. He made his descriptions go hand in hand in harmony as a master artist could play an aria in perfect harmony.[8]

At the heart of Dr. Ford's style of delivering lectures and of his understanding of anatomy as an object-based science stood one of North America's premier collections in anatomical specimens and models (fig. 43).

As Peter M. McIsaac discusses elsewhere in this volume, interest in the Anatomical Museum seemed to lessen after Dr. Ford's death and the turn of the twentieth century. The record keeping and cataloging of the collection were poor, and because the Department of Anatomy underwent multiple name and location changes, much of the collection was either misplaced or destroyed. As new methods of medical training developed, the importance

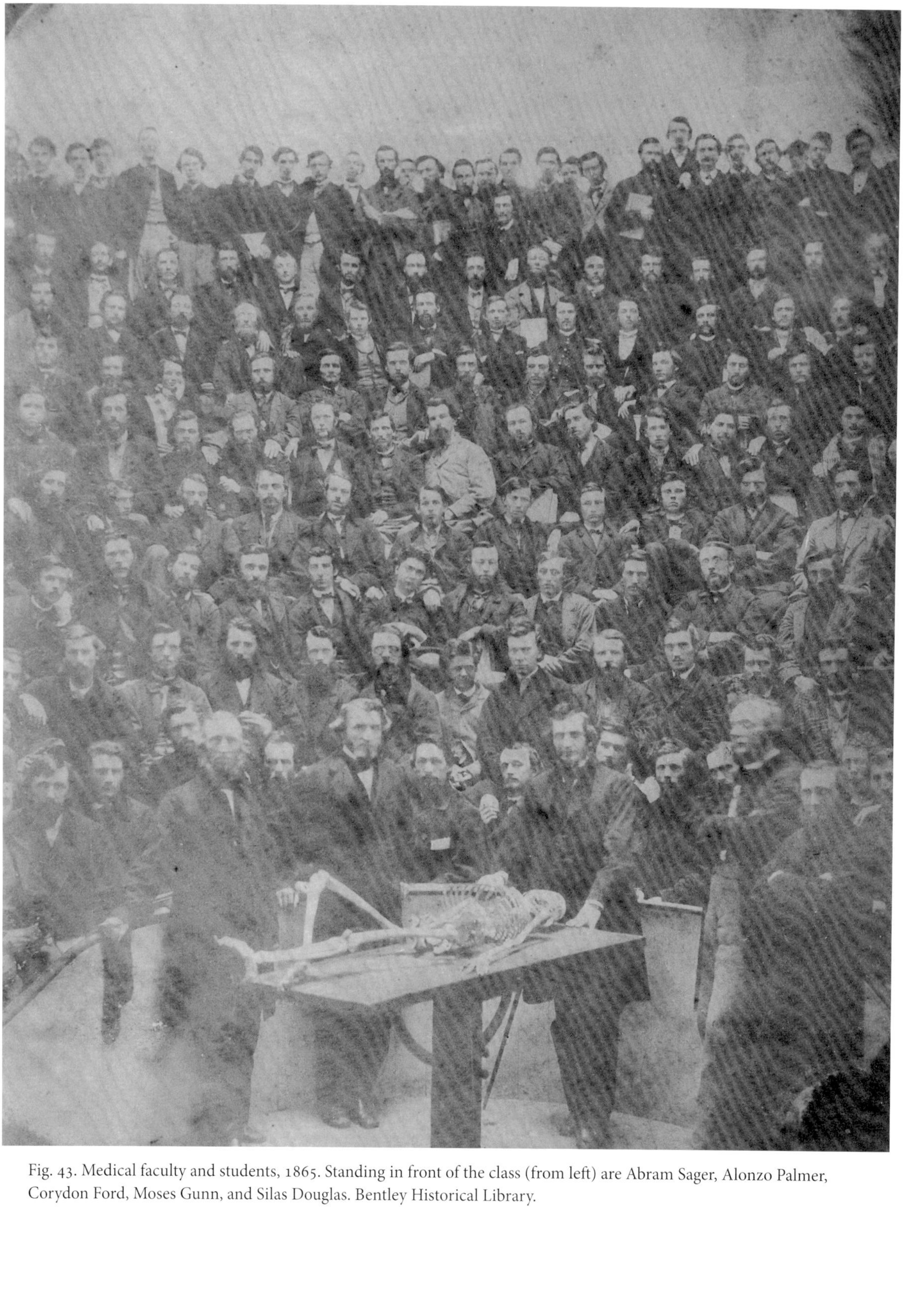

Fig. 43. Medical faculty and students, 1865. Standing in front of the class (from left) are Abram Sager, Alonzo Palmer, Corydon Ford, Moses Gunn, and Silas Douglas. Bentley Historical Library.

of the museum's collections declined. Many items once on display, including Dr. Ford's cane and various other memorabilia from his early years in the Medical School, were lost. In the Corydon Ford papers at the Bentley Historical Library, the handwritten bequest of a collection of human skulls and teeth to the School of Dentistry can be found, including a history of the collection and a list of 199 specimens and some plates.[9] Parts of this specific anatomical collection have been preserved and in 1996 were donated from the School of Dentistry to the U-M Museum of Anthropological Archaeology. The museum has documented the collection in a digital database accessible through the University of Michigan Libraries. Because of this database, students today can access some of the fruits of Corydon Ford's labor at any time, day or night, from around the globe.

Notes

1. Donald F. Huelke, "The History of the Department of Anatomy—The University of Michigan. Part I. 1850–1894," *Medical Bulletin (University of Michigan Medical Center journal)* 27 (1961), 1–27.

2. University of Michigan Bentley Historical Library: General University Timeline. http://bentley.umich.edu/legacy-support/umtimeline/general.php. Retrieved September 10, 2015.

3. Huelke, 7.

4. Ibid., 10.

5. Ibid.

6. Victor C. Vaughan and Martin Luther D'Ooge, *Memorial Discourses on the Life and Services of Corydon L. Ford Delivered in University Hall by Request of the Senate, June 10, 1894*, Ann Arbor: University of Michigan Press, 1894.

7. James Tobin, "Such Horrible Business," *The Heritage Project*, University of Michigan. The practice of raiding graveyards ended in 1880 when U-M demonstrator in anatomy William James Herdiman pressed for a state law that would grant the Medical School access to "bodies of the deceased" of poor people who had relied on state welfare and whose bodies were not claimed by relatives. http://heritage.umich.edu/story/such-horrible-business/#A-Necessary-Inhumanity. Retrieved April 4, 2015.

8. Vaughn and D'Ooge, 14.

9. Corydon L. Ford papers, Bentley Historical Library.

JOSEPH BEAL STEERE (1842–1940)

Naturalist, Explorer, and Curator

Kerstin Barndt

When Joseph Beal Steere returned from his first five-year-long world expedition in 1875, the University of Michigan bestowed upon him the first honorary PhD in its history and welcomed him back as a national celebrity.[1] During his travels through South America, the Pacific Islands, and East Asia, Steere had collected and purchased seashells (about 100,000), ancient and modern pottery, musical instruments, insects, bird and mammal skins, fishes, rocks, corals, and plants, including many specimens new to science. To house these rich collections, the university built a new museum on State Street and appointed Steere its first head curator. Steere curated the exhibit of Michigan's fauna for the 1893 World's Columbian Exposition in Chicago, and he served the university as professor of zoology until he was forced to retire in 1894.

Fig. 44. Joseph Beal Steere, 1880s. Bentley Historical Library.

Steere (fig. 44) was a nineteenth-century polymath and naturalist in ways both typical and unlikely. Throughout his life, unknown lands and "unbroken wilderness" beckoned to him.[2] Inspired by leading scientists of his day—such as Alexander and Wilhelm von Humboldt, Charles Darwin, Alfred Russel Wallace, Louis Agassiz, and Asa Gray— Steere repeatedly visited Brazil, Taiwan, and the Philippines. Throughout these voyages, he made collections, securing for the University Museum over seventy species of birds new to science, among many other things.[3] Steere's writings and collections cover a wide range of fields—from ethnography to linguistics, from zoology and geology to paleontology and botany. He was an avid writer, producing extensive field notes, scientific essays, and even short stories. He also wrote about political and moral questions of his day, advocated for women's education, mobilized humanitarian aid for China, and took a stance in favor of prohibition in Michigan.[4] Indeed, Steere's outspokenness against alcohol in Ann Arbor met with opposition and was one of the reasons the regents asked him to retire early. As a devout Christian, he believed in the civilizing force of religion and sought to reconcile the theory of evolution with natural theology.

Born in the township of Rollin in Michigan's Lenawee County in 1842, Steere spent most of his childhood close to relatives from his mother's side, the Beal family. When Joseph was eleven years old, the family moved to the mining settlement of Bloomer Township. His father, William Milhouse Steere, was a Quaker and farmer who had come to Michigan via Virginia and Ohio. When his health declined, the township elected him justice of the peace. His mother, Elisabeth Beal Steere, belonged to the Methodist church.[5] Due to his father's opposition, Joseph did not fight in the Civil War and moved instead to Ann Arbor, where he attended classes at Ann Arbor High School to prepare himself for the college entry exam. Mary H. Clark, an avid botanist and educator, and the physician Dr. George W. Pray, one of the first graduates from the University of Michigan, mentored and supported Joseph in his scholarly pursuits as well.[6] While Clark introduced him to naturalist fieldwork, Pray taught him Latin and Greek. In 1864, Steere enrolled in the University of Michigan earning his BA in 1868 and a law degree in 1870. His passions lay elsewhere, however, and while in law school, Steere also pursued his interests in the natural sciences. He joined Alexander Winchell's efforts to organize the University Museum's collections and together with his classmate Mark W. Harrington labeled and arranged the insect collection and other specimens.[7] He also began to contribute Michigan insects, shells, and birds to the collections, a practice he continued throughout his life.

WORLD TOUR, 1870–75

Steere's decision to embark on a world expedition must have occurred during the last years of his studies. In March of 1870, the regents noted his plans and pledged institutional support.[8] From the outset, Steere's travel was meant to contribute to the University Museum. While affiliation with the university eased shipping expenses and provided Steere with contacts to US consuls in Brazil and the Philippines, the global tour was cofinanced by his mother's cousin Rice Aner Beal, the owner and editor of Ann Arbor's weekly *Peninsular Courier and Family Visitant*. As a service to the newspaper, Steere agreed to send regular travel reports—almost one hundred over the course of the five years. Together with his detailed field notes held at the Bentley Historical Library, these articles grant insight into his scientific inquiries, modes of travel, and encounters with people, cities, and landscapes.

On September 20, 1870, Steere sailed from New York, arriving in Brazil seventy days later. He spent the following twelve months close to the Amazon and along the river's tributaries and the second year of his journey in Peru and Ecuador. He traveled by steamer, horseback, or raft, and crossed the Andes on foot. More than once he traveled alone, at other times securing the help of local guides and the support of European colonists, missionaries, and US consuls. Along the way he visited the old cities of Moyobamba, Chachapoyas, and Cajamarca, as well as Para, Lima, and Quito.[9] From Lima, he crossed the Pacific to China, where he traveled to Hong Kong and Canton (Guangzhou) before embarking on an extensive exploration of the not yet colonized interior parts of Formosa (Taiwan). Afterward, he visited the Philippines where he collected for almost a year. In 1875, he began to make his way back to Michigan, with shorter stays in Singapore and on the islands of Java, Ternate, and Macassar. As documented in his journals, Steere had weathered excruciating circumstances and feared for his life but never once entertained the thought of cutting his travels short. His bravery was noted by his contemporaries: "Dr. Steere will receive the hearty applause of all naturalists for the vigorous way in which he combatted the difficulties of climate and personal danger . . . , and reaped such an abundant harvest in the face of these trials."[10]

While Steere was traveling, he shipped specimens back to Ann Arbor. On March 4, 1871, Winchell and Harrington recorded the arrival of the first two boxes of specimens, from Para, Brazil, in the journal of the University Museum. They noted that the boxes included birds, shells, and crustaceans.[11] Over sixty boxes would follow; together, their contents served as the nucleus for continued global collecting on campus. Steere particularly boosted the collections of birds, mollusks, insects, and fishes; he also contributed valuable fern specimens from South America, Formosa, and the Philippines to the Herbarium. The anthropological collections received over five hundred pieces of ancient and modern pottery from Peru and the Amazon basin, and Taiwanese

Fig. 45. Steere (in center) with University of Michigan students, 1887 Philippine Expedition. Bentley Historical Library.

carpenter tools and manuscripts, household items, and clothing.[12]

With the eyes and ear of a naturalist attuned to variability among and within species and across geographical divides, Steere imbued his travel accounts with precise descriptions of flora and fauna—as well as of social, economic, gender, and racial divisions. He noted Brazilians' tolerance for interracial marriage as well as the country's reliance on slave labor.[13] From his home base in Para (Belém) on November 25, 1870, he commented on the citizens' religious practices, housing, and food. At the same time, he described the abundance of seashells and oysters he found on the beach and planned collecting forays into the rainforest.

During his extensive travels along the Amazon tributary, the Purus River, he extended his studies to indigenous languages and music.[14] In his field notes, Steere recorded lists of words to compare the basic vocabulary of several languages he encountered.[15] He continued his ethnographic and linguistic studies in Taiwan, leading to one of his first publications in 1874.[16] Following earlier comparative linguistic studies by naturalists Peter Simon Pallas and Wilhelm von Humboldt, Steere compiled word lists and notations of songs and also general descriptions and portraits of the island's population that show distinguishing tattoo art. He delineated the island's inhabitants hierarchically into three groups: (1) the colonial Chinese; (2) the Pepo-whan, "the ancient inhabitants of the great plains to the west, who possessed some of the elements of civilization, and who, from their exposed situation, soon fell under the dominion of the Chinese, when they entered the island two hundred years ago"; and (3) the indigenous "wild and savage" tribes in the mountainous parts of the island, among them the headhunting Bon-whan.[17] In his ethnographic travel writing, Steere's tone varied between sober description, moral indignation, and recognition of difference, tinged by contemporary colonial-era modes of categorization.

In the Philippines, Steere turned his attention to the archipelago's diverse bird population. He found several species unknown to him and sent the specimens to Richard B. Sharpe, ornithologist at the British Museum, for identification. Sharpe was able to confirm the discoveries and published on the new species on Steere's behalf, including on the Philippine azure-breasted pitta; in Steere's honor, the colorful bird was introduced into Western science as *pitta steerii*.[18]

With these discoveries, Steere followed in the footsteps of Darwin, Wallace, and Viscount Walden and contributed to the nascent field of biogeography, a field that supported the theory of evolution and the formation of continents and other landmasses through research into the geographical distribution of species. On his second trip to the Philippines in 1887, Steere continued his bird studies. Eventually, he exhibited the results at the Chicago Columbian Exhibition.[19]

MUSEUM CURATOR

The success of his global collecting tour brought Steere recognition abroad and at home. During his absence, the regents had already bestowed upon him an honorary doctoral degree; after Steere returned, they were eager to secure his expertise for the university and its museum. Steere accepted the position of assistant professor of paleontology and curator of the museum (later changed to assistant professor of zoology and paleontology). In 1879, he was promoted to professor of zoology, "the position of Curator in the Museum continuing with the advance in rank."[20] Steere's sponsor and cousin Rice A. Beal struck a deal with the university that transferred all of the specimens and artifacts that Steere had sent to campus during his travels to the museum as the newly named Beal-Steere collection.[21] With the collection and tenured professorship in hand, Steere married teacher and principal Helen F. Buzzard. The couple would have nine children.

The addition of several thousand new specimens strained the capacity of the original University Museum in the north wing of University Hall. President Angell and the regents successfully lobbied the state legislature and secured funds for a new museum building (see Barndt and Sinopoli, this volume). In 1881, six years after Steere's return from his world expedition, the museum opened and Steere became its first curator. Except for the Herbarium, which remained in University Hall, he now shared space with the many collections that he had helped build: zoology, ethnography, geology, paleontology, and archaeology. These collections remained united under the umbrella institution of "Museum of Natural History," and it was fitting that a naturalist like Steere—whose research bridged the gradually expanding divide between cultural and natural sciences—would take the lead. While the collections in mineralogy and geology were overseen by faculty in the corresponding fields, Steere curated the zoological, archaeological, and ethnographical collections.[22] The arrangement of the new display cases in the University Museum prominently exhibited Steere's collection of shells and many of his birds and mammals. The museum also showcased the anthropological collection, including burial urns, decorated food vessels, and vases from the Amazon delta and Southeast Asia.[23]

During Steere's curatorship, the University Museum also prepared the first public exhibition of the Chinese collection of porcelain and textiles that had been gifted to the university in 1885 by the Chinese Imperial Government after first being displayed at the 1884–85 New Orleans World Cotton Centennial Exposition.[24]

PHILIPPINE TAMARAW AT THE COLUMBIAN WORLD'S FAIR

To boost the University Museum's collections and his own research, Steere returned to the Amazon delta in 1879 and brought back "about a thousand botanical and zoological specimens" to the museum.[25] In 1887, he assembled a team of seven young men and local guides to continue his biogeographical studies of the Philippine Islands. Part of the team were U-M students Dean Conant Worcester (see Rice, this volume) and Frank Swift Bourns (fig. 45). Also joining the team was Mateo Francisco, a Filipino Visayan who had worked

Fig. 46. Tamaraw group display in University Museum, 1890s. Museum of Zoology.

for Steere during his first Philippine expedition and had accompanied him back to Ann Arbor in 1875.[26] Over twelve months, Steere, his students, and their guides identified fifty new bird species.[27] In one of his essays, Steere described the Philippine archipelago as a "naturalist's paradise" where he and his men "were rapidly adding species to [their] collection, and new to science."[28] This collecting fever resulted in a comprehensive study of Philippine birds and their distribution among the islands. Expanding on biogeographical research by Darwin, Wallace, and Philip Sclater, Steere formulated a new "law of distribution of non-migratory land birds of the Philippines." First presented at the American Association for the Advancement of Science in 1890, this paper contributed to the theory of evolution in its focus on "isolation" as the "first and the necessary step in the formation of species."[29]

Most of Steere's essays about the trip, however, were composed as travel narratives that closely described events of the journey alongside wide-ranging observations on modes of travel; weather conditions; the particularities of geological formations, flora, and fauna; and ethnographic descriptions of indigenous people and myths. On the island of Mindoro, for instance, the expedition set out to find the "savage beast" tamaraw (or tamarou), a small buffalo known as a legendary predator but not yet scientifically described. Francisco succeeded in killing a young bull and later two more animals.[30] Steere himself skinned the buffalos, prepared the skeletons, and shipped them back to Ann Arbor. He also scientifically described the species and sent his writings to Sclater, secretary of the Zoological Society of London, who published Steere's finding in the August 1888 issue of *Nature*. In his account of the tamaraw hunt, Steere reported that "the skins and skeletons were great curiosities to the people of the town."[31] In Ann Arbor this curiosity was replicated after museum taxidermist A. B. Convert mounted the buffalos (a bull, cow, and calf) for display. A photograph of the animal group (fig. 46) shows the family in peaceful assembly, tamed by the art of taxidermy. As type specimen, one of the buffalos is still part of today's U-M Museum of Zoology, albeit under secure cover because of prior applications of arsenic to repel insects.

After Steere's return to campus, President Angell asked him to serve on the committee in charge of the university's contributions to the 1893 Chicago Columbian Exposition. As part of the university showcase at the World's Fair, the University Museum displayed specimens from the Beal-Steere collection: the tamaraw group, corals and birds from East Asia, and a map showing the distribution of birds across the Philippines.[32] For the Michigan State Building, Steere assembled a comprehensive exhibit of Michigan fauna. Starting from the zoological collections of the university, he completed the series with a number of purchased specimens. The entire set returned to Ann Arbor after the fair's closing, positioning the museum more strongly than ever before as a distinct state institution.

RETIREMENT

While his research was being showcased in Chicago, Steere threw himself into a controversy about alcohol on campus that would cost him his professorship. He fought for prohibition on moral and religious grounds to "root out evil" and "protect boys and weak men."[33] In an open letter to "Christians of German descent," he tried to breach the divide on the issue that had developed between Ann Arbor immigrants from Germany and "Christians of older American families." On campus and in town his political activism was met with consternation rather than support. In June 1894, the regents officially asked Steere to resign. He refused and instead circulated pamphlets in town to protest, announcing his run for state senator on a prohibition ticket. The university stood its ground, citing economic reasons for Steere's termination.[34] In September, Steere was given the opportunity to appear at the meeting of the Board of Regents where he "was assured that the actions of the regents were not meant as a reflection of his character."[35] If not his character, letters by Steere's former student Dean Worcester criticized Steere's contributions as teacher and museum curator. According to Worcester, the mounts and specimens that had come back from Chicago had been left in disarray; in fact, the entire museum and its catalog were not in good shape. With equal disapproval, Worcester commented on Steere's teaching: "Bourns, or any one else of average intelligence and industry for that matter could take any good text-book and give a better course than Steere has been giving."[36]

Fig. 47. Steere engaged in field work, 1930s. Museum of Zoology.

Worcester's own ambitions may have been behind these critiques; however, one hint for why they took hold may come from notes taken by a student who attended a lecture course taught by Steere in 1891 on the history of zoology and evolution. The notes reveal a historiographical approach to the topic, which might have been out of step with new ideas and the turn to more laboratory-based inquiries in biology at the end of the nineteenth century.[37]

While it is difficult to ascertain how much Steere's Christian belief had influenced his university teaching (and provided the college with yet another bone of contention), he publicly discussed "Evolution from a Christian Standpoint" in a lecture series at Ann Arbor's First Methodist Episcopal Church immediately after his forced retirement.[38] More openly than in his previous writings he began to tackle questions regarding the compatibility between his scientific studies and religion. To his mind, the advances in science exposing evolution only confirmed his image of a Christian god as purposeful, loving, and with absolute generating power: "I taught and studied one branch of science for many years," he concluded, "and though I found many so-called scientific theories which conflicted with religion, I found no conflict of science itself with religion."[39]

Steere continued his scientific work and traveled to Brazil once more in 1901 to conduct fieldwork for the Smithsonian Institution. His field notes suggest that the collection of naturalia and anthropological artifacts was meant to complete a series on display at the Pan-American Exposition, a World's Fair held in Buffalo during the same year.[40] His published report resembles his earlier ethnographic writing in diction and detail, describing the social, economic, and environmental changes since his first encounter with the region in the 1870s, and including word lists, music, and songs. Small villages such as Manáos had seen immense growth into a "city of 30,000 with street cars and electric lights" and settlements of rubber gatherers lined the rivers.[41] Revisiting former settlements of indigenous tribes, he found Paumarí and Jamamadí greatly reduced in numbers and living in dire circumstances.[42]

With the passing of time and with new museum curators in place, Steere found his way back into the museum community on campus (fig. 47). Museum directors Alexander Ruthven, Frederick Gaige, and Carl Guthe reached out to Steere and acknowledged his pioneering work for the university museums. Steere attended the opening of the new Museum Building in 1928, which bears a bas relief of the naturalist on its north façade. On the occasion of Steere's ninetieth birthday in 1932, the museum journal *The Ark* dedicated an issue to his life's work and collections. Gaige reports that "hardly a year has passed that he [Steere] has not brought to the Museum interesting specimens and observations. . . . We look forward to his visits. Only a few days ago he came with specimens and told of explorations in a cave in Guimaras [Philippines] where he found the strange nest of the Edible-nest Swiftlet. He had the specimen in his hand."[43] Just thirteen months shy of his one-hundredth birthday, Steere died in Ann Arbor on December 7, 1940.

During his lifetime, Steere worked across disciplines and contributed to multiple collections that were subsequently dispersed as they entered separate museums. More recently, three museums that hold Steere-related collections (Zoology, Herbarium, and Paleontology) have joined forces to digitally reunite the naturalist's interdisciplinary collections.[44] Facilitating access and inspiring new research, this National Science Foundation supported project recognizes and cements Steere's lasting impact on the history of collecting at the University of Michigan.

Notes

1. This article is based on the Steere papers held at the Bentley Historical Library. William Morgan assisted me in working through this archive and I thank him for his thoughtful and thorough research notes.

2. Cf. Steere, "The Island of Mindoro," *The American Naturalist* 25 (December 1891), 1041–54, here: 1046.

3. The Thayer expedition had brought Agassiz to Brazil in 1846 and Steere sought to consult with the Harvard zoologist before his departure in 1870, albeit without success. Instead he met with geologist Charles Frederick Hart, who had accompanied Agassiz to Brazil. See Nelson Papavero, William Overal, Dante Teixeria, and Janet Hinshaw, "The Travels of Joseph Beal Steere in Brazil, Peru and Ecuador (1870–1873)," *Arquivos de Zoologica* 39.2 (2008), 87–269, here: 103–4.

4. See Steere, "Help China. Help America. Letter to the Editor," *Detroit News* September 3, 1931. For Steere's active role in the temperance movement, see endnote 33.

5. Steere, "The strange story of my father," handwritten autobiographical account, Steere Papers, Bentley Historical Library.

6. Steere, "Autobiography," typescript, Steere Papers, BHL.

7. Alexander Winchell, "Report to the Board of Regents," September 28, 1868.

8. Proceedings of the Board of Regents (1870–1876), March 25, 1870, 40.

9. See Papavero et al., 87.

10. Richard Bowdler Sharpe quoted in Gaige, "Joseph Beal Steere . . ." 1932, 352.

11. *Journal of The Museum of the University of Michigan in the Department of Geology, Zoölogy and Botany, and the Department of Archaeology and Relics*, Journal B (Roth), 1930 typescript of the original handwritten journal, Museum Library, University of Michigan, 19–20.

12. See descriptions of specimens collected by Steere for each division in the Museum of Zoology in *The Ark*, March 1932, 8–11.

13. Papavero et al., 108, 118.

14. On November 30, 1871, he translated a few lines of songs from the Hypurinas (Apurina) and Pamarys (Paumaris). Ibid., 168.

15. Steere, "Language and Music of a few Tribes of the River Purus, Brazil," and "Vocabularies of several of the Indian tribes of the river Purus in Brazil, and of Pebas and the river Huallaga in Peru, with some account of the different tribes and their customs, music, traditions," handwritten manuscripts, Steere Papers, Bentley Historical Library.

16. Steere, "Formosa," *Journal of the American Geographical Society of New York* 6 (1874), 302–34. Steere is credited as having been among the first naturalists to describe five important Austronesian tribes and languages of Formosa. See Paul Jen-kuei Li, "Forward" in *Joseph Beal Steere. Formosa and Its Inhabitants,* Institute for Taiwan History, 2002.

17. Ibid., 303–4.

18. Sharpe, "On the Birds collected by Prof. J. B. Steere in the Philippine Archipelago," read November 16, 1876. *Transactions of the Linnean Society of London*, vol. I, part 6, 1877; see also P. L. Sclater and Osbert Salvin, "On the Collection of Birds made by Prof. Steere in South America," *Proceedings of the Zoological Society of London* (February 1878), 135–42.

19. See Frederick Gaige, "Joseph Beal Steere. Naturalist, Explorer, Educator," in *Michigan Alumnus* 38 (February 1932), 345–46 and 352–53, here: 346; and Steere, "A list of the Birds and Mammals collected by the Steere Expedition to the Philippines, with localities and with brief preliminary descriptions of supposed new species," Ann Arbor: Courier Office, 1890.

20. Gaige, "Joseph Beal Steere," 353.

21. The deal was contested because Beal formulated his demands in the context of the Douglas-Rose controversy pertaining to the mismanagement of student fees in the chemistry laboratory. In this controversy, Beal pledged full support for Preston B. Rose and argued for the dismissal of long-time director of the laboratory Silas H. Douglas. See Lewis G. Vander Velde, "Douglas-Rose Controversy," *The University of Michigan, An Encyclopedic Survey*, part I, 208–13.

22. For the new bylaws governing the administrative organization of the university's collections, see *Proceedings of the Board of Regents*, October Meeting, 1882, 293.

23. See Harlan I. Smith, "The Study of Man. Anthropology at the University of Michigan," *Detroit Free Press*, November 12, 1893.

24. *Proceedings of the Board of Regents*, October 1886, 49.

25. Gaige, "Joseph Beal Steere," 353.

26. A photo of Mateo Francesco is included in Gaige, "Joseph Beal Steee," 352.

27. Steere, "A List of Birds and Mammals collected by the Steere expedition to the Phillipines" (see footnote 18).

28. Steere, "Observations Made in the Central Philippines," *The American Naturalist* 22 (July 1888), 622–26; here 622; and Steere, "A Month in the Eastern Philippines," *The American Naturalist* 23 (March 1889), 102–7; here 105.

29. Steere, "The Distribution of Genera and Species of Non-Migratory Land Birds in the Philippines," *The Auk* 11 (July 1894), 231–40; here 240.

30. Steere, "The Island of Mindoro," *The American Naturalist* 25 (December 1891), 1049–53.

31. Ibid., 1054.

32. *University Record* 3 (June 1893), 43.

33. Steere, "Appeal to Christian of German descent," undated typescript, Steere Papers, Bentley Historical Library. Steere was a member of the Michigan State Temperance Alliance and the U.S. Inter-Collegiate Prohibition Association.

34. Letter from Worcester to Reighard October 5, 1894, Reighard Papers, Bentley Historical Library; and "Regents Meet," *Ann Arbor Argus*, September 21, 1894.

35. Ibid.

36. Letter from Dean Worcester to Jacob Reighard, July 3, 1894; and Worcester to Reighard, July 9, 1895, Reighard Papers, Bentley Historical Library. In his letter to Reighard from February 27, 1895, Worcester reports that the laboratory course had become more popular among students since Steere's departure.

37. Arthur J. Tuttle, "Professor Steere: Course of Zoology. Development of Species and Theories of Classification," handwritten lecture notes; Arthur J. Tuttle Papers, Bentley Historical Library. On the history of biology, see Keith R. Benson, "From Museum Research to Laboratory Research. The Transformation of Natural History into Academic Biology," *The American Development of Biology*, ed. by Jane Maienschein et al., New Brunswick: Rutgers University Press, 1991, 49–83.

38. See note under "Local Brevities" in *Ann Arbor Argus*, November 23, 1894. At Ann Arbor's First Methodist Episcopal Church, Steere served as chairman of the Board of Stewards. See *Manual. First Methodist Episcopal Church*, Ann Arbor, 1893.

39. See Steere, "How the World Looks at Eighty. An Old Man's Confession of Faith," *The Michigan Christian Advocate*, January 21, 1925, 10 and 22. See also "This artificial World as Proof of the Existence of God," typescript, and "Life begins at Seventy-Five. An Address to the 500 Members of the Three-quarter-century club of Washtenaw County," *The Michigan Christian Advocate*, January 10, 1935, 9–10; Steere Papers, Bentley Historical Library.

40. "Joseph Beal Steere, Field Catalog, 1901," typescript, Smithsonian Institution, Division of Mammals Collection.

41. Steere, "Narrative of a Visit to the Indian Tribe of the Purus River," *Report of the United States National Museum for 1901*, 359–93; here: 364.

42. Ibid., 370.

43. Gaige, "Joseph Beal Steere. Master Naturalist," *The Ark*, March 1932, 7.

44. See http://sites.lsa.umich.edu/biomuseums-nsf-project.

WILBERT B. HINSDALE (1851–1944) AND THE BEGINNINGS OF MICHIGAN ARCHAEOLOGY

Richard I. Ford

Dr. Wilbert Hinsdale had long roots in Ohio (fig. 48). Born in Wadsworth, Ohio, on May 25, 1851, Wilbert B. Hinsdale was raised there until he went to Hiram College, receiving a bachelor of science degree in 1875. He then attended the Cleveland Homeopathic Medical College, where he received his doctor of medicine degree. Hinsdale returned to Hiram to earn a master's in biology before joining the faculty of the Cleveland Homeopathic Medical College in 1890. He left Cleveland in 1895 for the Homeopathic Medical College at the University of Michigan.

At the University of Michigan, Hinsdale had two unrelated but equally successful careers. The first was in the Homeopathic Medical College, where he quickly was promoted through the professorial ranks, becoming professor of materia medica, therapeutics, and clinical medicine and soon thereafter dean of the college. He published widely on homeopathic subjects and also on broader humanistic topics.[1] There was, however, longstanding tension between homeopathic treatment, which the regents of the University of Michigan had accepted in 1875, and the allopathic (contemporary) approach in the Medical School, which had been founded in the 1850s. The existence of two medical schools was regularly questioned by the state legislature and by the regents. Finally, in 1922, the Homeopathic Medical College was absorbed into the Medical School and "Doc" Hinsdale's term as dean ended with his first retirement.

Hinsdale had been a revered instructor and a greatly appreciated dean. At his retirement (and the closing of the hospital), the Homeopathic Hospital Guild, a women's charitable organization associated with the hospital decided to establish a scholarship in Hinsdale's honor. Rather than being dedicated to medicine, the scholarship focused on another of Hinsdale's interests: Michigan's Native peoples. Encouraged by Alexander Ruthven, members of the guild approached the university regents with the proposal to create the "Homeopathic Hospital Guild Scholarships in Michigan Ethnology," used to "aid sociological and ethnological studies with preference given to studies of the Indian population of Michigan," with the proviso that the funds would be administered by the newly created Museum of Anthropology[2] and that "the data secured by such studies shall be made available to the University and to philanthropic and social agencies interested therein."[3]

Hinsdale had longstanding interests in Michigan's Native communities. After moving to the state, the doctor and his wife, Estella Hinsdale, had begun vacationing each summer in northern Michigan, where "Doc" Hinsdale became familiar with the health and economic conditions of the Anishinaabe Indians. At the same time, Mrs. Hinsdale began to collect handmade black ash baskets (fig. 49) and Doc collected wooden artifacts ranging from bowls to kettle hooks. Together they developed one of the finest Michigan Indian ethnographic collections found in any private collection. In 1925, Estella Hinsdale donated a considerable portion of this collection to the new Museum of Anthropology; over the next few years she donated the remainder of the collection.

When the Homeopathic Hospital shut its doors in 1922, Doc was seventy years old. But he was in good health and ready to embark on a second career—as an archaeologist in the Museum of Anthropology. Because of his years of avocational study of Michigan archaeology and Indian ethnography, the regents' accepted his new position as custodian in charge of the collections in Michigan

Fig. 48. Wilbert B. Hinsdale at work, date unknown. Bentley Historical Library.

archaeology and as museum assistant to Museum of Zoology director Alexander Ruthven.

Although the Museum of Anthropology was approved by the regents in 1922, it did not become a reality until space was available; initially it was based in the Museum of Zoology (and wherever space for collections could be found) under Director Ruthven's supervision. The regents appointed Dr. Carl E. Guthe, a skilled and experienced archaeologist, to be in charge of the anthropological collections, with the initial title of associate director of anthropology (and later, director). However, Guthe immediately departed for the Philippines, where he excavated for three years (1922–25), leaving Hinsdale, without a degree in anthropology, to handle archaeological issues for the museum.

Hinsdale quickly embarked on an ambitious Michigan archaeology research program. For years Ruthven had received letters about archaeological sites in the state and offers of artifact collections for the museum. These Hinsdale now handled. With a legacy of already accumulated archaeology collections in the museum, combined with sheaves of Ruthven's correspondence about site locations, Hinsdale began to visit these locations and to meet with farmers and artifact collectors throughout the state. Hinsdale was a personable and popular ambassador for the museum and quickly won the confidence of these "citizen scientists." As his archaeological knowledge increased, Hinsdale gave lectures around the state about Michigan aboriginal archaeology, further expanding his contacts with interested citizens.

Ruthven encouraged Hinsdale to systematize this archaeology knowledge into an atlas. As Hinsdale conducted his statewide explorations, his confidence as a professional archaeologist had grown, and he synthesized his archaeological knowledge into a series of books[4] and professional papers[5] developed from topics in the evolving atlas. Still, although Hinsdale agreed with Ruthven about the importance of a comprehensive atlas,

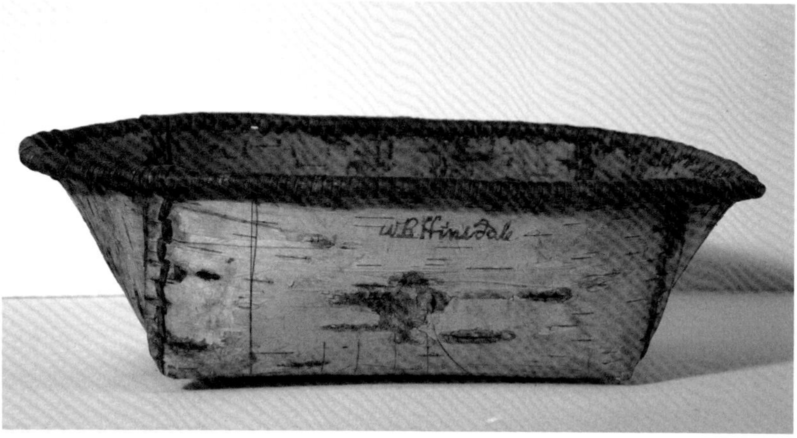

Fig. 49. Birch bark basket in the Estella Hinsdale Collection. Museum of Anthropological Archaeology UMMAA 5281.

he needed help to produce it—first, because he was traveling so much around a large and poorly paved state in a family car. This problem was solved by Henry Ford, who donated an automobile to the museum for Hinsdale's use. Second, Hinsdale was determined to map all the villages and mounds, trails, and portages onto individual county maps by township. To accomplish this, he needed a skilled cartographer, and so the artist Edward J. Stevens was hired. Hinsdale himself undertook to raise funds for Stevens's salary and the atlas, requesting donations from businessmen, especially in Kalamazoo and Grand Rapids, as well as from the public at his many lectures. It worked! The folio-size *Archaeological Atlas of Michigan*[6] was published in 1931 to much professional and national acclaim (fig. 50). It was the first publication to pull together Michigan site location information and, therefore, was valuable to the state's professional as well as amateur archaeologists. That same year, the U-M regents promoted Hinsdale to associate in charge of the Division of the Great Lakes.

Hinsdale's growing list of publications enabled him to become the face of anthropology at the university. He founded the Michigan Archaeological Society and was omnipresent at the annual meetings of the Michigan Academy of Science, Arts and Letters, where he created an anthropology section and frequently presented papers. He was elected president in 1932.[7] He occasionally attended the American Anthropological Association meetings to interact with other archaeologists and to present papers. However, he mostly attended the Central States Branch of the Anthropological Association, where he was a board officer. Well into his second career, he soon became recognized as "the father of Michigan archaeology."[8]

Although Hinsdale never taught a formal anthropology course at U-M, he was the only anthropologist on campus while Guthe was in the Philippines, and he made himself available to students who were interested in archaeology and in learning about Michigan's prehistoric past. One undergraduate in particular won his favor for his intelligence and enthusiasm: Emerson Frank Greenman, who often accompanied Hinsdale to explore sites. Greenman graduated with a bachelor of arts degree in 1923, and Hinsdale encouraged him to go for formal excavation training in G. G. MacCurdy's distinguished American School of Prehistoric Research in Europe. Greenman earned a diploma at Oxford in 1924 and returned to U-M, earning the university's first doctoral degree in anthropology. Because there was not yet an anthropology department or doctoral program, Greenman enrolled as a special student in the graduate school and was awarded his PhD in 1927. While still a student, he was hired as an assistant in the museum in 1926, under the supervision of the then-returned Guthe. He went to the

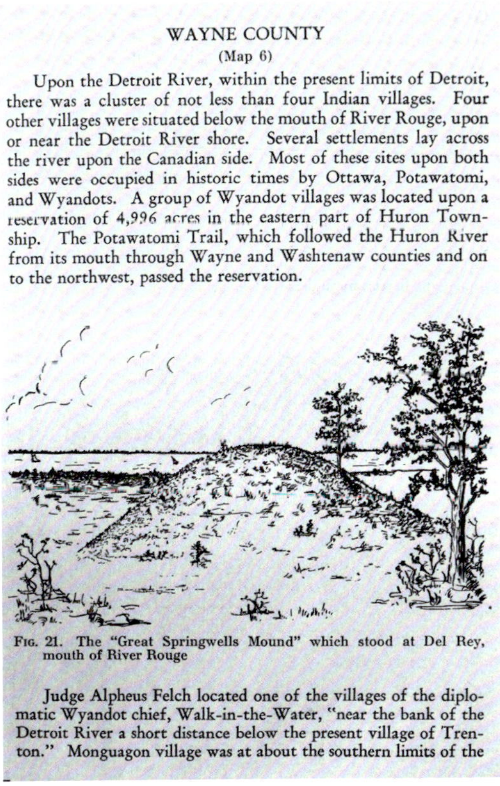

WAYNE COUNTY

(Map 6)

Upon the Detroit River, within the present limits of Detroit, there was a cluster of not less than four Indian villages. Four other villages were situated below the mouth of River Rouge, upon or near the Detroit River shore. Several settlements lay across the river upon the Canadian side. Most of these sites upon both sides were occupied in historic times by Ottawa, Potawatomi, and Wyandots. A group of Wyandot villages was located upon a reservation of 4,996 acres in the eastern part of Huron Township. The Potawatomi Trail, which followed the Huron River from its mouth through Wayne and Washtenaw counties and on to the northwest, passed the reservation.

Fig. 21. The "Great Springwells Mound" which stood at Del Rey, mouth of River Rouge

Judge Alpheus Felch located one of the villages of the diplomatic Wyandot chief, Walk-in-the-Water, "near the bank of the Detroit River a short distance below the present village of Trenton." Monguagon village was at about the southern limits of the

Fig. 50.Wayne Country entry in the *Archaeological Atlas of Michigan*. Photo by Carla Sinopoli.

Ohio Historical Society in 1928 for his first formal appointment. However, at Hinsdale's urging, Greenman returned to U-M in 1935 as a research associate in Great Lakes archaeology.

At eighty-four years old and after a prolonged hospitalization, Hinsdale no longer had the stamina to tour Michigan's archaeology sites. With Greenman in the Great Lakes Division in the Museum of Anthropology (by then it was a fully independent unit ensconced in its new space in the Museums Building), all field work became Greenman's responsibility, while Hinsdale spent the remainder of his professional career writing archaeological papers,[9] social commentary,[10] and fictionalized, autobiographical books about his youthful life in rural Ohio.[11] Then the oldest active professor at the University of Michigan, Hinsdale suffered a stroke on May 13, 1944, and died on July 25, 1944, leaving behind two marvelous legacies in two separate colleges at the University of Michigan.

Notes

1. Victor C. Vaughn and Wilbert B. Hinsdale (contributor), "The value of humanistic, particularly classical studies as a preparation for the study of medicine and of engineering, from the point of view of the professions," *Symposium from the Proceedings of the Classical Conference Held at Ann Arbor, Michigan, March 29, 1906*. Official publication. n.s. v. 8, no. 3, suppl. (Ann Arbor: University of Michigan, 1906).

2. The Museum of Anthropology was formally established as a separate division within the Museum of Zoology in 1922, the same year that the Homeopathic Hospital closed; see Museum of Anthropological Archaeology entry, this volume, and and the separate essays by Barndt and Sinopoli in part 1.

3. The original donation of $935 has grown considerably since 1922, and students continue to benefit from the Homeopathic Hospital Scholarship. Documents on file in the Museum of Anthropological Archaeology endowment records.

4. Wilbert B. Hinsdale, *Primitive Man in Michigan* (Ann Arbor: University of Michigan, 1925), *The Indians of Washtenaw County, Michigan* (Ann Arbor, MI: G. Wahr, 1927), *The First People of Michigan* (Ann Arbor, MI: G. Wahr, 1930).

5. Wilbert B. Hinsdale, "Indian Corn Culture in Michigan," *Papers of the Michigan Academy of Science, Arts and Letters* VII (1927): 31–49.

6. Wilbert B. Hinsdale, *Archaeological Atlas of Michigan* (Ann Arbor: University of Michigan Press, 1931).

7. Wilbert B. Hinsdale, "Tecumseh's Illusions: Presidential Address," *Papers of the Michigan Academy of Science, Arts and Letters* XVIII (1932): 31–52.

8. https://www.lsa.umich.edu/ummaa/about/history/wilbertbhinsdale_ci, accessed March 31, 2016.

9. Wilbert B. Hinsdale and Emerson Frank Greenman, *Perforated Indian Crania in Michigan* (Ann Arbor: University of Michigan Press, 1936).

10. Wilbert B. Hinsdale, "Property Rights and Indian Marriage: Two Cases in the Michigan Supreme Court," *Michigan Alumnus Quarterly Review* xx (1935): 154–61.

11. Wilbert B. Hinsdale, "Them Times: Memoirs of a Pioneer Community—I," *Michigan Alumnus Quarterly Review* (Autumn 1935): 619–32; "Them Times: Memoirs of a Pioneer Community—II," *Michigan Alumnus Quarterly Review* (Winter 1936): 49–62; "Them Times: Memoirs of a Pioneer Community—III," *Michigan Alumnus Quarterly Review* (Autumn 1936): 358–72.

FRANCIS W. KELSEY (1858–1927)

Shaping Classical Archeology

John Griffiths Pedley

Professor of Latin language and literature and chair of the Department of Latin from December 1889 until his death in 1927, Francis Kelsey (fig. 51) was a man of many parts: philologist, archaeologist, author, editor, administrator, fund raiser, campus planner, proponent of music and the fine arts, and advocate for World War I refugees.[1] At the center of developments in American education, he played a leading role in the growth of the University of Michigan and in the transformation of classical studies in North America.

Fig. 51. Francis W. Kelsey, 1920s. Kelsey Museum of Archaeology.

Born and raised in modest circumstances in New York State, Kelsey was educated at Lockport Union School and the University of Rochester. On graduation (1880), he was appointed instructor in classics at Lake Forest University, Illinois. During his nine years at Lake Forest, he published new editions of works by Cicero, Caesar, Lucretius, Ovid, and Xenophon that became the standard teaching texts of the time. By emphasizing archaeological and historical contexts and by the extensive use of lithography, line drawings, and photographs, Kelsey broke new ground. He spent the summer of 1883 in Europe familiarizing himself with the latest methods of archaeological research and a further full year (1884–85) acquainting himself with the German system of higher education. While at Lake Forest, he met and married Mary Isabelle Badger; they had three children: Ruth, Charlotte, and Easton.

Shortly after Kelsey arrived in Ann Arbor, he became chair of his department. While his innovative ways, such as asking the library to put books on reserve for his courses, stirred resistance, he knew how to resolve problems—Kelsey guaranteed to replace at his own expense any books lost. He brought the same zest for innovation to his leadership of the department. He introduced new courses and appointed new faculty, restructured the teaching of the Latin authors to include archaeology and history, and expanded the Roman archaeology courses (fig. 52). He encouraged the enrollment of graduate students so that by 1900 the department had awarded twenty-three master's degrees and six doctorates. Emphasizing his belief in graduate education and professional publication, he began the Michigan Humanistic Series, the precursor of the University of Michigan Press. Funded largely by supporters whose help Kelsey solicited personally, this initiative illuminates Kelsey's aptitude for scholarship, innovation, organizational skill, and hard work.

Fig. 52. Francis Kelsey in the classroom at the University of Michigan, c. 1900. Bentley Historical Library.

Instrumental in the appointment of William Warner Bishop, a former student, as university librarian, he was also influential in the establishment of the Department of Fine Arts in 1910, playing a similar role in establishing the Museum of Anthropology in 1922 and laying the groundwork for the Department of Anthropology (founded in 1928). He strove repeatedly for the establishment of a School of Music, urging the regents year after year to incorporate the Ann Arbor School of Music in the university. An enthusiastic musician himself, he served for over twenty-five years as president of the University Musical Society in which capacity he helped plan Hill Auditorium and bring artists as distinguished as Enrico Caruso and Ignacy Jan Paderewski to Ann Arbor.

His scholarly prominence, based on his collaboration with the German scholar August Mau in the publication of *Pompeii: Its Life and Art* (New York 1899, Leipzig 1900), supported his election as secretary (1906) and then president (1907) of the American Philological Association (since 2013 the Society for Classical Studies), and secretary (1902–7) and then president (1907–12) of the Archaeological Institute of America. He doubled the membership of the Archaeological Institute, extending its reach into Canada and the American West, and established the American School of Archaeology in Santa Fe. He worked to bring before Congress the bill for the preservation of antiquities, gave testimony himself, and lobbied then vice-president Theodore Roosevelt in person. The resulting American Antiquities Act of 1906 was used by President Roosevelt to protect intact natural environments including the Grand Canyon as well as American antiquities. A champion thus of archaeological conservation, Kelsey also vigorously promoted outreach, working tirelessly to spread news of archaeological discoveries and lecturing across the nation.

A leader in the development of American archaeological enterprises in Mediterranean lands, Kelsey emphasized the importance of international cooperation. In 1924, with Sir William Ramsay, he organized an expedition to Antioch in Turkey, and in 1924–25 directed the Franco-American work at Carthage in Tunisia. In 1924, he also began excavations at Karanis in Egypt. This program of fieldwork expanded beyond the examination of public monuments (at Antioch) and sanctuary studies (at Carthage) to the investigation of domestic contexts (at Karanis). Focusing on the recovery of every scrap of material evidence, the work at Karanis effectively revolutionized Greco-Roman archaeology in Egypt. Kelsey personally raised much of the funding for these expeditions from Detroit businessmen, including in 1923 an eye-watering gift from Horace H. Rackham of $100,000.

GROWING THE UNIVERSITY'S COLLECTIONS

A large part (approximately forty-six thousand objects) of the Kelsey Museum's holdings derives from the excavations at Karanis (1924–35). But well before this time, Kelsey was expanding the university's collections. In Tunisia in 1893, he acquired more than one hundred ancient lamps, pots, and building materials, and in Sicily, Naples, and Rome, many other examples of Roman daily life. He shipped back more than one thousand objects to Ann Arbor. When commitments held him in Ann Arbor, other faculty members acted for him. In 1898, Clarence Meader sent back more than three hundred artifacts from Rome, and in 1899 Duane Stuart dispatched thirty-four lamps acquired in Athens.

Latin Inscriptions and the de Criscio Collection

In 1897, Kelsey was informed by former student Walter Dennison that the parish priest of Puteoli (near Naples), Giuseppe de Criscio, was willing to sell his collection of antiquities, which included a large number of stone inscriptions. Kelsey immediately found a donor, Henry P. Glover of Ypsilanti, with the result that more than 250 of the inscriptions arrived in Michigan in 1899, to be followed in 1905 by nine more. All the de Criscio material comes from districts to the immediate west of Naples, and therefore offers new light on the social history of a specific area, providing important information about the Roman fleet at Misenum and the local *cursus honorum*. Together with more than one hundred inscriptions from various sites in Rome, acquired later (1909) from Walter Dennison's collection—most of them grave markers of slaves, freedmen, or soldiers—they form the largest collection of Latin inscriptions in the country.

The Hartwig-Kelsey Reliefs

Serving as professor at the American School of Classical Studies in Rome (1900–1), Kelsey continued to buy materials of educational and scholarly value. Of particular interest are five fragments of marble relief sculpture, including a head of Vespasian, identified (in the 1980s) as part of the Templum Gentis Flaviae, a Roman sanctuary honoring the three Flavian emperors (Vespasian, Titus, and Domitian). These Kelsey fragments have been linked with nine similar fragments donated to the Museo Nazionale Romano by Professor Paul Hartwig; all fourteen fragments are now shown to be from the same monument.

Brickstamps

Other acquisitions in this period included lamps, painted terracottas, lead water pipes with stamped inscriptions, stamped Arretine pottery, and numerous stamped bricks. As good examples of the different types of stamps, the bricks offer a wide range of consular dates showing that the Roman brick industry lasted from the end of the republic until the sixth century AD. They demonstrate Kelsey's good sense in putting together a small group of artifacts that is representative of the whole industry and thus ideal for teaching purposes. At the end of his stay in Rome, Kelsey shipped seventeen boxes to Ann Arbor, four of books and thirteen containing the brick stamps, stamped water pipes, and other antiquities.

The Dattari Collection of Coins

In 1909, Charles Lang Freer (a leading contributor to many of Kelsey's projects, known best for his gift of the Freer Art Gallery to the Smithsonian Institution) discovered that his fellow collector, Giovanni Dattari, wanted to find a home for his collection of Greek and Roman coins. Knowing that Kelsey was eager to build collections, Freer asked Dattari to consider the University of Michigan. As a result, Dattari donated part of his collection to U-M (other beneficiaries were Oxford and Toronto) in 1914. The Dattari coins that came to Ann Arbor are Greek and Roman issues spanning from the third century BC to the fourth century AD.

Papyri (Sliced, Flattened, and Inscribed Stalks of Papyrus)

In the late nineteenth century, two Oxford scholars, Bernard Grenfell and Arthur Hunt, were buying up papyri in Egypt. Seemingly worthless scraps of paper, papyri have in fact yielded much information about the social history of Graeco-Roman Egypt, as well as fragments of Greek literature (e.g., a fragment of Homer's *Iliad* in the U-M collection). When the supply of papyri from dealers began to dry up, Grenfell and Hunt turned to excavation, and by 1907 their work at Oxyrhynchus had produced half a million papyri. The significance of their discoveries was not lost on Kelsey, and in 1918 he began planning a postwar expedition, one of the objectives of which was to acquire papyri.

In Cairo, Kelsey met Bernard Grenfell and together they scoured the souks looking for papyri, Kelsey relying on Grenfell's expertise as to authenticity and value. They drove hard bargains. They also visited the Fayoum, a region fifty miles southwest of Cairo, where farmers digging for fertilizer had come across papyri. Here they looked for an American medical missionary, Dr. David Askren, who in the course of looking after the farmers' ailments had learned about their discoveries. Kelsey had struck up a friendship with Askren on a transatlantic journey in 1915, and the two reached an agreement whereby Askren looked out for papyri and manuscripts for him. For the rest of his life, Kelsey continued the purchase of papyri either alone or on behalf of the university, often working through Askren or in partnership with the British Museum and other institutions. Since it had become clear that excavation could provide an alternative source of papyri, Kelsey searched for a suitable site. The excavations at Karanis begun in 1924 swiftly became the major source for enlarging the collection with the result that the University of Michigan now boasts the largest collection of papyri in America. The Karanis papyri offer a fount of information valuable for social history—on agricultural production, demographics, personal concerns, marriage contracts, and social hierarchies.

Kelsey had a larger objective in mind too: the recovery of all the material evidence from the site, his purpose being the simultaneous study of papyri, the buildings, and their contents. Modifying the system of stratigraphic recovery and sequence dating introduced by Flinders Petrie, the Karanis excavators retrieved a vast number of objects, many of them now in the Kelsey Museum: coins; furniture and furnishings; kitchen equipment and products; crockery; living room utensils; equipment for agricultural work, for the workshop, and for the home; personal items including shoes, hair combs, pins, jewelry, cosmetic jars, and application sticks; and playthings and toys. Such material, along with the papyri, yields a full picture of daily life in a Greco-Roman provincial town.

Manuscripts

Expanding his attention to the photography, study, and acquisition of manuscripts, in September 1919 Kelsey made arrangements for the photography of manuscripts at the British Museum, including the famous Beatus manuscript (Beatus was an eighth-century monk who wrote a commentary on the Book of Revelation), and the following year spent a month on the Greek island of Patmos studying lectionaries and manuscripts in the Monastery of St. John.

Passing through Constantinople later that year, he bought fragments of several lectionaries and a Greek manuscript, subsequently purchasing another lectionary and three other manuscripts in

Paris. When in 1922 he learned of a sale in London of manuscripts from the library of Baroness Burdett Coutts, including Greek biblical manuscripts acquired by her circa 1870 in Ioannina (then Albania, now Greece), he determined to bid on some items. But Charles Freer had died. In search of a new benefactor, his enquiries in Detroit led him to Horace H. Rackham. Receptive to Kelsey's proposal, Rackham wrote out a check at once, allowing Kelsey to acquire many of the manuscripts for the library. In the same year, more manuscripts arrived from Paris.

In 1924, an antiquities dealer in Cairo, Maurice Nahman, was offering for sale Islamic and Persian manuscripts from the collection of Sultan Abdulhamid II. Kelsey conferred with Sir Frederic Kenyon, director of the British Museum, asking if the museum might wish to share the acquisition and if Edward Edwards of the museum staff, an expert on Persian manuscripts, could go to Cairo to examine the collection. At Nahman's house in Cairo, Kelsey looked through the material, which he described as "in part calligraphic, in part containing miniatures—Persian, Turkish, Arabic" and attractive. The more expert view of Edward Edwards after a three-day scrutiny was that four hundred of the manuscripts were "desirable." Nahman asked for £6,400 but settled for £4,000. In London, Edwards later reported that the collection surpassed expectations and a division was made: 270 manuscripts to Ann Arbor, 67 to the British Museum, and 54 to the collection of Alfred Chester Beatty, now in Dublin.[2]

Photographs, Books, and Musical Instruments

Whenever and wherever he could, Kelsey acquired photographs of archaeological objects, monuments, and sites. Included among these was an almost complete set of the Parker collection of Roman photographs. Numbering over three thousand, these photographs were prepared in the 1870s for John Henry Parker, the Oxford publisher and historian. Kelsey also employed G. R. Swain, assistant in photography in the Latin department (from 1913), to photograph every site and monument they visited on the overland journey from London to Egypt in 1919–20, and later employed him to document the archaeological work at Antioch, Carthage, and Karanis (1924–26). Swain's collected archaeological prints and negatives—of which there are more than twelve thousand in the Kelsey Museum—are as important now to the history of photography as they are to archaeological and historical research.

Kelsey was constantly involved in acquiring books for the library. Among the more notable were volumes from Thomas Spencer Jerome's library in Italy. By the terms of Jerome's will, his library was to be divided between the University of Michigan and the American Academy in Rome. After Jerome's death in 1914, Kelsey was asked by the executors to go to Italy to settle the will, deal with Jerome's papers and unfinished manuscripts, complete the sale of his villa, and see to the division of the library. The U-M library thus acquired fifteen hundred important books.

When Henry Frieze—longtime chair of the Department of Latin, three times interim president of the university, and a founder member of the University Musical Society—died in 1889, it fell to Kelsey and professor of music Albert Stanley to identify a suitable memorial for Frieze. After wide consultation the decision was taken to acquire the great organ from the World's Columbian Exposition in Chicago (1893) and transport it to Ann Arbor. Kelsey raised the necessary funds, supervised the acquisition of the organ, and saw to its removal and installation in University Hall. Known as the Frieze Memorial Organ, it was moved to Hill Auditorium in 1913. Kelsey and Stanley were also involved in the acquisition of Frederick Kimball Stearns' rich collection of over two thousand musical instruments that came to the university in 1899.

The Marburg Vases

In 1923, with funds provided by Paul Gray of Detroit, Kelsey purchased a group of Greek, Cypriot, Etruscan, and South Italian vases from a German dealer. These vases were in the care of the University of Marburg. The 130 vases provide

Fig. 53. Francis Kelsey (center) with a group of fine arts students from the American Academy in Rome, including four architects, visiting Pompeii in 1921. Photo by Jack Skinner. Bentley Historical Library.

examples of Greek, South Italian, and Corinthian styles with a comprehensive range of shapes and decorations. They range in date from the Late Bronze Age to the Early Hellenistic and include a Geometric pyxis, an Etruscan bucchero chalice, and a South Italian (Paestum) red figure krater. As examples of major styles of Greek pottery, they provide an excellent study and teaching collection.

The Barosso Watercolors

One spectacular addition to the collections was the acquisition of replicas of the wall paintings in Room 5 of the Villa of the Mysteries at Pompeii (fig. 53). Uncovered in 1909, this fresco cycle represents women enacting enigmatic rituals generally agreed to be Dionysiac in nature. Kelsey, realizing that the colors would inevitably fade, in 1924 commissioned the artist and archaeologist Maria Barosso to make replicas of all the walls of the room. Work in progress on her twenty-one almost life-size (executed at 5/6 scale) watercolors was seen by Kelsey in 1925 and again in 1926 when the paintings were unrolled and mounted in Room 5 itself to compare replica with original and to ensure that Barosso's colors matched those of the originals. Subsequently the replicas were exhibited in Rome at the Villa Borghese before coming to Ann Arbor. A re-creation of the whole room may now be seen at the museum that bears Kelsey's name, part of the remarkable legacy that Francis W. Kelsey left to the University of Michigan.

Notes

1. For an extensive biography of Francis W. Kelsey, see John G Pedley, *The Life and Work of Francis Willey Kelsey: Archaeology, Antiquity, and the Arts*. Ann Arbor: University of Michigan Press, 2011.

2. Kelsey was also instrumental in acquiring the Yahuda collection of mostly Arabic manuscripts; approved by Edwards, they came to the university in 1926.

WILLIAM L. CLEMENTS (1861–1934) AND HIS EARLY AMERICANA COLLECTION

J. Kevin Graffagnino

The nineteenth and early twentieth centuries were a golden age for collecting antiquarian American books, manuscripts, maps, and other paper-based primary sources. As Americans' pride in their young nation grew, interest in acquiring original materials relating to the early heritage of the United States became widespread. With prices low, few institutions competing in the chase, and a growing circle of dealers and auctioneers eager to help prospective customers fill their shelves, book-loving Victorian and Edwardian Americans of means gathered impressive personal collections of North American history. The most enthusiastic and persistent—John Carter Brown, James Lenox, Peter Force, George Brinley, Hubert Howe Bancroft, and Edward E. Ayer—did so at levels that succeeding generations of bibliophiles could only envy. Many of these great early Americana collections dispersed or sold at auction after their owners' deaths, but those that went intact to academic and other institutional libraries became the basis of some of this nation's greatest historical research centers.[1]

Fig. 54. William Lawrence Clements, oil portrait by Hermann Hanatschek, 1928. William L. Clements Library.

William L. Clements (fig. 54) was one of the outstanding American book collectors of his or any other generation. In just three decades of activity, from 1904 to his death in 1934, Clements more than held his own in the passionate pursuit of early Americana. Well-to-do by Michigan standards but never close to commanding the resources that contemporaries like Henry Ford, William Randolph Hearst, and Henry E. Huntington applied to their collecting interests, Clements assembled an extraordinary library of rarities on early American exploration and discovery, eighteenth-century military history, colonial government and politics, and the creation and development of the United States. By his own estimate, Clements over the years devoted more than half his fortune to the hobby that became an obsession. When he gave his collection to the University of Michigan in 1923, he established one of the first great American history libraries at a public university and ensured a lasting personal legacy as a significant American bibliophile.[2]

William L. Clements was born in Ann Arbor, Michigan, on April 1, 1861, the youngest child of James and Agnes Clements. James was a civil engineer who installed Ann Arbor's first gas system, and William attended the University of Michigan as an engineering major. Soon after his graduation in 1882 he went north to Bay City, Michigan, where James had a share of the Bay City Industrial Works. William became manager of the firm, and in the decades that followed he built the company into a leading manufacturer of railroad cranes, derricks, steam shovels, and other heavy equipment. Clements obtained patents for several of his improvements in machinery design, and the Industrial Works production capacity grew from 12-ton cranes in the 1880s to 200-ton cranes in the 1920s. Some of the steam shovels that dug the Panama Canal came from the Industrial Works factory. As the company prospered, Clements acquired the trappings of growing wealth and influence—a twenty-room mansion in Bay City designed by Albert Kahn; principal ownership of the Bay City Bank; and election in 1909 to the U-M Board of Regents. By the time he turned fifty, Clements had achieved noteworthy success in his business and public careers.

Clements became serious about book collecting later in life than most bibliophiles. He filled his first Bay City home with new and old books, but the assortment of literature sets, early printed books, modern nonfiction, and some American history lacked cohesion and focus. His evolution into a serious collector of early Americana began in 1903 when he purchased the one thousand-volume library of his friend Aaron J. Cooke, a Bay City dry goods merchant who had bought heavily at the 1870s–1890s dispersal auctions of leading Americanists William Menzies, George Brinley, and Samuel L. M. Barlow.[3] Acquiring the Cooke collection as a base spurred Clements to keep going. By 1912 he was spending as much as $17,500 to buy 140 seventeenth- and eighteenth-century rarities from the distinguished collection of Newbold Edgar. In 1913 he hired his first full-time librarian to tend to his shelves, and the following year he issued a printed catalog of the highlights of his three thousand American books. In Clements's eyes, the outstanding entry in *Uncommon, Scarce and Rare Books Relating to American History during the Discovery and Colonial Periods, together with other Americana from the Library of William L. Clements, Bay City, Michigan* was the London 1588 edition of Thomas Hariot's *A Briefe and True Report of the New Found Land of Virginia*, which he purchased in February 1914 for $7,500 and described as "the star of all Americana" (fig. 55).[4] By now Clements was hooked; "I am adding to my Americana as fast as opportunity offers," he wrote to a friend in October 1916, "I do not know what I would do if I did not have this interest."[5]

The Clements collection continued to grow rapidly. In 1916 he began assembling full sets of the exploration chronicles of Theodor de Bry and Levinus Hulsius, complicated bibliographical challenges for any early Americanist. Clements bought more than $60,000 in books in 1919, bringing his total library expenditures since 1903 close to half a million dollars (it is worth keeping in mind that $60,000 in 1919 would be equal to approximately $820,000 today). August 1920 brought the publication of *A Checklist Preliminary to a General Catalogue of Books Published Before 1700 Belong-*

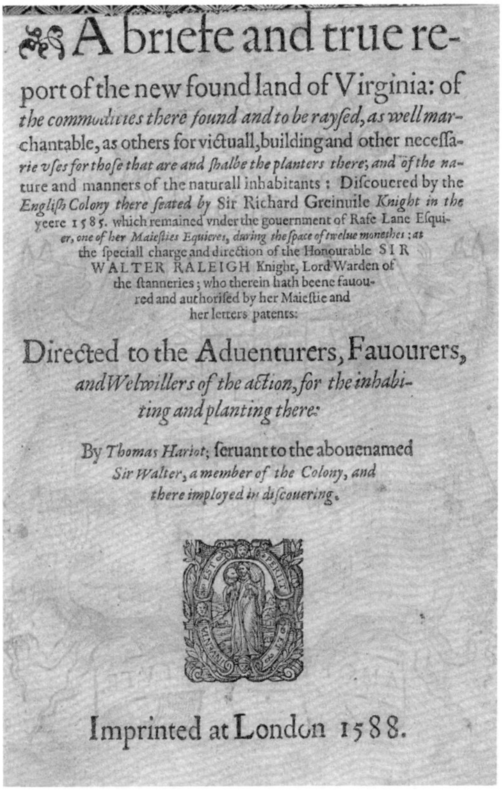

A briefe and true re-
port of the new found land of Virginia: of
the commodities there found and to be rayſed, as well mar-
chantable, as others for victuall, building and other neceſſa-
rie vſes for thoſe that are and ſhalbe the planters there; and *of the na-*
ture and manners of the naturall inhabitants: Diſcouered by the
Engliſh Colony there ſeated by Sir Richard Greinuile *Knight in the*
yeere 1585. which remained vnder the gouernment of Rafe Lane Eſqui-
er, one of her Maieſties Equieres, during the ſpace of twelue monethes: at
the ſpeciall charge and direction of the Honourable SIR
WALTER RALEIGH Knight, Lord Warden of
the ſtanneries; who therein hath beene fauou-
red and authoriſed by her Maieſtie and
her letters patents:

Directed to the Aduenturers, Fauourers,
and Welwillers of the action, for the inhabi-
ting and planting there:

By *Thomas Hariot*; ſeruant to the abouenamed
Sir Walter, a member of the Colony, and
there imployed in diſcouering.

Imprinted at London 1588.

Fig. 55. Thomas Hariot's *A briefe and true report of the new found land of Virginia* (London, 1588) recounted the story of the "lost colony" of Roanoke. This volume came to the University with William L. Clements's original gift. William L. Clements Library.

ing to the Library of William L. Clements, Bay City, Michigan, a listing of 630 rarities that indicated the dramatic expansion of the library in just six years. Clements's pace might have slowed at that point, but in July 1921, while on a trip to England, he made a fateful decision to expand his collecting into manuscript material. At a Sotheby's auction in London, he bought 170 bound volumes of the papers of William Petty, second Earl of Shelburne and prime minister of England from 1782 to 1783, for $10,000. The Shelburne archive was rich in manuscripts relating to England's colonial empire and the American Revolution, and the acquisition drove Clements to make the era of the Revolution a principal focus of his library. For the remaining thirteen years of his life, searching out and acquiring the papers of key British and American participants in 1764–83 Anglo-American affairs dominated his collecting activities. The costs were substantial ($70,000 in 1921 alone), but Clements found the thrill of the chase exhilarating and the growing reputation of his library a gratifying return on his investment.

By 1920 Clements had plans for the future of his collection. As early as 1916 he made a tour of the Northeast to visit the American Antiquarian Society, the John Carter Brown Library, the Yale Elizabethan Club, Harvard University's libraries, and Henry E. Huntington's library in New York City to gather ideas and inspiration for creating a permanent home for his treasures. At the September 1919 meeting of the U-M regents, he informally offered his library to the university, and the following month he took the board's Library Committee and university librarian William W. Bishop east to visit the institutions that had most impressed him. In February 1920, one month after the dedication of U-M's new library building that included an exhibition of some sixty-five of Clements's rarest books, he made a formal proposal to the regents: he would give his collection and pay for a building to house it if the university would guarantee an annual appropriation of $25,000 for operations and acquisitions. No doubt recognizing the opportunity to vault the university's American history holdings from inconsequential to first-rate, the regents quickly accepted. It took more than two years to work out the details, in part because near the end of the process, Clements introduced the idea of having the new library's director report to a five-person Committee of Management headed by U-M's president rather than to the university librarian. Both the university president, Marion L. Burton, and librarian Bishop questioned that, but Clements stubbornly insisted that he would not proceed with his gift if the university made his collection part of the U-M library system. Burton and Bishop acquiesced, and in December 1922 the regents approved the detailed gift agreement.[6]

Construction on the William L. Clements Library got under way well before the debate over governance ended. Clements helped lay the cornerstone of the new building on March 31, 1922, "in a driving rain and under the most inauspicious conditions."[7] Clements's position as regent and head of the Buildings Committee, not to mention his strong personality, contributed to the prime location next to the president's house. It is somewhat ironic that the university tore down one of the oldest structures on campus, constructed in 1840 as a residence for faculty and in 1922 known as the Old Engineering Building, to put up a history library. Clements selected Albert Kahn—architect for Hill Auditorium, Burton Memorial Tower, and the Graduate Library at U-M—to design his library and asked him to adapt the layout of the John Carter Brown Library at Brown University for it. Kahn followed the spirit of that instruction but chose a 1587 casino on the grounds of the Villa Farnese in Caprarola, Italy, as his model. Construction took fourteen months, and the results were impressive and elegant. Kahn's design led visitors up a brick front walk and two tiers of marble steps to massive bronze doors that opened into a two-story great room featuring oak paneling, glass-front bookcases, and an ornately painted barrel-vault ceiling. Staff and collections spaces in the rest of the building added function to form. Clements was pleased, and for the rest of his life Kahn designated the library building as his favorite among the hundreds he designed.[8]

The university dedicated the William L. Clements Library on June 15, 1923. In formal ceremonies at Hill Auditorium, noted scholar J. Franklin Jameson delivered a keynote address on "The American Historian's Raw Materials." Clements spoke briefly, transferring his interest in the library, "with its work and development," to his alma mater. He expressed pride in placing his collection at the university, but warned that he did not want U-M students to use the collections; rather, he said, "It is primarily a library for advanced research on the part of scholars already well equipped."[9] Yet although the Clements Library was open, it had no director. Clements agonized over the selection, and it was not until August that he settled on Randolph G. Adams, a thirty-year-old assistant professor of history at Trinity College (now Duke University) who came to Clements's attention through George Parker Winship, head of the Widener Library at Harvard. Adams arrived on campus in October 1923 to begin a twenty-eight-year tenure as director. For his part, Clements returned to Bay City pleased with what had transpired, but sad to see his collection gone. "I have returned home to a house empty of nearly all books," he wrote to historian and editor Worthington C. Ford, "so it is needless to tell you how totally lost I am."[10]

Clements dealt with his sense of disorientation in classic collector fashion, by plunging back into the search for more antiquarian treasures. "This day and hour mark the conclusion of a book-collector's career," he told the audience at the Clements Library dedication ceremonies, but he could not live up to that prediction. Instead, he filled the gap in his life and on his shelves by renewing the hunt for manuscript collections relating to the American Revolution. In 1925 he bought ten linear feet of the papers of American general Nathanael Greene, one of George Washington's most capable military subordinates, for $35,000. In December of the same year he purchased ninety feet of manuscripts of Sir Henry Clinton, commander in chief of the British army in North America during the Revolution, for $88,500 from a Clinton descendant in England. September 1927 brought an expenditure of $25,000 for a collection of the papers of Lord George Germain, secretary of state for America in the British government from 1775 to 1782. When a New York gallery offered Benjamin West's monumental five-by-eight-foot painting, *The Death of General Wolfe*, depicting Wolfe's heroic death after defeating Montcalm at the Battle of Quebec in 1759, dedicated Anglophile Clements could not resist paying $7,500 to acquire it for display in the great room. The final great purchase in this five-year flurry of buying came in 1930 when Clements and his agents outmaneuvered Henry E. Huntington to acquire seventy linear feet of manuscript material of Thomas Gage, commander in chief of British military forces in North America from 1764 to 1775. That $100,000 expense ran Clements's 1925–30 acquisitions total to somewhere north of $300,000. Clements kept all the Revolutionary War papers at Bay City, where he enjoyed studying them, but he consistently said that he would bequeath them to the university when he died.[11]

Clements lived eleven years beyond the 1923 opening of his library. He remained a constant presence and influence there—often to the consternation of Randolph G. Adams, who agreed with Clements about barring undergraduate and graduate students from using the collections but chafed at the founder's frequent suggestions and orders. The Great Depression had a severe impact on Clements's finances and his last years. His Bay City bank closed and business at the Industrial Works declined, cutting deeply into his income. In March 1932, he added a codicil to his will stating that U-M would have to purchase the Revolutionary War manuscript collections that he had always intended to donate; Clements told Adams about the change but swore him to secrecy, a request that Adams honored. The 1933 fall elections resulted in a Democratic landslide in Michigan, ending Clements's long tenure on the U-M Board of Regents. The university marked his contributions and leadership by awarding Clements an honorary degree in June 1934, but by then illness and age had weakened him and he died in Bay City on November 6, 1934, at the age of seventy-three.[12]

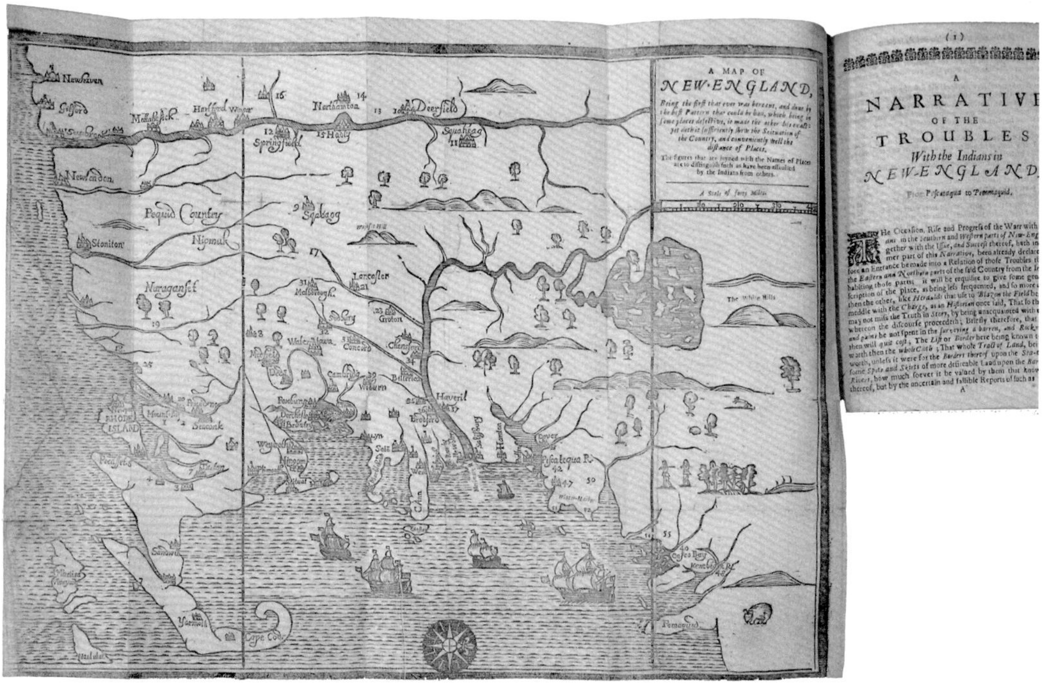

Fig. 56. William Hubbard's *A Narrative of the Trouble with the Indians in New-England* (Boston, 1677) included the earliest map printed in North Americ. William L. Clements Library.

Following Clements's death, the news that the university would not receive the Revolutionary War collections as expected generated considerable publicity. Some regents insisted that U-M should not buy the manuscripts, and it took until March 1937 to hammer out an agreement under which the university would pay $300,000 over fifteen years to acquire the papers. During and after that process, Adams continued to run the Clements Library as Clements had wanted, admitting no students and few others as readers. The collections grew steadily and in impressive fashion, but access did not. Throughout Adams's long tenure, which ended with his death in 1951 at the age of fifty-eight, the Clements Library remained much as its founder had envisioned—an exclusive depository that was *at* the University of Michigan but not *of* the university and certainly not accessible to most U-M faculty or students.[13]

The gradual process of changing the Clements from forbidding storehouse of priceless rare books to historical research center began in the early 1950s. In keeping with a national trend toward greater accessibility at American history museums, libraries, and sites, U-M administrators and Clements staff changed course to a more open and welcoming philosophy. In 1953, Howard H. Peckham, who had been the library's curator of manuscripts from 1936 to 1945 before leaving to head the Indiana Historical Bureau, succeeded Adams as director; during his twenty-four years at the helm, the Clements Library opened its doors to far more researchers. Third director John C. Dann continued that trend in his three decades of leadership (1977–2007), significantly expanding both the scope of the collections and the number of users. Michigan faculty began bringing their classes to the library to introduce them to sources not available elsewhere; students proved that they could use antiquarian books, manuscripts, and other papers respectfully and safely; and the Clements Library became much more a part of the campus that surrounds it. Today, as the library nears its 2023 centennial, the chronological parameters of the holdings run from 1492 to 1900 and U-M undergraduates make up more than half the twelve hundred readers who conduct research in the Avenir Foundation Room every year.

In part because, in the past ninety-four years, the philosophy at the Clements Library he created has moved away from an elitist vision typical of his Edwardian generation, William L. Clements's legacy at the University of Michigan and in American book-collecting history is secure. His early Americana collection ranks among the finest ever assembled by any individual (fig. 56), and his accumulation of manuscripts from the era of the American Revolution remains unequaled. In designating his alma mater as home for his library, Clements broke with the tradition of East Coast, private institutions as repositories for significant American history source materials to establish one at a Midwestern public university. In so doing, Clements took U-M from mediocre library holdings in American history to an extraordinary collection that has become a prime destination for scholars from around the world. In 1923, Clements expressed his hope that he had created "special facilities for historical research" at "the greatest and oldest Middle West university." The library that bears his name fulfilled that promise. In that accomplishment, William L. Clements epitomizes the wisdom of Henry E. Huntington's observation: "The ownership of a fine library is the surest and swiftest way to immortality."[14]

Notes

1. Carl L. Cannon, *American Book Collectors and Collecting from Colonial Times to the Present*. New York: The H. W. Wilson Company, 1941; Donald C. Dickinson, *Dictionary of American Antiquarian Book Collectors*. Westport, CT: Greenwood Press, 1998; Robert A. Shaddy, *"A Mad World, My Masters!": Book Collecting in America, 1890–1930*. PhD diss., University of Missouri-Columbia, 1990.

2. The best biographical source on Clements is Margaret Maxwell, *Shaping a Library: William L. Clements as Collector*. Amsterdam: Nico Israel, 1973.

3. Joseph Sabin, comp., *Catalogue of the Books, Manuscripts and Engravings Belonging to William Menzies of New York*. Albany, NY: Press of J. Munsell, 1875; *Catalogue of the American Library of the Late Mr. George Brinley, of Hartford, Conn*. Hartford, CT: Press of the Case, Lockwood & Brainard Company, 1878–93), 3 vols.; James Osborne Wright, comp., *Catalogue of the American Library of Samuel Latham Mitchill Barlow*. New York: Douglas Taylor, 1889.

4. William L. Clements to Randolph G. Adams, September 30, 1925, Adams Papers, Clements Library, University of Michigan.

5. William L. Clements to Clarence S. Brigham, October 26, 1916, Clements Papers, Clements Library.

6. Maxwell, *Shaping a Library*, chapters 8 9.

7. William L. Clements to Lathrop C. Harper, March 31, 1922, Clements Papers.

8. Maxwell, *Shaping a Library*, chapter 9.

9. *The Dedication of the William L. Clements Library of Americana at the University of Michigan*. Ann Arbor: University of Michigan, 1923, 11–15. The publication provides a detailed description of the Clements Library collections at its opening.

10. William L. Clements to Worthington C. Ford, June 27, 1923, Clements Papers.

11. Maxwell, *Shaping a Library*, chapter 15; Clayton A. Lewis, "William L. Clements and *The Death of General Wolfe*," in Carole McNamara, *Benjamin West: General Wolfe and the Art of Empire*. Ann Arbor: University of Michigan Museum of Art, 2012, 26–33.

12. Maxwell, *Shaping a Library*, chapter 16.

13. Clements remained adamant about limiting access. On July 18, 1934, four months before his death, he wrote to Randolph G. Adams, "If I understand from your letter that the President asked you to turn over in your mind, making the Library more available to members of the Staff and graduate students, this request alarms me more than anything in recent years. From the very thought of the foundation of a rare historical library and the incidental money expenditure and thought necessary to form it, the idea that such a Library would ever be used by under-graduate students or graduate students, as of the ordinary kind, was the most serious objection, and, at times, frequently upset the whole idea. . . . I should indeed be very sorry to see any unqualified person mauling over the papers so precious now and more so in the future of American History. It never will be done with my consent." Clements Papers.

14. Clements remarks at 1923 opening, in *Dedication*, pp. 11–15. James Thorpe, *Henry Huntington: A Biography*. Berkeley: University of California Press, 1994.

DEAN CONANT WORCESTER (1866–1924)

Colonizing and Collecting the Philippines

Mark Rice

Although his fame has faded over the years, there was a time when Dean Conant Worcester (1866–1924) was a very influential—and equally controversial—individual (fig. 57). A physically imposing man with an abundance of energy and self-confidence, Worcester's career path went from scientist to colonial administrator to successful businessman. He was one of the many players who helped usher the United States into its position of global dominance in the twentieth century. His representations of the Philippines in writing and images were instrumental in the debates within the United States about the future of its most far-flung colony. Indeed, as one biographer has written, Worcester "was the foremost creator and popularizer of American notions about the Philippines" in the first two decades of the twentieth century.[1]

Fig. 57. Dean Conant Worcester. University of Michigan Special Collections Library.

Worcester was an avid photographer and a prolific writer, and his records serve as a rich source of historical details about the early years of US control of the Philippines. There are multiple archives of Worcester materials at the University of Michigan. The Dean C. Worcester photographic collection at the Museum of Anthropological Archaeology contains nearly five thousand original glass plate negatives as well as hundreds of lantern slides made from Worcester's photographs. This collection may be the most extensive collection of original negatives from the US colonial era in the Philippines. It has proven to be an invaluable resource for scholars interested in the history of American anthropology, the history of ethnological photography, and American colonial history in the Philippines.[2]

Other important collections of Worcester materials at the university include zoological specimens (now in the Museum of Zoology) and ethnographic and archaeological objects (now part of the Museum of Anthropological Archaeology's Philippine Expedition Collection) that he collected. The Dean C. Worcester Papers in the Special Collections Library include twenty-one volumes of letters, reports, manuscripts, and other written materials that focus primarily on the fourteen years (1899–1913) that Worcester was involved in the US colonial regime in the Philippines. In addition, the library has nearly nine hundred photographs (taken between 1890 and 1907 and donated by Worcester in 1914) that show various "people, places, and activities in the Philippines."[3] These photographs have been digitized and are available through the library's Philippine Photographs Digital Archive. At the Bentley Historical Library, the Dean C. Worcester Papers, 1887–1925 (donated in 1957 by Worcester's daughter, Alice Worcester Day), include photographs, correspondence, lecture notes, newspaper clippings, and other documents related primarily to Worcester's time as a colonial administrator.

Dean Worcester was born on October 1, 1866, in the small town of Thetford, Vermont, the seventh of eight children of Ezra and Ellen Conant Worcester. Ezra was a country doctor whose medical practice was not very lucrative. With eight children to feed, the family did not have a lot of extra money to spend on education. Consequently, Dean turned to his much older brother, William, a successful physician who specialized in mental illness, to help him pay for his college expenses. With William's support, Worcester headed west in 1884 to attend the University of Michigan, where he majored in zoology. Shortly before he received his degree in 1889, "he was appointed to the university' staff as an assistant in botany, although he had no undergraduate preparation in that field, an appointment that reflected the indistinct boundaries separating many subdisciplines of science at that time."[4]

Worcester's first exposure to the Philippines came while he was an undergraduate. In 1887, he accompanied his professor, J. B. Steere, on a zoological expedition to the Philippines, returning to Ann Arbor in 1888. After his graduation, Worcester desired both additional adventure and additional opportunity for scientific exploration, and he believed that much zoological work was left to be done in the Philippines. Consequently, he returned to the archipelago in September 1890, together with his classmate and friend Frank Swift Bourns (who had also taken part in the Steere Expedition), as part of the Menage Expedition (fig 58). While Minneapolis-based real estate tycoon Louis F. Menage was the main sponsor, the expedition was also supported by the Minnesota Academy of Natural Sciences. For two and a half years, Worcester and Bourns traveled through much of the central and southern Philippines. They acquired thousands of zoological specimens and cultural artifacts from some of the ethnolinguistic groups they encountered there—in particular the Mangyans of Mindoro and Tagbanuas of Palawan—and they took an unknown number of photographs. The Museum of Anthropological Archeology has more than one hundred glass plate negatives from the Menage Expedition in its collection and the Museum of Zoology houses some of the biological specimens.

In late 1892, Worcester and Bourns separated in order to maximize the territory they could cover in the late stages of the Menage Expedition. Bourns traveled to Borneo to hunt orangutan, while Worcester continued working in the Philippines. Although he had planned to travel to various sites for several more months, Worcester had to cut short his time in the Philippines due to a severe case of typhoid fever. After convalescing in Manila for a few weeks, Worcester left the Philippines in February 1893 and arrived back in the United States in April.

Within days of his arrival back in the United States, Worcester traveled to Los Angeles to marry Nanon Fay Leas, a fellow alumnus of the University of Michigan. The small ceremony took place in the Leas home on April 27. The couple settled in Ann Arbor where Worcester was hired as curator of zoological and anthropological collections and custodian of all collections in the University Museum,[5] and to teach zoology courses at the university. Their daughter, Alice, was born in 1896, and their son, Frederick, in 1898.

From 1893 to 1898, Worcester focused his energies primarily on his family life and on his work as a scientist and instructor. He also spent some time in Minneapolis cataloging the specimens collected during the Menage Expedition. However, much of that collection was broken up due to unexpected financial straits that the economic depression placed on the Minnesota Academy of Natural Sciences. In 1894, however, Worcester gave a lecture about the Menage Expedition before the Minnesota Academy, a lecture that he illustrated with his own photographs. This may have been the moment when he began to understand the potential power of his illustrated lectures. Accord-

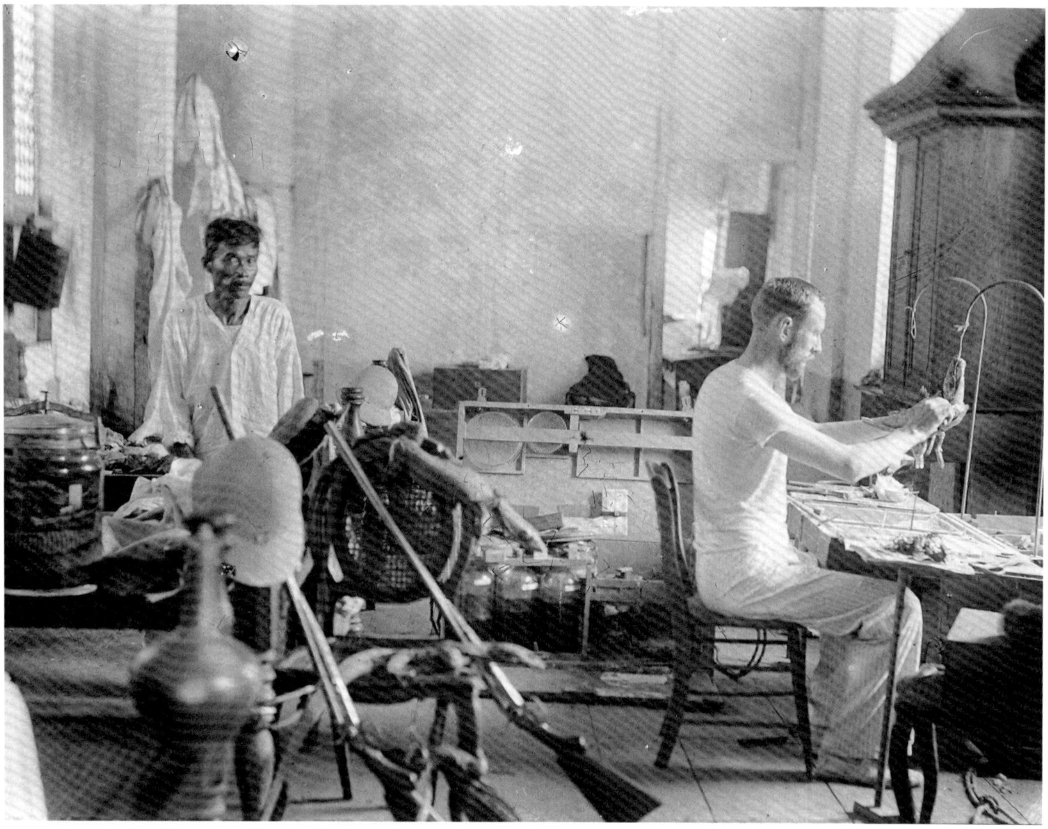

Fig. 58. Dean Worcester and assistant processing specimens for the Menage Expedition, Capiz, Panay, October 1890. University of Michigan Museum of Anthropological Archaeology.

ing to the Minneapolis *Tribune*, his lecture "proved as interesting as anything which has been heard in Minneapolis for many days."[6]

The next year, in 1895, Worcester, along with millions of other Americans, took careful note of the Cuban War of Independence. Although he remained focused primarily on zoology, it appears that that war began to shift his thinking about what he had learned while in the Philippines. That shift continued in 1896 with the start of the Philippine Revolution. By the end of that year, Worcester coauthored (with Bourns) an article about societal conditions in the Philippines. That article, "Spanish Rule in the Philippines," appeared in the October 1897 issue of *Cosmopolitan* magazine and was an attack against "Spanish misrule" that drew parallels between the conditions in the Philippines and Cuba: "For months the deadly struggle in unhappy Cuba has held the attention of the American people. On the other side of the globe another struggle is in progress, no less tragic, though infinitely more hopeless, so far as the human eye can see, than that at our own door."[7]

Also in 1897, Worcester presented two papers at the meeting of the American Association for the Advancement of Science, held in August in Detroit. Those papers, described in one report as "elaborate," were "illustrated by photographs" and dealt with two ethnolinguistic minority groups in the Philippines, the Tagbanua and the Mangyan.[8] Worcester was invited to give those same talks later that month in Toronto at the British Association for the Advancement of Science, signaling his emerging interest in describing and categorizing various minority groups in the Philippines. On October 18 of the same year, he gave an illustrated lecture about Spanish rule and conditions leading up to the Philippine Revolution to the Unity Club of Ann Arbor.

The culmination of Worcester's shifting interest in the Philippines and the event that opened the door for his long colonial career was his 1898 book, *The Philippine Islands and Their People*. The book recounted his two expeditions to the Philippines and blended his personal observations and experiences with broader historical and geographical information about the Philippines that he gleaned from other sources, particularly John Foreman's 1890 book, *The Philippine Islands*. Worcester argued that the Philippines was not fit for self-government but that with the right guidance its rich natural resources could be developed more effectively than they had been under Spanish rule. Worcester's book, illustrated with photographs taken during the Menage Expedition, quickly became a bestseller, received generally positive reviews, and was republished multiple times in just a few months, American interest in the Philippines having been whetted by the Spanish-American War.

Based in large measure on the success of his book, Worcester was invited to meet with President McKinley in January 1899 to talk about his experiences in the Philippines. Shortly after their meeting, McKinley appointed him as the sole civilian member of the Schurman Commission, which was charged with investigating conditions in the Philippines and making recommendations for what the United States ought to do next there. Worcester arrived back in the Philippines on March 4, one month after the start of the Philippine-American War. He reestablished contact with some of the friends he had made during the Menage Expedition, most of whom were Spanish landowners, and he passed along information provided to him by these friends to US military authorities, in order to assist in the American fight against Philippine independence.

The Schurman Commission completed its work later that same year and returned to the United States to file its report. The next year, Worcester was back in the Philippines as part of the Taft Commission, the only member of the Schurman Commission to be appointed to the new commission. With the establishment of a civil government in the Philippines in 1901, Worcester was appointed to the position of secretary of the interior, a post he held until 1913. His wife and children relocated to the Philippines to be with him. The family lived mostly in Manila but also built a summer house in the mountain town of Baguio, which served as the summer capital for the new colonial government.

Two dimensions to Worcester's political career in the Philippines merit highlighting. The first is that Worcester served longer than any other commissioner in the Philippines. Not only was he the only person to serve on both the first and the second Philippine commissions, but his thirteen-year career as secretary of the interior was unmatched by any other colonial administrator. The second dimension is the scope of his responsibilities. As historian Rodney Sullivan writes: "The major areas for which Worcester bore responsibility for most of his long tenure as secretary of the interior were agriculture, health, science, lands, tribal peoples, and forestry."[9] Of these, Worcester devoted most of his energy to the tribal peoples.

Every year that he was secretary of the interior, Worcester made one or more "inspection tours" through northern Luzon, the Visayas, and Mindanao, where he oversaw American efforts to "civilize" the non-Christian populations. These efforts included building schools, establishing peaceful relations between previously hostile groups, and introducing Western-style clothing. Worcester brought his camera with him on these tours and photographed a wide range of ethnolinguistic minority groups. He suggested that these groups comprised a larger percentage of the islands' population than they actually did. He also presented them as more "authentic" representations of the islands than were the Hispanicized and Catholic majority of the country, thereby bolstering his argument that the United States needed to retain control of the Philippines for a long time in order to guarantee that American-style civilization would grow deep roots in the colony. Between

Fig. 59. "Party of Ifugaos come to welcome party of Secretary of the Interior" (original caption). 1908. Nueva Vizcaya. University of Michigan Museum of Anthropological Archaeology, Worchester Collection.

his own photographs and the photographs made by government photographers whose work he oversaw, Worcester built up an archive of more than fifteen thousand photographs (fig. 59). These photographs were routinely included in his books, articles, and government reports, as well as in the articles and books of other people reporting on the activities of the US colonial regime in the Philippines.

One important and influential venue for Worcester to publicize his ideas about the benevolence of US control of the Philippines was *National Geographic Magazine*. His first article in that magazine, published in the June 1898 issue, was titled "Notes on Some Primitive Philippine Tribes"; following articles were "Field Sports Among the Wild Men of Northern Luzon" (March 1911), "Taal Volcano and Its Recent Destructive Eruption" (April 1912), "Head-Hunters of Northern Luzon" (September 1912), and "The Non-Christian Peoples of the Philippine Islands: With an Account of What Has Been Done for Them Under American Rule" (November 1913). The last two articles encompassed entire issues of the magazine, which was rarely done. They were heavily illustrated with Worcester's photographs, some of which were hand-colored to heighten the apparent exoticism and primitivism of the Filipinos. The last of the articles appeared after Worcester was back in the United States preparing for an intensive campaign against Philippine independence.

Worcester's reliance on photographs to promote US activities in the Philippines was not without its critics. In the 1912 book, *The American*

Occupation of the Philippines, the anti-imperialist James Blount called Worcester "the direst calamity that has befallen the Filipinos since the American occupation, neither war, pestilence, famine, reconcentration, nor tariff-wrought poverty excepted."[10] Discussing Worcester's representations of Filipinos, Blount referred to Worcester as the "P.T. Barnum of the 'non-Christian tribe' industry," and he criticized Worcester for "humbugging the American people into the belief that the Islands must be retained until the three hundred thousand or so Negritos, Igorrotes, and other primitive wild peoples sprinkled throughout the archipelago are 'reconstructed.'"[11] He also criticized "the click of the Worcester kodak" seen in the Bureau of Insular Affairs photographs lent out for publication in popular American magazines, resulting in what he called a "libelous panorama" of images of the Philippines.[12]

Partly in response to Blount's book, partly in response to the anti-imperial platform of the newly elected President Woodrow Wilson, and partly to secure his place in history, Worcester published his two-volume opus, *The Philippines Past and Present*, in 1914. The book was a comprehensive overview of the effects of US control of the Philippines since 1898, painted in a thoroughly positive light. For much of the twentieth century it served as the standard text on the history of US-Philippine relations.

Also in 1914, Worcester embarked on a lecture series in dozens of cities, mostly in the Eastern and Midwestern states. His lectures were illustrated with lantern slides and he sometimes showed films that he had made of Philippine minority groups. These lectures, intended to promote indefinite US retention of the Philippines, received a great deal of attention in newspapers, which would frequently include lengthy excerpts from the lectures, thereby expanding their reach beyond the audiences that routinely packed the ballrooms and lecture halls where Worcester spoke. In December 1914, in what was essentially the finale of his political career, Worcester gave one of his lectures before a US Senate committee debating the Jones Bill, which some congressional Democrats hoped would grant autonomy to the Philippines. An amended, the Jones Bill was signed into law in August 1916. While it granted increased legislative authority to elected Filipinos, executive power remained firmly in the hands of Americans, and no timetable was set for Philippine independence.

In 1915, Worcester and his family returned to the Philippines, Worcester having been appointed vice president of the Philippine-American Company in late 1913. For the next nine years, Worcester oversaw a number of business interests in the islands, including cattle ranching on the island of Mindanao and the production of coconut oil in Cebu, where he lived with his family. He proved to be a capable manager who oversaw a successful business that paid and treated its employees well, including "wages at least 20% higher than elsewhere."[13] Nevertheless, he remained a polarizing figure in the Philippines until his death from heart disease on May 2, 1924.

While Worcester's time as a student and employee of the University of Michigan was short-lived, his impact on its museum and archival collections, and to enduring university ties to the Philippines, was profound. It is in no small part due to his efforts that U-M houses some of the United States' most extensive and important assemblages of biological, archaeological, photographic, and archival collections from the Philippines.

Notes

1. Rodney J. Sullivan, *Exemplar of Americanism: The Philippine Career of Dean C. Worcester*. Ann Arbor: University of Michigan Center for South and Southeast Asian Studies, 1991, 234.

2. See, for example, Mark Rice, *Dean Worcester's Fantasy Islands: Photography, Film, and the Colonial Philippines*. Ann Arbor: University of Michigan Press, 2014, and Manila: Ateneo de Manila University Press, 2015; Christopher Cappozola, "Photography and Power in the Colonial Philippines," *MIT Visualizing Cultures*, http://ocw.mit.edu/ans7870/21f/21f.027/photography_and_power/index.html; Paul Kramer, *The Blood of Government: Race, Empire, the United States, and the Philippines*. Chapel Hill: University of North Carolina

Press, 2006; and Rodney J. Sullivan, *Exemplar of Americanism: The Philippine Career of Dean C. Worcester.* Ann Arbor: Center of South and Southeast Asian Studies at the University of Michigan, 1991.

3. Dean Conant Worcester papers, Special Collections Library, University of Michigan; see http://www.lib.umich.edu/special-collections-library/dean-c-worcester-papers-1834-1915 (accessed on September 4, 2015).

4. Sullivan, *Exemplar of Americanism*, 19.

5. Carl E. Guthe, "The Museums at Michigan," *The Quarterly Review of the Michigan Alumnus* 42 (1935), 472.

6. "The Menage Expedition," *Minneapolis Tribune*, April 21, 1894.

7. Dean C. Worcester and Frank S. Bourns, "Spanish Rule in the Philippines," *Cosmopolitan*, October 1897, 587.

8. Anita Newcomb, McGee, "Anthropology at the American Association for the Advancement of Science," *Science* (October 1, 1897), 512.

9. Sullivan, *Exemplar of Americanism*, 98.

10. James Blount, *The American Occupation of the Philippines, 1898–1912*. New York: Putnam, 1912, 571.

11. Ibid., 578.

12. Ibid., 579.

13. Resil B. Mojares, "Worcester in Cebu: Filipino Response to American Business, 1915–1924," *Philippine Quarterly of Culture and Society* 13:1 (1985), 7.

MARGARET WATSON PARKER (1867–1936)

A Private Passion and a Public Legacy

Carole McNamara

The quality and breadth of any museum's collections are developed through the expertise of the institution's staff in identifying acquisitions—both purchases and gift cultivation—as well as through the generosity of collectors and supporters who become donors. For the University of Michigan Museum of Art (UMMA), the collector-donor whose gift has done more to shape the direction of the museum's collecting than any other individual is Margaret Watson Parker (Vermont 1867–Detroit 1936).[1] Her sizeable gift of 679 works of art, plus the funding her husband provided in her memory, enabled decades of collecting, particularly in Asian art, and has been the defining influence on the character of the Museum of Art's collection.

Fig. 60. Dorothy Jarvis, portrait of Margaret Watson Parker, undated photograph. University of Michigan Museum of Art.

Margaret Selkirk Parker (fig. 60), née Watson, was raised in Evanston, Illinois, the youngest of three daughters of prominent Chicago lawyer and real estate investor George Watson and Julia M. Dickinson. She and her sister Janet were among the first women to attend Northwestern University, where Margaret was enrolled as a student from 1883 to 1887, although she left before graduating. As a young woman she began collecting the works of James McNeill Whistler and Japanese *ukiyo-e* woodblock prints, competing in the art market with some of the most important collectors of the early twentieth century. Her collecting interests brought her into contact with two people whose collections and areas of interest were to help shape her own direction: Charles Morse, an engineer from her hometown of Evanston, and Charles Lang Freer, the Detroit industrialist whose collection was gifted to the Smithsonian Institution in Washington, DC. Both Morse and Freer collected Japanese art and the work of Whistler.

Charles J. Morse (1852–1911) was born in Oregon and graduated with a degree in engineering from the Sheffield Scientific School at Yale University. In 1889, he moved to Chicago as a representative of an Eastern iron firm engaged to construct the Manufacturers' Building at the World's Columbian Exposition that opened in 1893. During the World's Fair, Morse was so taken with the art in the Japan exhibit that he retired from his job to spend a year of study and research in Japan. Although we don't really know how Morse and Watson became acquainted, they clearly shared an enthusiasm for Japanese art. Their interest in the aesthetics of Asian art also led them both to the art of Whistler, whose paintings and prints indicate a tremendous sympathy with the art of Asia, particularly of Japan. Following Morse's death in 1911, his widow retained his important collection of Asian art; when it was sold at auction in 1957, the sale featured works by Masanobu, Toyonobu, Harunobu, and Utamaro and the collection was celebrated as one that was—as had been earlier expressed in Morse's obituary in the *Chicago Daily Tribune*—"internationally famous."[2]

The other collector who mentored Margaret Watson amassed a collection of such stature that it eclipsed those of Morse and Watson: Charles Lang Freer (1854–1919). Freer's astute eye and almost voracious desire to create a collection of lasting breadth and importance is evidenced by the gift of his collection to the Smithsonian Institution. His friendship with Whistler meant that Freer had direct access to the artist; Whistler, in turn, hoped that with Freer's patronage, he and Freer together might create the definitive collection of his work. As Whistler wrote to Freer on July 29, 1899, "I wish you to have a fine collection of Whistlers!!—perhaps *The* collection."[3] We do not know the circumstances of how Freer met Morse but it may have been through an exhibition of Japanese art in New York in 1896 to which Morse was a lender. We do know that Freer and Morse were corresponding by 1896.[4]

The fact that Morse and Freer each collected both Japanese woodblock prints (and eventually Asian art more broadly) and the work of James McNeill Whistler was bound to influence their young protégé. This pairing of interests was indicative of an aesthetic synergy between a recognized American master artist and one of his major sources of inspiration. Many collections formed in the early twentieth century included both components regardless whether the initial interest was grounded in the art of Asia, as in the case of Morse, or in that of Whistler. However, by the time she met Freer in 1900, Watson had already been collecting both Japanese prints and the prints of Whistler.[5] The museum's documentary files contain a receipt for Watson's purchase, in November of 1900, of eight Japanese prints from Morse's own collection, including works by Haronobu, Hokusai, and Hiroshige. Many of those prints, now in UMMA's collection, retain their original mount with Morse's signature on the face mat. The numerous receipts and correspondence with dealers that came with the gift also reveal that as early as 1894—years before she met Freer—Watson had already acquired impressions of Whistler's Venice etchings, such as *Little Venice* and *The Palaces*, from the Chicago dealer Frederick Keppel.

The first decade of the twentieth century was one of explosive activity in the market for Whistler's art.[6] His death in 1903 was followed by the sales of some of the earliest and finest holdings of his works, and Watson, Morse, and Freer were among the collectors hoping to secure exceptional impressions of Whistler's works from dealers and at auctions. In addition to a natural competitiveness, there was also a spirit of camaraderie among collectors. For instance, when Freer wrote to dealers such as Obach & Company in London or Knoedler's in New York, he frequently had prints sent on approval for both Morse and Watson as well as for himself. Works acquired by the other two would be paid for by Freer along with those that he was retaining for his own collection. Watson and Morse would in turn reimburse Freer for their purchases. This cooperative approach was discernable with other collectors as well. A letter from Knoedler's in the Parker files at UMMA dated November 27, 1906, alludes to another pair of Whistler collectors working together: "Mr. [Howard] Mansfield was in again this morning and was anxious to get some more for Mr. [Clarence] Buckingham. He is also anxious to get some more for his own collection." [7] Freer provided introductions to the major Whistler dealers for Watson, lauding her discernment and high standards. To Gustav Mayer at Obach & Company he wrote,

> Miss Watson is a lady of very refined tastes and already owns a considerable number of fine things. She wishes to add from time to time to her collection of Whistler prints and I have suggested that she communicate with you directly concerning future purchases. . . . I would add for your information that Miss Watson is likely to be very particular in the quality of prints she buys. I do

Fig. 61. James McNeill Whistler, America, 1834–1903, *Nocturne*, from *The First Venice Set*, 1879–1880, etching and drypoint printed in warm black ink, Margaret Watson Parker Art Collection. University of Michigan Museum of Art 1954/1.372.

> not think that she expects to attempt to make anything like a complete collection of Mr. Whistler's etchings and she will be attracted only by the finest plates in the best states and condition.[8]

The years 1905 to 1907 were a period of great activity and frequent correspondence between Freer and Watson as they discussed works coming up for sale, such as the Whistler holdings from Queen Victoria's collection at Windsor Castle—a group of impressions that Whistler had personally selected for the monarch. Between Freer, Watson, and Morse, letters were exchanged nearly every week when important collections came up for sale. In addition to the continuing interest in Japanese art and the work of Whistler, it was during this period that Watson began to branch out into other areas of interest to include ceramics from the Near East, such as the Syrian Rakka ware chalice that she bought in 1905 from Marie Nordlinger and Paris dealer Siegfried Bing.

The relationship between Freer and Watson is known largely through the letters that Freer sent her; very little of her responses exist, although we can see in Freer's side of the correspondence where the relationship began to change. In March of 1905, Freer wrote to Watson clearly in response to a suggestion in an earlier letter of hers that she be allowed to deal directly with Obach regarding works she might wish to purchase rather than routing both the shipment of the works and the billing through Freer. He replied,

> I will be very pleased to write Mr. Mayer when I return the present shipment, telling him of your interest in Mr. Whistler's work and asking him to treat you exactly the same as he does me, and I wish you to particularly feel that there is not the slightest probability of any misunderstanding or jealousy ever arising between you and me.

Freer expresses the hope that they can meet at

Obach's later in the spring and review the Whistler prints there together. He concludes by observing how collectors help avoid senseless bidding up of prices at auction by one or the other party dropping out, thus saving the winning bid from being prohibitive and evidence of "friendly interest felt in each others collections."[9]

Freer continued to applaud Watson's ambitions to create a less extensive, but still outstanding, collection. When an unusually beautiful impression of the highly atmospheric *Nocturne* from Whistler's *First Venice Set* came on the market at Knoedler's in New York (fig. 61), Freer immediately wrote to Parker to tell her of the breathtaking quality of the print, adding that "the price seems to me almost prohibitive—$1500.00. Still, it is so superb I thought I must tell you about it." As a further indication of the exceptional quality of this impression, Freer continued, "Mr. Petty [at Knoedler's] tells me that a number of collectors have taken their own proofs, of the same plate, to the Exhibition, for the purpose of comparison and that the opinion seems to be unanimous that the Knoedler impression is the finest one which has thus far been submitted to the test."[10] Watson did buy this impression from Knoedler's and it is one of the most outstanding works she ever acquired.

In 1907, Freer made his second trip to Japan where he met up with Mr. and Mrs. Morse and Miss Watson. During this rendezvous in Japan, something occurred that changed Watson's estimation of Freer, and a gradual cooling in their relationship began. It was in December of that year that forty-year-old Watson married the eminent University of Michigan professor of ophthalmology Walter Robert Parker (1866–1955). Together they lived in Grosse Pointe, and although Freer and the new Mrs. Parker were both active in Detroit arts organizations, it seems that they saw little of one another. The last indication of correspondence between them was a brief letter, still discussing their shared interest in Whistler, from Parker to Freer in November of 1908.

Mrs. Parker joined the Detroit Society of Arts and Crafts in 1909 and through that society became involved with the Detroit-based firm of Pewabic Pottery, becoming friends with Mary Chase Perry Stratton, who had founded Pewabic in 1903. She owned several Pewabic works by 1917 and in 1932 the Parkers acquired a large number more.[11] Following Mrs. Parker's death in 1936, Dr. Parker asked Stratton to review the collection in anticipation of transferring the works to the University of Michigan; she provided a handwritten annotated description of the Pewabic examples in the Parker collection.

Other Detroit arts organizations enjoyed the patronage of the Parkers. The Detroit Museum of Art[12] received an anonymous loan from Parker in 1918 that included some of the best works in her collection: the unusual early seascape by Whistler, *Sea and Rain,* as well as a selection of both Japanese (fig. 62) and Whistler prints; paintings by Alfred Stevens and by George Inness; a selection of Pewabic Pottery works; and sculptures by the artist Edward Kemeys. In 1930, Benjamin March of the Detroit Institute of Arts wrote Parker to thank her for her funding that allowed the museum to acquire a Chinese celadon rabbit.

Throughout this time, Parker continued to acquire works that reflected her interests. She bought exceptional Whistler prints as they came available; as late as 1932 she still found exceedingly rare impressions of the late etchings executed in Paris, such as *Café Corazza* and *Rue Vauvillers,* which are known in only a handful of impressions, and an extraordinary complete *French Set.* Parker's approach to collecting was that of a connoisseur, and she was armed with a sizeable library of reference volumes that came to the University of Michigan as part of her bequest. She made lists and records of works she had seen, even making notes on the back of a list from Knoedler in which she questioned the dealer's information, correcting the cross listings to the Wedmore catalog *raisonné* of Whistler's prints.

Parker's interest in Asian art extended to Central and South Asia. She became acquainted with Walter Koelz (1895–1989), whose research as a zoologist and whose anthropological collections have been important contributions to the University of Michigan's holdings (see Sinopoli,

this volume). In 1932, University of Michigan President Ruthven wrote to thank the Parkers for their $500 donation toward Koelz's expenses for an expedition to Tibet. For the next several years, Parker received lengthy letters from Koelz describing his journeys to the Punjab region of India as well as his travels to Tibet. In 1936, Koelz and Josselyn Van Tyne published a list of seven new birds they had identified in the Punjab; one of them, in the partridge and grouse family, was named for Parker.[13]

Margaret Watson Parker died suddenly in 1936 while she and her husband were traveling in Hot Springs, Virginia. Her will stipulated that her husband retain a life interest in her collection but that the entire art collection be given to the University of Michigan upon Dr. Parker's death. In the years following her death, Dr. Parker began to transfer portions of her collections to the university. Upon his death in 1955, the remainder of the collection, including the Whistler prints, came to the university. An additional gift in Dr. Parker's will stated that half of his residual estate should be added to his wife's bequest to be spent in "acquiring additions to the Margaret Watson Parker Art Collection, or to be added to a fund for the erection of a suitable art gallery to house the Margaret Watson Parker Art Collection and other Art Collections or for both purposes."[14]

The Museum of Art, established as a separate unit of the university only in 1946, was the natural recipient of the bequest gift. A draft regents' communication of October 8, 1957, stated that the executive committee agreed that a major acquisitions program should be undertaken as soon as possible and asserted the primacy of learning from original works of art, which this bequest enabled. For many years, the fund allowed the Museum of Art to acquire important works that helped to place it as one of the leading collections of Asian art in any university museum. Among the many

Fig. 62. Kitagawa Utamaro, Japan, 1754–1806, *Courtesan in Procession*, 1793–1795, ink, white pigment, and color on paper, Margaret Watson Parker Art Collection. University of Michigan Museum of Art 1962/1.104.

works added to the collection were Chinese and Japanese paintings in all formats (hanging scrolls, folding screens, album leaves), Asian sculpture, and decorative arts, including works from India. In more recent years, the fund also allowed the purchase of Whistler prints that Parker had hoped to buy but had not secured during her lifetime, such as the *Guitar Player* and *The Limeburner*. The last of the funds were spent during the expansion and renovation of the museum in 2007–2009. The upgrading of the galleries where Mrs. Parker's works are featured included all new display furniture and climate system upgrades, as well as extensive research and reinterpretation of the collection, all in accordance with Dr. Parker's cash bequest gift.

We can only surmise whether Freer's example of a collection gift to a public institution figured in her thinking, but Parker's bequest to the university has had a foundational importance to the Museum of Art. The collection remains the most important and valuable single gift to the museum, even after decades of growth. Parker's collecting interests may have been fueled and shaped by her two early mentors, but given her independence, it is not surprising that her collection branched well beyond the interests of Asian art and Whistler. At the time of her death, her collection included a rich selection of Old Master prints and drawings by artists such as Rembrandt and Millet; nineteenth-century paintings by artists such as Alfred Stevens, George Inness, and Henri Fantin-Latour; and European and Asian textiles and metalwork. Past and present curators and directors have built upon the strengths of the Parker collection to amass internationally renowned holdings in Asian art and the graphic works of James McNeill Whistler. To assemble her collection now would be impossible. Her belief in the compelling presence of original works of art ensured that her gift to the University of Michigan would enable the growth of a first-class collection available to students and scholars. The strength of UMMA's collections is a direct result of the collection of Margaret Watson Parker and her generous bequest.

Notes

1. *Margaret Watson Parker, A Collector's Legacy*. Exhibition catalogue, Ann Arbor: University of Michigan Museum of Art, 1982; *UMMA—A Beacon for Art*, Ann Arbor: University of Michigan Museum of Art, 2009.

2. *Chicago Daily Tribune*, December 18, 1911, 20.

3. Letter in Freer Gallery Art Archives, Washington, DC., quoted in Linda Merrill, editor, *With Kindest Regards: The Correspondence of Charles Lang Freer and James McNeill Whistler*. Washington, DC: Freer Gallery of Art, Smithsonian Institution, 1995, 122–23.

4. Confirmation of the date marking the beginning of the Freer-Morse correspondence as well as the suggestion regarding the meeting of Freer and Charles Morse through an exhibition in New York are contained in an email from Lee Glazer, Associate Curator, American Art, Freer and Sackler Galleries, Washington, DC.

5. UMMA documentary files, letter of Charles Lang Freer to Charles J. Morse, February 6, 1901. Charles Lang Freer Papers, Freer Gallery of Art and Arthur M. Sackler Archives, Smithsonian Institution. Morse introduced Miss Watson to Freer while visiting the New York dealer Rufus E. Moore in 1900 or early 1901. My thanks to David Hogge, Head of Archives, Freer Gallery of Art and Arthur M. Sackler Gallery, Smithsonian Institution for locating and confirming Watson and Freer's initial meeting.

6. See discussion in Carole McNamara and John Siewert, *Whistler: Prosaic Views, Poetic Vision* (Exhibition catalogue), New York: Thames and Hudson, 1994.

7. Like Freer, Howard Mansfield (1849–1938) was another collector who knew Whistler and with the artist's help assembled one of the most important Whistler print collections (and Japanese art) in private hands. Margaret Watson acquired several works from Mansfield's collection when they went on sale at Knoedler's in 1907. Clarence Buckingham (1855–1913) was an important Chicago financier and collector whose holdings comprised not only Whistler prints and Japanese woodblock prints but also an extensive collection of Old Master prints, many of which are now part of the collection of the Art Institute of Chicago.

8. UMMA documentary files, letter of Charles Lang Freer to Gustav Mayer, Obach & Company, March 21, 1905. Charles Lang Freer Papers, Freer Gallery of Art and Arthur M. Sackler Archives, Smithsonian Institution.

Margaret Watson was acknowledged by her friends to be a very poor correspondent. Hence, most of what we know about her relationship to Freer is based on his correspondence. The Freer Gallery of Art has very kindly provided the Museum of Art with copies of Freer's letters to Miss Watson, thus augmenting our understanding of Freer's formative role in guiding Watson.

9. UMMA documentary files, letter of Charles Lang Freer to Margaret Selkirk Watson, March 9, 1905. Charles Lang Freer Papers, Freer Gallery of Art and Arthur M. Sackler Archives, Smithsonian Institution.

10. UMMA documentary files, letter of Charles Lang Freer to Margaret Selkirk Watson, March 26, 1906. Charles Lang Freer Papers, Freer Gallery of Art and Arthur M. Sackler Archives, Smithsonian Institution.

11. Thomas W. Brunk, *Painting with Fire: Pewabic Vessels in the Margaret Watson Parker Collection* (Exhibition catalogue), Ann Arbor: University of Michigan Museum of Art, 1995.

12. Renamed the Detroit Institute of Arts in 1919.

13. Josselyn Van Tyne and Walter Koelz, "Seven New Birds from the Punjab," *Occasional Papers of the Museum of Zoology*, no. 334, May 27, 1936. The new subspecies *Francolinus francolinus parkerae* was "named in honor of Mrs. Walter Parker of Detroit, in grateful recognition of her active interest in the junior author's researches in India."

14. UMMA documentary files.

AGNES INGLIS (1870–1952) AND THE BIRTH OF A RADICAL ARCHIVE

Julie Herrada

When Agnes Inglis (fig. 63) began volunteering in the University of Michigan's Labadie Collection in Ann Arbor in 1924, she was fifty-two years old, and she thought it was a pretty good job. "I really don't want pay for I am so much freer doing it the way I do. I could not be tied to either routine or hours," she wrote in 1927.[1] It was not until 1929, five years after she commenced her work, that the library began paying her a nominal weekly salary.[2]

Born in Detroit into a prominent and strait-laced Presbyterian home in 1870, Inglis's early life had been a sheltered one. Her father died when she was only four. Although she had four siblings, she was the oldest girl, so she was duty bound to care for her widowed mother until she was in her late twenties. After her mother passed away, Inglis left behind her privileged existence, determined to experience the world. She was about to be transformed.

Fig. 63. Agnes Inglis at barn, July 1928. Labadie Photograph Collection, University of Michigan Library.

For the next twenty years, Inglis lived her life as she pleased—reading, traveling, and working with social justice causes. In 1899, she enrolled in history and English classes at U-M, but she never graduated. Although she found academia intellectually stimulating, she craved more real world experience, so she left the university to volunteer in settlement houses in Chicago and Detroit. Witnessing the disparities in the class structure and the struggles of the poor and working class, she organized rallies in support of better pay and working conditions for women, advocated for the legal and economic rights of prostitutes, and briefly joined the Socialist Party.

After meeting Emma Goldman in 1912, Inglis joined the anarchists. In 1913, she became an atheist. She arranged lectures, rallied support for labor and civil liberties causes, organized strikes and boycotts, and contributed bail money for political prisoners and World War I sedition law violators. She never married, so she made her own decisions about how to spend her time and money and was always generous with both (although her brother eventually reduced her allowance to a modest living stipend so that she would not squander her inheritance on radical causes).

As a volunteer in the U-M Labadie Collection, which had no on-site supervisor, Inglis assigned herself the first task to carefully unpack and examine the gift that had arrived from Joseph (Jo) Antoine Labadie over a decade earlier. Labadie, an anarchist and atheist printer and publisher, with the help of his wife, Sophie, a devout Catholic,

Fig. 64. Agnes Inglis in the Labadie Collection, August 1929, photo by Judson Grenel. Labadie Photograph Collection, University of Michigan Library.

had collected and carefully preserved a houseful of radical literature, some of which was produced on his own printing press. It was a unique and valuable archive.

Labadie had chosen to donate his collection to U-M for sentimental as well as practical reasons—his family had resided in Michigan for two hundred years; one of his daughters had attended the university; and, thinking he could visit, he wanted it to be geographically close to him. Thus, the shipment of twenty boxes arrived at the library in September 1912.[3] Although the Board of Regents had accepted the gift, the library administration was perplexed about what to do with this radical trove of literature.[4] And so nothing was done with them. None of the correspondence, books, pamphlets, by-laws, proceedings, newspapers, newsletters, membership cards, photographs, broadsides, palm cards, and badges relating to labor, anarchism, socialism, cooperatives, and monetary reform movements of the Gilded Age were even unpacked. The whole archive was placed in an out-of-the-way corner of the library, and inquiring patrons would be given a key and sent into a locked cage, left alone with boxes of unprocessed and uncatalogued material. If anyone *had* unpacked it before Inglis arrived on the scene, it is very possible that the Labadie Collection would not exist today, certainly not in the form it eventually took by the end of the 1940s.

When she commenced her work, Inglis received no library or archival training and there was no one to guide her. This afforded her almost total freedom in the ways she would describe, organize, preserve, and make accessible the archive (fig. 64). With no operating budget, she gleaned furniture and office supplies from other parts of the library. Promotion of the collection came only through word of mouth.

Despite her maturity and experience, Inglis still had curiosity and conviction—and a temperament that engendered confidence, allowing her to ferret out family papers and stories. She also possessed a deep sense of responsibility for the materials and well-honed skills as an organizer and social justice activist. Although she was not a trained librarian, archivist, or historian, she was concerned about the materials and wrote to Labadie, "It is true someone who cared had to begin. It is a big job to get started."[5] Inglis thought she would spend a few months "getting it started." She once wrote to her friend, the famous anarchist Emma Goldman, "It's no joke to take all that mass of material and fix it up so students can really use it. It is not a work everyone can do. One has to know the material. People don't appreciate that."[6]

Inglis noticed that there were gaps in the holdings, so it wasn't long before she decided to build on the original donation, tapping into her large network of activists; the fact that she was there was enough to build a solid foundation of credibility among those who were typically dubious of institutions. And because no one in the administration was really paying attention—and the few who were, were at least acquiescent—Inglis was able to establish her own collection development program, as long as she could acquire materials and manage it by herself, which she did very successfully for nearly three decades.

She began corresponding, questioning, and probing the lives of the people she was tracing even after they had slowed down or retired from the radical movement. She wrote directly to them or to their families if they were deceased. She was intensely interested in the personal as well as the political lives of not only the most prominent but also the lesser-known radicals, who were just as relevant to her. She was always curious to know what became of people.

It was obvious to Labadie that Inglis had brought his collection into the forefront of radical archives and that it would stay there at least as long as she did. This archive—created by Labadie out of a love for labor and the underdog—became Inglis's labor of love. She was dedicated to preserving history and creating a valuable labor library, as well as adding to her own enlightenment, and so she carefully tended to the items in the collection. Her many connections and her commitment gave her access to a constant flow of information and documentation.

Inglis strived to obtain full sets of periodicals and newspapers by writing to everyone she knew. Her letters are full of details describing the holdings and what was still missing. There was no acquisitions budget and no resources for managing the archive, so she relied almost completely on donations. She had the sympathetic help of associate librarian Francis Goodrich, who arranged for binding newspapers in the Labadie Collection after Inglis had filled in and collated the sets. Goodrich left the library in 1930, but by then Inglis was settled in and pleased with the way things were going. At Inglis's urging, head librarian William Warner Bishop arranged for custom bookplates to be made for books, pamphlets, and scrapbooks, and was himself dedicated to the preservation and access of the materials.

Jo Labadie died in 1933 and is buried next to his wife in an unmarked grave (his wish) at Parkview Memorial Cemetery in Livonia, Michigan. The remainder of his collection was a bequest to the library through his three children. His son Laurance maintained the Labadie printing press and continued to print radical literature. He kept in touch with Inglis; their correspondence, along with many others, is contained in the Agnes Inglis Papers (Laurance's own archive was a bequest to the Labadie Collection through his niece Carlotta Anderson after his death in 1978).

After Jo Labadie's death, Inglis continued to build on the original collection. During her tenure, she increased its size dramatically. Part of her time was spent writing notes on historical events and biographical information on many activists and writers. These notes are a valuable resource, as are all the print and visual materials she collected and painstakingly preserved, bound, referenced, analyzed, and cataloged, in her own idiosyncratic way (fig. 65).

Inglis remained active throughout her life in political circles, writing letters, visiting prisoners, and organizing lectures and fundraising events for the radical movement. Her extensive research into people and events uncovered more sources for materials. She wrote to living radicals such as Emma Goldman, Ralph Chaplin, Granville Hicks, Elizabeth Gurley Flynn, Joseph Ishill, Ben Reitman, Mary Gallagher, Louis Adamic, Ammon Hennacy, and Augustin Souchy, and to families of

Students! Comrades!

We Appeal To You

You are the youth in the schools. We are the youth in the factories, in the fields, in the mines. We all are the youth of the land. The great war taught us that the world depends upon us. We left the schools and universities, the factories, fields and mines. Together we marched to the trenches, or else together we maintained the great industries. We felled the spruce. We mined the copper, the iron. We built the ships. We harvested the grain. We loaded and ran the cars and the trains. We loaded and manned the ships.

They said fight or work. And we fought and worked, we, the boys and girls—we, the youth! We know now that we can. We realize our potential powers. The old men, our Masters, the old men, our Professors, told US to make the world safe for democracy. They were too old. So we fought and worked together for Freedom. We were the youth of many nations, of different colors and tongues, boys and girls. All were needed. They said all—all of us were needed. And thus we discovered our potential powers and possibilities. We learned our **individual** power. We learned, too, our power **organized.** We learned our collective power.

We fought to make the world safe for Democracy. We fought to win Freedom.

DID WE DO IT!

NO!

WE in the factories are NOT free. We in the fields and the mines are NOT free.

STUDENTS! COMRADES! We appeal to you.

We are all back from the trenches now.

The class rooms are yours. The libraries are yours. The

I wrote this but did not distribute my own work! as it were, I was not bold enough. Agnes Inglis.

Fig. 65. Students! Comrades! broadside. The handwritten note at the bottom says "I wrote this but did not distribute my own work! as it were, I was not bold enough. Agnes Inglis." Labadie Collection, University of Michigan Library.

deceased ones, as evidenced in her copious notes. By doing so, she was able to preserve history that probably would have otherwise been lost forever. As a result of her efforts, letters, diaries, journals, artwork, autobiographies, photographs, posters, scrapbooks, badges, and buttons, as well as hundreds of books, journals, and newspapers, came into the collection.

Although there is some evidence that Inglis loaned out materials, by 1950 she seems to have changed that policy. When Jean Levey, executor of the meager estate of Emma Goldman, wrote to Inglis asking to borrow some of Goldman's letters for her research on a biography, Inglis replied back, "I sent Emma her letters when she was writing her book. But I can't send them out to anyone again. They are too rare and too fragile. . . . This is the way libraries do. This correspondence now belongs to the Library and to History. I have to treat it as I must all documents in my charge."[7]

In addition to the students and faculty who used the Labadie Collection for their research, Inglis assisted many people with their books, including Emma Goldman and Ralph Chaplin with their autobiographies (*Living My Life* and *Wobbly*, respectively), Henry David with the seminal *The Haymarket Tragedy*, and James J. Martin with *Men Against the State*. But her letters reveal that she was not partial to big name radicals. She always made time for visitors to the library. Since there was no one else to rely on, she felt an innate responsibility to make the materials accessible to anyone who was interested. In a letter to Belle Chaapel, she described "two young people from NYC who came out to spend a day in the Collection. I love to have people come like that but as I grow older I get tired after 3 hours. These young people came at 10 a.m. and left at 5 p.m. Literature does not tire me but people do, in case they stay too long."[8]

A demonstration of Inglis's continuing legacy is the story of E. F. Doree, an organizer for the radical labor union, the Industrial Workers of the World, who was sentenced to twenty years for violating the federal Espionage Act in 1917. Doree was an appreciative recipient of Inglis's financial support (in the form of bail money), long before she came to work in the Labadie Collection. Doree later had a daughter, Ellen Doree Rosen, who—in 1997, forty-five years after Inglis's death, but ever grateful to her for her generosity to her parents so long ago—donated her father's valuable collection of prison letters and other documents to the collection. These letters are now used frequently in an undergraduate history course.

By 1950, Inglis still walked daily to and from her rented rooms near the library, but had slowed down considerably. Although she always made time for researchers, in a letter to Jeanne Levey, she wrote, "I am obliged to tell you I have not been very well, having lost a good many lbs. this last year. So I have to limit myself to supplying the material and information but then pretty well leaving you to the work itself and not being so sociable as I used to be. I go home and am a very poor hostess."[9] She passed away on January 30, 1952, less than two months after her eighty-first birthday. Her ashes are buried in the Inglis family plot at the historic Elmwood Cemetery in Detroit.

Despite the promise of Dr. Warner Rice, the new head librarian, that the library would continue to add to the Labadie Collection, its sole employee was not replaced for eight years.[10] Hence, the collection was neglected and its fastidiously organized and fragile contents were left in unskilled hands. No cataloging or collecting was done. It was a period of dark days for the collection, as well as for its researchers who had no one to help them navigate the labyrinth of Inglis's intricate filing and shelving system. Indeed, there was no one to whom she could have passed her knowledge. There is almost no evidence of activity in the Labadie Collection in 1951, the year before she died. Afterward, the person answering queries from patrons did not seem to appreciate her system of organization. A letter from librarian Alfred Harris, dated 1953, states, "I am sorry to have to inform you that there may be a delay of a couple of months in supplying the exact information you have requested. The reason for this is that it has been found necessary to revise the pamphlet section of the

Labadie Collection due to the death of Miss Agnes Inglis, the former Curator, who as a philosophical anarchist had her own individualist, or anarchistic 'methods' for handling these materials."[11]

After the relocation and integration into the newly built Special Collections Library in the south building of Hatcher Library in 1970, Inglis's precise filing and locating system was lost forever. Her index cards, mostly handwritten with notes and analytics, are a wealth of institutional memory. They still survive and are filed in a card catalog on the seventh floor of Hatcher Library. A Mellon grant in 2012 allowed for the transcription of several hundred of these cards into a searchable, public database as part of the Editors' Notes Project.[12] The entire holdings of the Labadie Collection, now many times larger, are almost fully cataloged and accessible through the online public catalog and various public online databases.

Despite the temporary lack of attention to such a valuable treasure, there are numerous reminders of Inglis's contributions. Her inscriptions can be found inside many of the books and pamphlets (fig. 65), and her notes and letters remain among the library's most heavily used materials. A curator's notes and inscriptions help document the source of materials in an archive, but Inglis's rich correspondence with and about these people is part of the record that weaves their stories together.

The tradition of collection building initiated by Inglis continues, although its methods vary. Today, it is the fastest growing and most heavily used archive within the Special Collections Library and throughout the entire campus. "It is pretty well assured a place for all time," Inglis told Labadie in 1931.[13] That statement sums up the level of satisfaction she obtained from her life's work, and it is with gratitude that many people still benefit from it. It is to U-M's credit that the ingenuity and fervor of Agnes Inglis—although mostly unappreciated at the time, as the making of history often is—has been allowed to thrive and grow a century later.

Notes

1. Letter from Agnes Inglis to Joseph Labadie, May 12, 1927. Joseph Labadie Papers, Box 4. Joseph A. Labadie Collection, University of Michigan Library.
2. Letter from Agnes Inglis to Joseph Labadie, May 3, 1929. Joseph Labadie Papers, Box 4. Joseph A. Labadie Collection.
3. Letter from Theodore W. Koch to Joseph Labadie, September 3, 1912, Joseph Labadie Papers, Box 4. Joseph A. Labadie Collection.
4. Letter from University of Michigan to Joseph Labadie, November 22, 1911. Joseph Labadie Papers, Box 4. Joseph A. Labadie Collection.
5. Letter from Agnes Inglis to Joseph Labadie, November 24, 1925. Joseph Labadie Papers, Box 4. Joseph A. Labadie Collection.
6. Letter from Agnes Inglis to Emma Goldman, March 19, 1925. Emma Goldman Papers. Original at International Insitute of Social History, Amsterdam, The Netherlands. Joseph A. Labadie Collection.
7. Letter from Agnes Inglis to Jean Levey, January 14, 1950, Agnes Inglis Papers, Box 13. Joseph A. Labadie Collection.
8. Letter from Agnes Inglis to Belle Chaapel, November 29, 1946. Agnes Inglis Papers, Box 4. Joseph A. Labadie Collection.
9. Letter from Agnes Inglis to Jean Levey, April 5, 1950. Agnes Inglis Papers, Box 13. Joseph A. Labadie Collection.
10. Edward Weber, a social sciences reference librarian, was appointed curator in 1960.
11. Alfred Harris to Malcolm Young, February 23, 1953. Edward Weber Papers, Box 37. Joseph A. Labadie Collection.
12. www.editorsnotes.org
13. Letter from Agnes Inglis to Joseph Labadie, May 1, 1931. Joseph Labadie Papers, Box 4. Joseph A. Labadie Collection.

ERMINE COWLES CASE (1871–1953)

Pioneer of Paleontology

Philip D. Gingerich

Ermine Cowles Case, commonly known as E. C. Case, was the first twentieth-century paleontologist to join the University of Michigan faculty. His 1907 appointment initiated the development of broad and successful teaching and research programs that brought both national and international distinction to the university. Whereas nineteenth-century paleontological research at U-M had largely been focused on the state of Michigan (which has a rich Paleozoic invertebrate fossil record that has been important for interpreting its geological history) and was often carried out in collaboration with the Michigan Geological Survey, Case was a vertebrate paleontologist trained in biology and comparative anatomy as well as geology. Thus, he represented a new and different tradition.

Fig. 66. Ermine Cowles Case, 1934. Bentley Historical Library.

E. C. Case (fig. 66) was born on September 11, 1872, in Kansas City, Missouri. The youngest of three children, he was named for his paternal grandfather (Ermine Case) and grandmother (Mary Ann Cowles). Case's father, trained as a physician, rose to a quartermaster and colonel in the Union Army during the Civil War and was later active in business in Kansas City, eventually being appointed postmaster. Case's mother, daughter of Kansas City's first mayor, died shortly after he was born.

Case entered the University of Kansas at Lawrence in 1889, at the tender age of seventeen. Samuel Wendell Williston arrived from Yale a year later to serve as professor of geology, paleontology, anatomy, and medicine. Williston had trained with Benjamin Franklin Mudge at Kansas State Agricultural College (now Kansas State University) at Manhattan before moving to Yale University to work for fourteen years as a research assistant for paleontologist and professor Othniel Charles Marsh. Williston had a big influence on Case while he was an undergraduate at Kansas. Case was the sole student accompanying Williston on an 1891 field expedition to western Kansas, and his first scientific publication, "Kansas Mosasaurs," in 1892, was coauthored with Williston.[1] Although Case graduated in 1893 with an AB degree in chemistry, Williston considered him to be his first paleontology student, and Case was possibly Williston's most successful student. Case worked for Williston for another year and received an AM degree from Kansas in 1894.

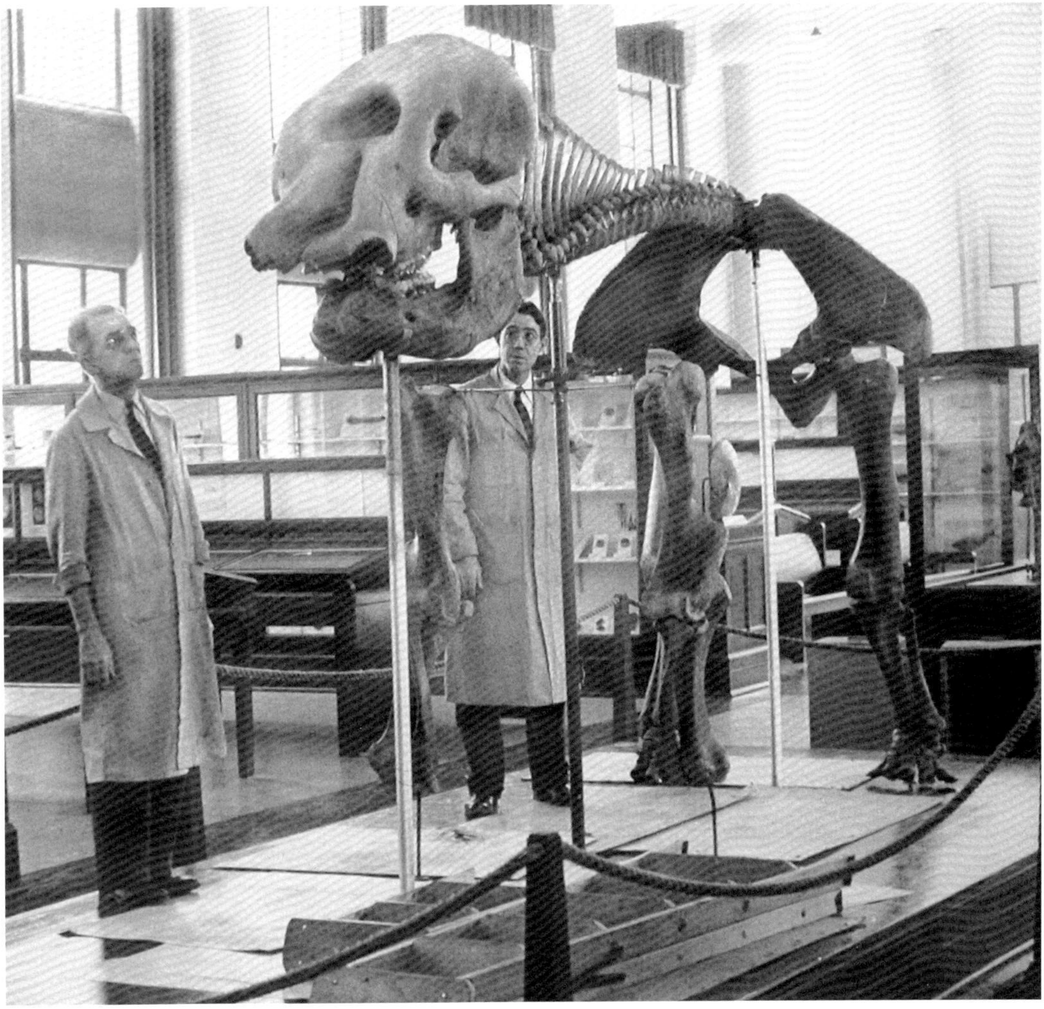

Fig. 67. E. Case with the Owosso Mastodon in the Hall of Evolution, photogravure, c.1947. Museum of Natural History.

From Kansas, Case moved to Cornell University in Ithaca, New York, where in 1895 he received a master's degree in geology for his thesis "The Drumlin Region of Central New York, together with Certain Experiments in Ice Action." He completed his PhD in 1896, at the age of twenty-four, working in the Department of Paleontology at the University of Chicago under the direction of another former O. C. Marsh assistant, George Baur; his dissertation, "On the Osteology and Relationships of *Protostega*," analyzed the marine turtle from the Cretaceous Niobrara Chalk of western Kansas. After George Baur died in 1898, Williston moved from Kansas to the University of Chicago in 1902 to succeed him as professor of paleontology.

Case spent 1897 to 1907 teaching in the Department of Geology and Physical Geography at the Wisconsin State Normal School in Milwaukee. In 1898, he married Mary Snow, daughter of one of the first professors and then chancellor at the University of Kansas. Their two children, Francis (professor of chemistry at Temple University) and Theodore (physician in Chicago), were born in 1899 and 1900, respectively. Case's years in Milwaukee were very productive. He published a series of didactic studies for students, covering all of the major vertebrate groups; a book on the geology and physical geography of Wisconsin; and two major monographs on pelycosaurs. He also received a grant from the Carnegie Institution of Washington in 1903 that enabled him to visit many museums in Europe to further his research on fossil vertebrates. In 1907, William H. Hobbs, the chair of geology at the University of Michigan, offered Case an assistant professorship to teach historical geology and paleontology and to organize and augment long-neglected collections of specimens relevant to these subjects.

Weaving the threads of Case's, Mudge's, Williston's, and Baur's careers together, it is clear that when Case reached U-M, he brought a field-oriented Yale tradition of empirical vertebrate paleontology with him, a tradition reflected eastward from Kansas to Chicago and finally to Michigan. With expansion of the faculty following Case's arrival and the opening of the Museums Building in 1928, geology and paleontology at U-M were considered to have a faculty and facilities comparable to any institution in the country—and Professor Case played a major role in the growth and national visibility of both disciplines.

Case was most active in research in the years following completion of his PhD. He made a transition from the study of Mesozoic marine vertebrates like the mosasaurs and turtles of his native Kansas to the older and more difficult late Paleozoic aquatic and terrestrial vertebrates of Texas, initially focusing on the pelycosaurs or "sail-lizards." Some of this research was carried out with his advisors George Baur and Samuel Wendall Williston, but Case himself was responsible for the fieldwork and for most of the analysis and writing. George Gaylord Simpson summed up this work:

> Professor Case's first major research interest was in the land life of the Permo-Carboniferous of North America. He envisioned and accomplished not only many individual contributions but also a synthesis of the whole of that subject.[2]

From 1899 through 1926, Case published nine monographs totaling 1,360 printed pages in the *Transactions of the American Philosophical Society* or as *Publications of the Carnegie Institution of Washington* (in 1946, he published a tenth monograph).[3] Many shorter reports were published as journal articles, but the monographs made Case the authority of his time on the composition, morphology, and environmental setting of North American late Paleozoic and early Mesozoic vertebrate faunas.

At U-M, Case's original research was principally on Paleozoic reptiles, but he also wrote on fish, amphibians, birds, and mammals of the Mesozoic and Cenozoic. Case was active throughout his career collecting fossils in the field, investigating them in the museum, and teaching about prehistoric life from all three of the richly fossiliferous geological eras. He explained his motivation for an active field program:

> The paleontological collections in the Museum . . . have been gathered to illustrate the history of the earth and the origin and development of life. . . . One of the most important steps was the institution of expeditions by members of the department, directed to definite localities with the object of procuring material to fill gaps in the series and obtain material for student use. No university can hope to compete with the large museums of the country in obtaining and displaying large and expensive specimens, nor is it desirable that they should attempt this, but every institution engaged in teaching and investigation must have an ample supply of material to illustrate the . . . history of life upon the earth. Fossils are the relics of the past and are used for two purposes: first, to record the passage of geological time; and second, to record the evolution of life.[4]

For many decades, the most tangible way to appreciate Professor Case's contributions to vertebrate paleontology at U-M has been through a visit to the Hall of Evolution in the Museum of Natural History in the Ruthven Museums Building (fig. 67).[5] Case's Museum of Paleontol-

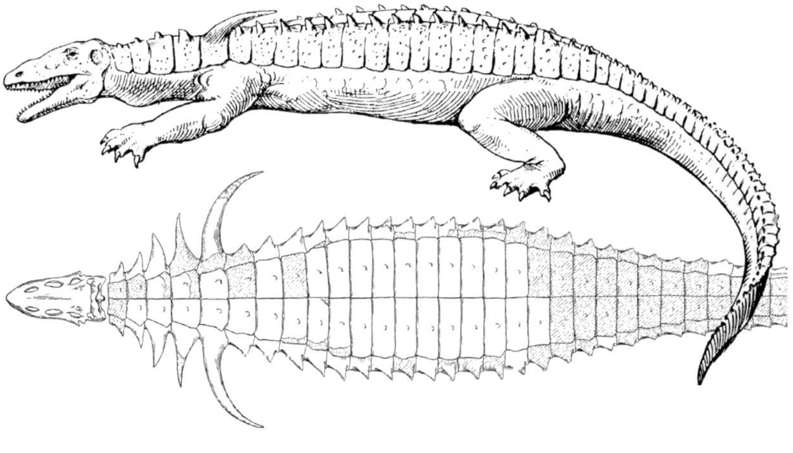

Fig. 68. E. Case's restoration and type skeleton of the ten-foot long Triassic-age armored reptile *Desmatosuchus spurensis*. Museum of Paleontology.

ogy specimens document Paleozoic stages of the transition from Devonian lobe-finned fishes (e.g., *Eusthenopteron*) to diverse Permian amphibians (e.g., *Trimerorachis*, *Eryops*, and *Diplocaulus*) and Permian stem reptiles (e.g., *Captorhinus*) and synapsid reptiles (e.g., *Dimetrodon* and *Edaphosaurus*). Case's Triassic specimens include a beautiful skull of *Phytosaurus* and a skeleton of the horned and armored *Desmatosuchus* (fig. 68). There are sauropod dinosaur specimens from the Jurassic, and a popular skeleton of the ornithopod *Edmontosaurus* from the Cretaceous. Case's Cenozoic mammal specimens include skulls of true and false sabre-tooth cats from the Oligocene and skeletons of some of their better-known plant-eating prey. There is an emphasis on faunas and paleoecology that reflects Case's interest in both throughout his career.

Professor Case had an international profile and traveled to Africa, Asia, Australia, and Europe in connection with his work. During a trip to England, he was able to purchase the Richard Lydekker library of paleontology books and reprints, filling existing gaps in the paleontology literature available at U-M. Mrs. Case accompanied him on a 1923 trip to Africa; she became ill and died while they were in Cape Town, a tragedy that helps explain a transition that Case soon made in his career. Before 1923, his energy was given to hands-on research. From 1926 to his retirement, Case's focus shifted to administration of the then new Museum of Paleontology research and teaching institute, and to development of paleontology exhibits for the Visual Education Department's displays in the Museums Building.[6] When Case retired from the faculty in 1941, he was simultaneously professor of historical geology and paleontology, chairman of the Department of Geology, director of the Museum of Paleontology, and curator of fossil vertebrates in the Museum of Paleontology.[7] Professor Case trained two doctoral students, Claude W. Hibbard and John A. Wilson, who received their PhDs in 1941 and went on to distinguished academic careers at U-M and the University of Texas, respectively.

Professor Case was active in scientific societies and received many honors. He was a research associate of the Carnegie Institution of Washington, which sponsored many of his monographic publications; a fellow of the Geological Society of America; and a member of the American Association for the Advancement of Science, American Society of Mammalogists, American Society of Naturalists, Michigan Academy of Sciences, Sigma Xi, Society of Vertebrate Paleontology, and Washington Academy of Sciences. Case was elected president of the Michigan Academy of Science in 1912 and president of the Paleontological Society in 1929, was honored with membership in the American Philosophical Society in 1931, and was U-M's Henry Russel Lecturer in 1934. When Case died in 1953, the Board of Regents established an endowed Ermine Cowles Case Memorial Lecture and an Ermine Cowles Case Student Award. These endowments have brought distinguished lecturers to campus and have rewarded deserving students for more than sixty years, both living testaments of the enduring influence E. C. Case has had on the science of paleontology at the university.[8]

Case was known for his self-deprecating humor. In his later years, he published tongue-in-cheek essays including "My Monument" (1942), "The Chipmunk, the Professor, and the Tom Cat" (1944), and "The Torch and the Tin Can" (1950).[9] In the latter, he lamented that ten years of retirement might not be enough to correct the scientific mistakes of his fifty-year career. For all his imagination, Case probably never expected to be named, as he was posthumously in 1974, one of a cohort of notable women in the history of American science![10]

Humor aside, Case had a distinguished career, and through his research and teaching brought national and international distinction to the university. His specimens and exhibits are a living legacy as they continue to supplement the course work of University of Michigan students and provide educational outreach to many middle and high school students. The Museum of Natural History's Hall of Evolution in the Ruthven Museums Building, featuring many skulls and skeletons of fossils collected by Case, has long been a source of enjoyment and lifetime learning for visitors and residents of southeastern Michigan and will be central to exhibitions in the new museum in the Biological Science Building expected to open in 2019. The Museum of Paleontology that Professor Case nurtured has continued to flourish as a research and teaching institution, and its outstanding collections underpin ongoing progress in understanding the history of life, the process of its evolution, the earth's changing environments, and the interdependence of all three.

Notes

1. Willeston, S.W., and Case, Ermine C., "A Kansas Mosasaurs," *Kansas University Quarterly* 8 (1892), 15–32.

2. Simpson, G. G., "Ermine Cowles Case (1871–1953)," *Yearbook of the American Philosophical Society* (1953), 329–33, here 331.

3. These include: Case, *Ermine C., Revision of the Amphibia and Pisces of the Permian of North America* (1911), *The Permo-Carboniferous Red Reds of North America and their Vertebrate Fauna* (1915), *The Environment of Vertebrate Life in the Late Paleozoic in North America, a Paleographic Study* (1919), and *Environment of Tetrapod Life in the Late Paleozoic of Regions Other than North America* (1926), all published by the Carnegie Institution, Washington, DC.; see also Case, "Modern Developments in Paleontology" (Henry Russel Lecture for 1934), *Michigan Michigan Alumnus Quarterly Review* 41 (1935), 512–22; for a complete list of Case's publications, see Kellum, L. B., "Memorial to Ermine Cowles Case," *Proceedings of the Geological Society of America* (1955), 93–102.

4. Case, Ermine C., "Something about the Paleontological Collections in the University," *Michigan Alumnus Quarterly Review* 27 (1921), 292–300, here: 292.

5. The Museum of Natural History, currently in the Ruthven Museums Building, will close in 2018 to make way for a new science and nature museum in the Biological Sciences Building. Many of Case's specimens will continue to be featured in the new museum.

6. See Museum of Natural History entry, this volume.

7. See Case, "The Dilemma of the Paleontologist," *Contributions from the Museum of Paleontology, University of Michigan* 9 (1951), 173–215.

8. A list of the Ermine Cowles Case lectures can be found at http://www.lsa.umich.edu/UMICH/paleontology/Home/Events/Case%20Lecture/Case%20lecture%20list%202014.pdf

9. Case, Ermine C., "My Monument," *Michigan Alumnus Quarterly Review* 48 (1942), 233–42; Case, "The Chipmunk, the Professor, and the Tom Cat," *Michigan Alumnus Quarterly Review* 50 (1944), 343–45; Case, "Reminiscences and Impressions, or the Torch and the Tin Can," *Biologist* 33 (1950), 1–9.

10. Rossiter, M. W., "Women Scientists in America before 1920," *American Scientist* 62 (1974), 312–23; and Case, E. C. [Women in paleontology]. *Michigan Daily* 92 (1982), 1.

ALEXANDER G. RUTHVEN (1882–1971)

Envisioning the Modern Research Museum

Rachel Miller

In his forty-eight years at the University of Michigan, Alexander Grant Ruthven was a graduate student, faculty member, museum curator and director, and university president. Trained as a field biologist and herpetologist with a zest for snakes, Ruthven retained his commitment to scholarly research throughout his career. This passion—combined with outstanding administrative and leadership skills and a belief in the university museum as a cornerstone of academic work—made him one of the University of Michigan's most influential twentieth-century figures.

EARLY LIFE

Alexander Grant Ruthven (fig. 69) was born in 1882 in northern Iowa in a town named for his grandfather, a Scottish emigrant who first earned a living as a railroad contractor and was followed by his sons into farming, banking, and business. As a farm boy in the rural Midwest, Ruthven's childhood allowed him to explore early interests in animal life, and his family was also prosperous enough to send him to Morningside College, a small Methodist school outside Sioux City, Iowa. Ruthven had been set on a career as a biologist from an early age—he later cited Charles Darwin's *Origin of Species*, a Christmas gift at age twelve, as early inspiration. He took summer classes at the University of Chicago as an undergraduate, where he met the zoologist Charles C. Adams.[1] Adams was a critical early influence on Ruthven, offering professional guidance and providing a model for mentoring that Ruthven emulated. Most significantly, Adams imparted to Ruthven a guiding vision of the university museum as a dynamic research institution. When Adams moved from Chicago to Michigan in 1903 to assume a curatorship, Ruthven followed and accepted a graduate fellowship in zoology, earning his PhD in 1906.

Fig. 69. Alexander Ruthven, 1940s. Bentley Historical Library.

RUTHVEN AS SCHOLAR

The foundation of Ruthven's work as an administrator and museum professional was his commitment to scholarly research. Although trained as a herpetologist, his work included reptiles and amphibians and was important to a wide range of scholars across fields such as geography, ecology, and zoology. When Ruthven arrived at Michigan, zoology was in its first phase of institutionaliza-

Fig. 70. Detroit Nature Club in Portage, 1917. Ruthven (second from right) in front row with Bryant Walker (second from left), and other members of the club, including curators of the Museum of Zoology Norman Wood (second row behind Ruthven), and F. M. Gaige (third row, first from left). Bentley Historical Library.

tion as a field of study separate from the more amorphous "natural history." Borrowing techniques from European universities, zoologists of Ruthven's generation considered themselves more secular, specialized, and utilitarian than their predecessors, and they advocated for new forms of pedagogy and research that prioritized hands-on engagement and collaboration in the field and laboratory.[2]

The published version of Ruthven's dissertation, *Variations and Genetic Relations of the Garter Snake*, remains a classic in the field thanks to Ruthven's "major contribution of systematically presenting geographic variation in meristic [countable] characters, as well as his thoughtful and lengthy discussions of character variation and the origin and evolution of the four species group he recognized."[3]

Following this early success, Ruthven had a highly productive two decades, leading eighteen expeditions under the aegis of the Museum of Zoology (UMMZ) and publishing 128 scientific articles in twenty-eight journals between 1906 and 1930. In this same period of time, he named five new genera and thirty-four new species.[4] The breadth and depth of his research was due in large part to his variety of field sites. While he conducted much of his early research close to home, in Michigan's Porcupine Islands, Saginaw Bay, Isle Royale, and the Sturgeon River (fig. 70), he also led expeditions to Nevada, North Carolina, Texas, Columbia, Mexico, and British Guiana. Ruthven's involvement in scholarly publishing extended beyond his own research. He established an Occasional Papers of the Museum of Zoology series to publicize the efforts of his research team, as well as the Miscellaneous Papers series on collecting and preserving museum specimens.[5]

RUTHVEN AS MUSEUM MAN

Among Ruthven's most significant contributions to the University of Michigan was his development of the role of the scientific researcher in the museum. After completing his PhD, Ruthven succeeded Charles Adams as curator of zoology within the University Museum, with additional responsibilities for the anthropology collections and only two additional staff members.[6] At the time, the museum consisted of the mineralogy, geology, botany, and zoology collections, with the professor at the head of each department serving as director of its respective collection. Ruthven thus worked under Dr. Jacob Reighard, the director of the Zoological Laboratory and Museum.

After only a few years as curator, Ruthven began to work to alter this leadership structure, which he felt lacked cohesion and did not adequately support collections development, training, fundraising, and scholarly publications. In 1913, he convinced the regents to rename the University Museum the Museum of Zoology, arguing that the new name better reflected the institution's strengths—and to appoint him its director.[7] As Ruthven explained, this change "distinctly separates the museum from all other departments in the university, placing it upon the same footing as the General Library, and definitely fixes the responsibility for its development upon the person directly in charge."[8] Ruthven saw his museum as a research department that ran parallel to but independent of academic departments, complete with its own facilities, funding streams, staff, and graduate students.[9] In the end, his zealous campaign to make the zoology collections a crown jewel in the university protected both the collection's supremacy and the legacy of its fiercest champion. However, Ruthven's new organizational structure also sidelined the other museum disciplines. The mineralogical and geological specimens were accessioned to the Department of Geology, eventually joining the Herbarium in the Natural Science Building, while the anthropology collections remained under zoology and without an independent presence until 1922.

The administrative reorganization of the collections in the University Museum only partially addressed the fact that the building was now bursting at the seams. Even before Ruthven was named director, he had begun pressing the regents for a new, larger, and more modern facility for the collections.[10] He intensified his requests after 1913, actively and repeatedly making his case to the university president and members of the Board of Regents. The museum's collections, laboratories, and exhibits competed for a limited amount of space; curators were forced to limit their research, restrict the number of admitted graduate students, and manage collections across multiple buildings.[11] Ruthven initially advocated for his Museum of Zoology. However, as he made his case for a new building over more than a decade, his vision became increasingly expansive, eventually accommodating all of the university's separate natural history collections (i.e., herbarium, paleontology, and anthropology) in his conception of a new museums building.

The regents, in response, proposed merging the scientific collections scattered across the campus under a single head. Ruthven countered with a director in charge of the new building, but with each collection as a separate museum with financial and administrative autonomy. "This scheme will maintain the independence of each museum," he wrote to Bryant Walker, "and it will at the same time give the University a man in the position of Director of the museums who will have the responsibility of all the collections housed in the building."[12] After more than a decade of negotiations with both the regents and the state legislature, in 1925 funds were approved for a new museum with a floor plan more than eight times larger than the old museum (see Sinopoli, this volume). Designed by Albert Kahn, the University Museums Building opened on Geddes Avenue in 1928, with Ruthven as director.

Ruthven's multidecade experience in an evolving museum structure allowed him to develop his vision of the role of the museum curator in an academic setting. While, he acknowledged, some museums were correct to orient themselves toward popular instruction, Ruthven believed that a university museum's priority was research. He was in frequent conversation about the future of the profession with his counterparts at the Field Museum, the Museum of Comparative Zoology at Harvard, and the American Museum of Natural History (AMNH). Ruthven's key goal was to "formulate a little more exactly the legitimate fields of museum activity," he explained to the AMNH director. "Just like teachers, museum men tend to follow well marked trails. Someone should look over the whole field of museum work and present tentative programs for the consideration of directors and curators."[13] He also argued that museum training should not occur separately from a specific field of knowledge, and that it was better understood as a technique or method of scholarship rather than an object of study in itself. "Directors and curators should be scholars," he wrote, "not technologists."[14]

AN ADMINISTRATOR IN TRAINING

Ruthven's zoological research at U-M was made possible by the relationships he forged with naturalists outside the academy, many of whom he first met at the Nature Club of Detroit. Some of these connections with avocational scientists had been initiated by his predecessor, Dr. Adams, but it was Ruthven who made them formal—as honorary curators, trusted advisors, and donors. The shoestring budget of the Museum of Zoology barely covered staff salaries and collections maintenance; in 1909, for example, salaries accounted for 80 percent of the budget, which Ruthven argued "renders impossible any extensive acquirement of specimens by purchase or expeditions."[15] Keenly aware of the importance of new research to establishing the museum's reputation, Ruthven turned to Detroit professional men who pursued zoology as a hobby and had the autonomy and financial security both to support and to participate in university expeditions. Ruthven's closest partner was Bryant Walker, a Detroit attorney and mollusk expert, who funded the museum's expeditions to the Upper Peninsula, Iowa, South and Central America, and Mexico, and after whom Ruthven named his first child.

Ruthven was careful to establish these relations as mutually beneficial, publishing the work of his honorary curators, involving them in expeditions, giving them a stake in the future of the museum, and positioning the Museum of Zoology's research program as an ideal site for civic engagement.[16] Ruthven also leveraged the financial and professional support of his honorary curators when negotiating with university presidents and regents, as both the regents and the museum's patrons (some of whom were Michigan alumni) came from the same Detroit professional networks. "The interest which you have shown in the work has an excellent effect on the museum authorities," he wrote to nature preservationist and photographer of Michigan's fauna George Shiras. "I am trying to educate the Regents and administrative officers, and I find that whenever I show them that this work is approved by such men as yourself they quickly see the value of it themselves."[17]

Ruthven's expansive vision for UMMZ required an expanded staff as well as financial support. By 1922, Ruthven's team was extensive enough to support *The Ark*, a light-hearted staff newsletter that tied together the far-flung network. In addition to his many graduate students, to whom Ruthven was a responsive and generous advisor, the herpetologist also relied on the expertise of other specialists. By 1930, the Museum of Zoology had a staff of forty, the majority of whom were paid staff. It was barely recognizable as the institution Ruthven first encountered in 1903.

A MICHIGAN FOCUS

In addition to its robust research capacities, the zoology museum, as envisioned by Ruthven, was a key site in the documentation and conservation of Michigan's natural history, and in the general biological education of the state's populace. The

Fig. 71. Alexander Ruthven at his desk in the University Museums Building, 1930s. Bentley Historical Library.

university had been centrally involved in mapping and categorizing the fledgling state since the State Geological Survey's inception in 1837. In 1905, at Ruthven's urging, the legislature authorized a biological division of the survey and named the herpetologist as chief naturalist (fig. 71).[18]

As a complement to his role in the State Geological Survey, Ruthven imagined the Museum of Zoology as a state institution that primarily focused on its immediate territory, while being involved in an extensive network of exchange with peer institutions. Such a vision required multiple nodes of specialized research, allowing for both local field-based research and global breadth through scholarly exchange; it was a vision first possible in the era of museum development that Ruthven was crucial in constructing. "We will never attempt to build up a big general museum here," he explained, "but do hope to build up a large state institution, with as permanent a staff as possible, going outside of the state only to get material which will permit the members of the staff to develop."[19] Ruthven's policy was to balance comprehensive Michigan knowledge of Michigan biology—the most efficient option available—with institutional expertise in particular groups (reptiles) and geographic locations (Mexico). This two-pronged approach was necessary both to establish a national reputation and to recruit the most talented scholars to work in the museum.[20]

Also in keeping with his role as the state's chief naturalist, Ruthven believed that the museum's educational mission extended beyond the university's students, and that his institution should provide a service "to the people of the state, particularly to teachers and local naturalists."[21] For while Ruthven's ideal museum was primarily a research collection and not an exhibition space, he viewed himself as engaged with the wider world. The museum would thus serve the public in four ways: limited exhibits, as space permitted; loan collections to schools; receiving and determining specimens; and serving as a general source of information.[22] An informed and participatory citizenry would only enhance the museum's scholarly research, for while "field parties of trained men" was the preferred mode of collecting, specimens from "interested persons" were also highly appreciated because they often came with the local knowledge of the collector.[23]

UNIVERSITY PRESIDENCY

The successful funding and construction of the new museum building increased Ruthven's stature on campus, and he was named dean of administration by President C. C. Little in 1927. After his many years of successful interactions with the regents, Ruthven was a clear choice to succeed Little as president in 1929, a position he held for twenty-two years.[24] He shepherded the university through the Great Depression and World War Two, advocating first for additional funding from the legislature and then against the conscription of college students.[25] Ruthven also supported the preservation of institutional memory, backing the establishment of a university archive and the centennial celebration in 1937. As president, Ruthven was frequently caught in political snares; he tangled frequently with the *Michigan Daily* and had several students expelled for "radical" behavior during the 1930s, and his partnership with the Congress of Industrial Organization to offer a "Workers' Education Service" for the state's industrial workers was effectively thwarted by management.[26]

Ruthven's primary presidential legacy, however, was a new era of administrative delegation and corporatization. Drawing from organizational structures pioneered in the private sector, Ruthven's "Michigan System" distributed authority through a council of vice presidents, deans, and directors who could more effectively deal with the complex operations of finance, instruction, public relations, and alumnae affairs. When Ruthven retired in 1951, the student population had more than doubled to almost thirty thousand; the faculty, the budget, and the physical plant had grown in tandem. As scholar, museum administrator, and university president, Alexander Grant Ruthven dedicated his life to building the University of Michigan into a model modern research institution, and his efforts continue to shape the structure and function of the university today.

Notes

1. Peter E. Van de Water, *Alexander Grant Ruthven of Michigan: Biography of a University President*. Grand Rapids, MI: Eerdmans, 1977.

2. Keith R. Benson, "From Museum Research to Laboratory Research: The Transformation of Natural History into Academic Biology," in *The American Development of Biology*, ed. Ronald Rainger, Keith R. Benson, and Jane Maienschein. New Brunswick, NJ: Rutgers University Press, 1991, 49–77.

3. Douglas A. Rossman, Neil B. Ford, and Richard A. Seigel, *The Garter Snakes: Evolution and Ecology*, Norman: University of Oklahoma Press, 1996, 9.

4. Frederick Gaige, "Dr. Ruthven as a Writer," *Michigan Alumnus* 36 (May 1930), 563. For his contribution to the fauna of Michigan, see Ruthven, Crystal Thompson, and Helen Thompson, *The Herpotology of Michigan*, Lansing, MI: Wynkoop Hallenbeck Crawford, State Printers, 1912.

5. The occasional papers series was funded by Walker and Swales. "Report of the Director of the Museum of Zoology to the Board of Regents, 1917–1918," University of Michigan, University Museum, 1918, 19–20.

6. In 1895, a new position was created for a curator of zoology and anthropology who would act as custodian for all other collections and work separately from the Department of Zoology. This was the position Ruthven assumed in 1906, a role tied more closely to collections development and custodianship than to the classroom.

7. "I have had entire charge, made up my own budget," Ruthven wrote to Regent Harry Bulkey, and "Reighard freely admits this and entirely approves of the change in care and the transference of the directorship to me." Ruthven to Harry Bulkey, December 2, 1912. Alexander Ruthven Papers, Bentley Historical Library..

8. "Report of the Head Curator of the Museum of Natural History to the Board of Regents, 1911–1912," University of Michigan, 1912, 8.

9. Ruthven and Reighard often locked horns over whether the department or the museum would be the centerpiece of zoology at Michigan. When the museum acquired a new set of fish specimens, Ruthven wrote to Bryant Walker that "Reighard will kick back like hell, but I am used to that [. . .] He says we are absorbing, or rather dominating, the Department of Zoology." Ruthven to Walker, November 26, 1919. Alexander Ruthven Papers, Bentley Historical Library.

10. Originally described as a "Science Hall," Shirley Smith (secretary to the regents) to Alexander Ruthven, January 24, 1911; Museum Papers, Bentley Historical Library.

11. In 1925, part of the museum's collections were housed in Morris Hall, Angell Hall, University Hall, and in the Natural Science Building. Ruthven to Shirley

Smith, January 20, 1925. Alexander Ruthven Papers, Bentley Historical Library.

12. Ruthven to Bryant Walker, May 8, 1922. Alexander Ruthven Papers, Bentley Historical Library.

13. Ruthven to F. Lucas, January 23, 1925. Museum of Zoology records, Bentley Historical Library.

14. Alexander Ruthven, *Naturalist in a University Museum*, 29.

15. In the 1909–10 Report of the Curator, Ruthven reported a budget of $3,500 with only $700 left after staff salaries. By 1914, the university had allocated $5,353, with $4,500 going toward salaries, and Ruthven's additional coup of $1,300 in gifts. "Report of the Curator, 1909–1910," 13; "Report of the Curator, 1913–1914," 6.

16. Robert E. Kohler, *All Creatures: Naturalists, Collectors, and Biodiversity, 1850–1950*. Princeton: Princeton University Press, 2006, 91–134.

17. Ruthven to George Shiras, January 4, 1913. Alexander Ruthven Papers, Bentley Historical Library

18. Robert E. Kohler, 101–2.

19. Ruthven to George Shiras, April 10, 1916. Alexander Ruthven Papers, Bentley Historical Library.

20. "In order to retain people who are competent to work on state material we must be able to offer them something more and this is the way we do it." Ruthven to George Shiras, April 10, 1916. Alexander Ruthven Papers, Bentley Historical Library.

21. "Report of the Head Curator of the Museum of Natural History, 1910–11, 7–8

22. "Report of the Director of the Museum of Zoology to the Board of Regents, 1913–1914," University of Michigan, 1914, 7.

23. "Report of the Head Curator of the Museum of Natural History, 1910–11," 7–8.

24. Van de Water, *Alexander Grant Ruthven of Michigan*.

25. For a detailed account of Ruthven's presidency during the Depression and World War II, see Howard Peckham, *The Making of the University of Michigan, 1817–1967*. Ann Arbor: University of Michigan Press, 1967, 169–217.

26. Peter E. Van de Water, "The Ruthven Administration," *The University of Michigan, An Encyclopedic Survey, Vol. V, The University of Michigan 1940–1975*, ed. Ferol Brinkman. Ann Arbor: University of Michigan. Bentley Historical Library, 1977, 1–7.

HARLEY HARRIS BARTLETT (1886–1960)

Visionary Botanical Scholar

David Michener and Anton Reznicek

Harley H. Bartlett's childhood interests and travels foreshadowed only part of the trajectory of one of the University of Michigan's distinguished multidisciplinary scholars, collectors, and self-taught intellectuals. Bartlett (fig. 72) was born on March 9, 1886, in Anaconda, Montana, to Jonathan Hodgkin and Harriet Amanda (Potter) Bartlett during a mining boom. By 1899 the family moved to Indianapolis, at his uncle's invitation, so that the children could have better schooling. At Shortridge High School, Harley's love of geology, botany, and chemistry became evident. An 1899 photograph captures him engrossed in collecting botanical samples from a log in a semiabandoned Midwestern field, apparently oblivious to the photographer.[1] His schooling was interrupted for one year when he traveled with his father to Georgia on an ill-fated goldmining venture, but he soon returned to Indiana and earned his high school degree.[2] After graduation he was hired by the school to continue first as an assistant in botany and then in chemistry.

Fig. 72. Harley Harris Bartlett, 1938. University of Michigan Herbarium.

In 1904 Bartlett entered Harvard College where he was an undergraduate assistant to botanists Benjamin L. Robinson and Merritt L. Fernald, who were then revising *Gray's Manual of Botany*. Bartlett was given responsibility for the treatment of *Juncus*, a challenging group of grass-like plants. This was only one of the ten scientific publications he wrote as an undergraduate in academic English and German, supplemented by the lucid botanical Latin for which he became a well-respected authority. For all his botanical interests and his work with specimens brought back from faculty expeditions to Mexico and Guatemala, he took only one course in botany and graduated cum laude with an AB in chemistry in 1908.

At Harvard, distinguished chemistry professor (and future Nobel Prize winner) Theodore W. Richards invited Bartlett to join him for graduate study. Bartlett declined, apparently in part from intolerance for academic bureaucracies[3]—a character trait that persisted at U-M, where he cultivated "bland methods of cutting red tape."[4] Instead, Bartlett chose to begin his career as an obscure chemical biologist in the United States Department of Agriculture's division of Plant Industry, which oversaw research on drugs and poisonous plants and the chemistry of physiology and fermentation. Bartlett's love of field botany brought him to join the Washington Biologists'

Field Club in 1911, a membership that continued to his death.[5] His years in Washington, DC, led to multiple publications; in addition to his official work he conducted genetic research on evening primroses (*Oenothera* of the Onagraceae) on his own time. Some of his results still stand: four of the *Oenothera* species he genetically differentiated and named remain recognized.[6] His research brought him to the attention of the great Dutch plant geneticist, Dr. Hugo de Vries, as well as faculty in Ann Arbor.

Bartlett came to Ann Arbor in 1915 as an acting assistant professor, hired by botany chair Frederick Charles Newcombe. Almost immediately, he began further research on *Oenothera* at the new Botanical Gardens, then on Iroquois Street. His intellectual prowess was soon established: he became a full professor in 1921 and served as chair of the Department of Botany (succeeding Newcombe) from 1922 to 1947.

From 1919 to 1955 Bartlett was director of the Botanical Gardens, a role he clearly relished, perhaps because he delegated many administrative responsibilities. Bartlett had accepted the directorship on the condition that his doctoral student on *Oenothera* genetics, Freida Cobb, become the assistant director, a position she held until after his retirement.[7]

Bartlett retired in 1956 and was made a professor and director emeritus by the regents; his friends and past students started the Harley Harris Bartlett Plant Exploration Fund.[8] On February 21, 1960, Bartlett died of a heart attack.

Although he had chaired thirty-two doctoral committees, had been an exchange professor at the University of the Philippines, and had at least 179 publications to his credit, Bartlett had only briefly pursued graduate training and had declined all honorary degrees offered. His successor as Botanical Gardens director, Edward Voss, noted that "he seemed to enjoy that he never earned nor accepted"[9] a graduate degree. Frequent references to him as "Dr." Bartlett are erroneous, based on posthumous assumptions that such a distinguished professor must have had a PhD.

Bartlett was an inveterate collector of anything that piqued his intellectual curiosity—plants, books, references, anthropological materials, stamps. In his scientific endeavors, he was a polymath, publishing papers on a huge diversity of topics. However, one of the major threads that ran through all his work was that it was based on collections. And for plants, even on trips that were nonbotanical, he often found ways to make collections—even pressing botanical specimens under the mattress of his hotel room.[10]

Bartlett made the first of many significant expeditions to Sumatra in 1918 for the US Rubber Company, rubber being an essential war material. He continued conducting fieldwork until his retirement, visiting tropical destinations often in the context of studying rubber-producing plants or antimalarials, in expeditions sponsored by the University of Michigan, Carnegie Institution of Washington, Smithsonian Institution, University of the Philippines, US Rubber Company, and United States Department of Agriculture (USDA) Office of Rubber Plant Investigations. His expeditionary absences could be lengthy. In 1940, the USDA, concerned that the rubber plantations of Southeast Asia would fall under Japanese control, assigned Bartlett to personally oversee the transfer of 4,800 rubber plants from the Goodyear Rubber Company plantation in Mindanao to Haiti. This was successfully achieved in 1941.[11]

Bartlett's botanical collections came to the Herbarium after his death (he was never formally part of the Herbarium faculty). They total more than two hundred thousand sheets and include both his own collections and those of collectors he trained or hired.[12] Today, his collections remain major resources for researchers working in Argentina, Belize, Chile, Guatemala, Haiti, Mexico, Panama, the Philippines, Sumatra (fig. 73), Taiwan, Uruguay, and, of course, the United States and Canada. Bartlett never let lack of support interfere with his work and first-hand acquisition of collections; he was perennially broke, spending funds as quickly as he got them on books, travel, and generously helping others, and often having to borrow to pay bills.[13]

Bartlett never published a flora, a definitive

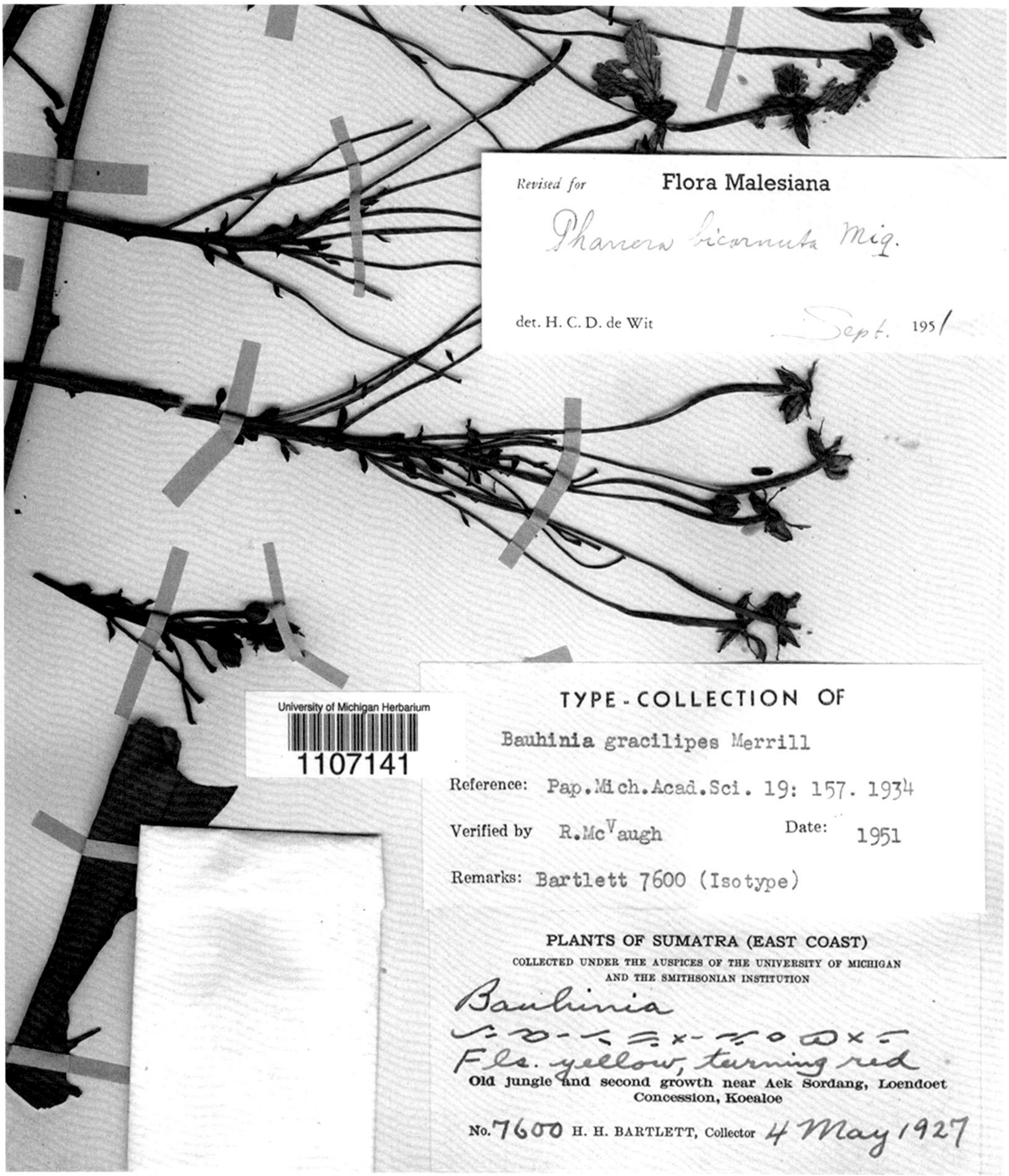

Fig. 73. Detail of Bartlett isotype specimen, collected in Sumatra on May 4, 1927, with Batak text on original label. University of Michigan Herbarium.

work describing plant species of a region. His studies of plants, however, were wide-ranging, and included laying the ground work for floras of various regions of the world, including parts of Mexico, Guatemala, the Philippines, Sumatra, and Taiwan. Though he collected little in Michigan, he was instrumental in founding the Michigan Flora project and the *Asa Gray Bulletin*, which transitioned into the *Michigan Botanist*.

He also had strong interests in a number of specific plant groups, publishing names for at least 130 new plants in over forty different families. But his main focal point for systematic studies remained the evening primroses, *Oenothera*, and, later in his career, palms (Arecaceae). His library of over two thousand items, given to the University Library after his death, was deep and extensive. Voss noted him saying, almost as an afterthought, after lending him an old book, "Take good care of it; there's only one other copy in the western hemisphere."[14] Bartlett was an advisor to Rachel McMasters Miller Hunt (Mrs. Roy Arthur Hunt) of Pittsburgh on her book collecting; her library later became the core of the Hunt Institute for Botanical Documentation at Carnegie-Mellon University. Bartlett's detailed annotations and documentation of an exhibit from her collections, held at the Clements Library in 1949, remains a useful book to botanical historians, and was first in a pair of influential exhibits on botanical Japanese books and manuscripts at the Clements Library of the University of Michigan.[15]

Bartlett was also a keen observer of human societies, observing that "[l]ocal customs and local languages are among the things that make the world interesting."[16] During his extended travels to Sumatra and the Philippines, he learned Batak, an Austronesian language. He wrote extensively about Batak culture, illustrating his writing with his many photographs. And he amassed important anthropological collections of writings (on bamboo and bark paper), textiles, tools, and toys,[17] for which he is renowned outside botanical circles.

In all areas of his research, Bartlett was intrepid, persistent, and passionate. Here he recounts a return trip he made to a Batak community in Sumatra to rebut a critic's statement that a particular structure and related cultural practice he had reported did not in fact exist:

> By 1927 so great had been the change that only one *parsoeroan* [ritual structure] was seen. It was in a little village reserve (Kampong Taratak) surrounded by the great Tindjoan oil-palm concession in Tano Djawa. The natives refused to tell where it was, but by hunting diligently, fording a deep river, and finally subsidizing a small boy, the writer found it. The fence was gone, but a hen had been sacrificed under the building within a day or two, showing that it was still used.[18]

Bartlett's wide range of publications spanned from plant genetics to paleobotany, from color terminology in Batak and Malay (perhaps not coincidentally he had researched the genetics of floral color) to the history of fire in primitive agriculture in the tropics, from contemporary postage stamps to the rarest of botanical books. This spectrum of interests helped him to mentor a range of graduate student projects and was a critical foundation to the Botanical Gardens' record of scholarship while at the Iroquois site.

Bartlett's contributions were acknowledged when the new botanical garden on Dixboro Road was dedicated in 1962, two years after Bartlett's death. The published program for the event, which was presided over by President Harlan Hatcher, noted:

> The subsequent history of this, the Iroquois site, is in great measure a record of his [Bartlett's] interests and activities through the 36-year period during which he served as Director.[19]

Although there is no physical tribute dedicated to Harley H. Bartlett at the Matthaei Botanical Gardens (as the institution on Dixboro Road is now known), and the Gardens published for only a short time a journal honorifically named *Bartlettia*, the interdisciplinary scope encompassed by the Gardens bears the indelible imprint of his vision and leadership.

BARTLETT COLLECTIONS AT U-M AND BEYOND

Harley Bartlett's collections were separated after his death.[20] His personal diaries and papers are primarily at the University of Michigan's Bentley Historical Library (11 linear feet), while his scientific collections are more broadly dispersed. Over two hundred thousand herbarium specimens were processed by the University Herbarium, and the University Library accepted his extensive personal botanical library of over two thousand items. His anthropological collections were divided among the U-M Museum of Anthropological Archaeology (where 155 Batak bark and bamboo manuscripts are featured in an online exhibit), the American Philosophical Society (approximately six hundred objects), and the Logan Museum of Anthropology of Beloit College (approximately three hundred objects).[21]

Notes

1. See Indiana University photography archives http://webapp1.dlib.indiana.edu/archivesphotos/results/item.do?itemId=P0037357 (accessed October 7, 2015).

2. See U-M Faculty History Project Memorial Minute http://www.lib.umich.edu/faculty-history/faculty/harley-harris-bartlett/memorial (accessed October 7, 2015).

3. Edward Voss published a meticulously detailed obituary of Bartlett with memorable firsthand observations and anecdotes. The two men had years of professional interaction, beginning when Voss was a student. First published as Edward G. Voss "Harley Harris Bartlett," *Bulletin of the Torrey Botanical Club* (1961): 47–56, it was reprinted in *Bartlettia* (February 1982): 38–47.

4. 1956 Review of Theodore P. Bank II, *Birthplace of the Winds. Michigan Alumnus Quarterly Review* 62 (21): 367–68. In this review, Volney Jones noted that Bartlett sponsored the arctic expedition as he had done for diverse students out of his personal funds.

5. Perry, M. C, C. S. Bond, and E. J. R. Lohnes. 2007. Washington Biologists' Field Club, USGS Patuxent Wildlife Research Center, Laurel, MD. Accessed October 7, 2015: http://www.pwrc.usgs.gov/resshow/perry/bios/wbfchome.htm

6. Species count is from Theplantlist.org. Bartlett and his coauthors named seventeen *Oenothera* taxa—as of 2015, four are still "accepted" and thirteen are "unresolved"—quite a feat for the early days of cytotaxonomy.

7. Shackman, 2009: AADL web-published history (accessed October 7, 2015): *The Botanical Gardens on Iroquois*, http://aaobserver.aadl.org/aaobserver/15253

8. The Harley Harris Bartlett Plant Exploration Fund was founded with $2,300, as noted in the September 1955 U-M regent's minutes; the last contribution, noted in the December 1965 minutes, brought the total contribution to $3,600. Since the fund no longer exists, it was presumably expended (several publications had been seen up to 1961 acknowledging its support).

9. Voss, Edward G. "Harley Harris Bartlett." *Bulletin of the Torrey Botanical Club* (1961): 47–56, it was reprinted as *Bartlettia* (February 1982): 38–47.

10. Personal communication to Tony Reznicek from Rogers McVaugh.

11. The web-published *Haitian Rubber Timeline 1903–1945* by Thomas Fennell can be found within the Haitian history website compiled by Bob Corbet, emeritus faculty member of Webster University, St. Louis, Missouri: http://faculty.webster.edu/corbetre/haiti/misctopic/leftover/rubber.htm (accessed October 7, 2015).

12. Jones, Kenneth Lester, *The Harley Harris Bartlett Diaries (1926–1959)*. Published and circulated privately by K. L. Jones, Ann Arbor, 1975.

13. Ibid., 5.

14. Voss, "Harley Harris Bartlett," 40.

15. Bartlett, *Fifty-five Rare Books from the Botanical Library of Mrs. Roy Arthur Hunt*. Clements Library Bulletin 57 (1949), 55 pp. See also Tsukamoto, Yotaro, and J. L. Creech, *Japanese Horticulture: Origins and History* (Yokahama-shi: Woods Press, Inc., 2015). This posthumous publication was "Edited by The Committee, for Publishing 'Japanese Horticulture." In the editors notes (178–79) both Bartlett's exhibit, here listed as Bartlett and Shohara, and the Clement's Library 1954 exhibit are contextualized for their roles in developing Dr. Creech's understanding of Japanese horticulture and the "enduring role of plants in Japan/United States relations" (167).

16. Voss, "Harley H. Bartlett," 40.

17. Including an important collection now housed in U-M's Museum of Anthropological Archaeology, http://webapps.lsa.umich.edu/umma/exhibits/Batak2009/index.html

18. Bartlett, *The sacred edifices of the Batak of Sumatra*. Ann Arbor: University of Michigan Press, 1934, 5.

19. The *Dedicatory Exercises* program for the Botanical Gardens on Dixboro, June 14, 1962, is filed at the Bentley Historical Library. A pdf-file of the document can be found at https://webapps.lsa.umich.edu/mbg/storeadmin/paperfiles/1008.pdf

20. Websites for institutions holding Harley H. Bartlett's materials include:

American Philosophical Society
https://www.amphilsoc.org
Logan Museum of Anthropology–Beloit College
https://www.beloit.edu/logan/
U-M Bentley Historical Library
http://bentley.umich.edu/
U-M Herbarium
https://www.lsa.umich.edu/herb/
U-M Museum of Anthropological Archaeology
https://www.lsa.umich.edu/ummaa/

21. The materials at the American Philosophical Society and Beloit College reflect his professional affiliation with the former since 1929, and his sister's (Hazel Bartlett) with the latter.

LEWIS G. VANDER VELDE (1890–1975)

Michigan's Archive Builder

Brian Williams

In June 1934, Lewis George Vander Velde (fig. 74), a forty-four-year-old assistant professor of history at the University of Michigan, applied for funding to launch a project to locate and collect historical material relating to the state of Michigan. Vander Velde was developing a seminar on Michigan history and was sorely disappointed by the lack of original source material available to support the research of his graduate students. His proposal went before the Faculty Research Fund administered by the Horace H. Rackham School of Graduate Studies and was approved for $700.[1] The modest funding was to be used for travel, field expenses, and the production of photostatic copies of manuscripts. The project progress report, written by Vander Velde in June 1935, indicated that a serious illness had prevented him from active work until April 1935. But by that spring, operating out of a small office in the University Press Building on Maynard Street, Vander Velde was diligently sending letters to the families of former governors and other leading Michigan citizens, while pursuing a number of leads throughout southeastern Michigan. The papers of former governor John J. Bagley (1873–77) were the first to arrive, and promises were in hand to receive the papers of former governors Alpheus Felch (1846–47) and Henry Howland Crapo (1865–69). Vander Velde also hired two graduate students, Arthur R. Kooker and Elizabeth Sparks, for research assistance and clerical support.[2] Among the expenses reported were $40 for a typewriter, stationery, and office furniture, and $25 paid to a janitor in Monroe who rescued, cleaned up, and turned over early county records destined for the pulp mill. The janitor saved examples of each type of record from more than two tons of records slated for destruction. Vander Velde concluded his report by noting, "[T]here is no doubt in my mind that the project is very worthy and very promising."

Fig. 74. Lewis G. VanderVelde in the archive, 1950s. Bentley Historical Library.

Vander Velde's "very promising" project was granted an extension and another $700 for 1935–36. The project also caught the attention of President Alexander Ruthven, who had earlier expressed interest in collecting materials relating to the history of the University of Michigan. Ruthven was very supportive of Vander Velde's fledgling program and agreed to provide budgetary assistance to enable the collection of printed and manuscript material pertaining to the history of the university. In September 1935, the regents authorized $2,000 to be added to the Department of History budget. Of that, $1,500 was to pay the salary of a research assistant, with the remainder used for expenses. The approaching centennial of the university's 1837 relocation from Detroit to Ann Arbor was fueling a growing appreciation of its history, and so the university history initiative was formalized in November 1935 when the regents established the Committee on University Archives, appointing Vander Velde as its executive secretary. Chaired by Frank Robbins, the assistant to the president, the committee also included William Warner Bishop, university librarian; Randolph G. Adams, director of the William L. Clements Library; Wilfred B. Shaw, head of the Alumni Association; Shirley Wheeler Smith, university secretary; and Mortimer Cooley, emeritus dean of engineering.[3] The Committee on University Archives promptly sent out circulars to campus personnel, asking them to report on historical material in their offices. Relatives of former regents and faculty were also solicited for early material. Field visits, surveys of attics and basements in campus buildings, and the actual collection of historical university records soon followed. A move into a room in the Clements, designated as "the Michigan History Room," provided a place to base the operations as well as the opportunity to learn valuable pointers from the staff of the Clements.

The newly collected records included the papers of former president James B. Angell, law professor Thomas M. Cooley, and professor of geology Alexander Winchell. The Winchell papers had originally been donated to the Minnesota Historical Society, but given the strong Michigan content of the papers, the Minnesota Historical Society, with the blessing of the original donors, agreed to transfer the collection to Michigan. Vander Velde sent a notice to the leading journal of history in the United States, the *Mississippi Valley Historical Review*, mentioning the transfer of the Winchell papers and simultaneously announcing the arrival of the new and ambitious collecting program launched at U-M. An edited version of the notice appeared in print in the March 1938 journal, omitting mention of the Winchell papers, but emphasizing that a principal objective of this new university archives program was the "collection of the papers of former regents, faculty members, and administrative officials" in conjunction with "records from the offices of the various divisions of the university; and the collection of newspapers and other non-campus printed sources relating to the university's history."[4]

As Vander Velde had secretly hoped, many of the papers of regents and others associated with the university included significant amounts of nonuniversity material. The papers of Regent Frank Fletcher, for instance, included extensive records of his lumbering business, reinforcing Vander Velde's notion that the collecting focus should be twofold—collecting both university and nonuniversity material on Michigan history. In February 1938, the "collections of historical books and manuscripts" under the care of Vander Velde were fully legitimized when the regents formally designated them as the Michigan Historical Collections (MHC). Along with that designation, the budget, which formerly had been a line item in the Department of History budget, was included directly in the main university budget. Vander Velde was named as director of the newly designated MHC, which was not a stretch: Vander Velde's administrative abilities had been recognized earlier by his appointment as assistant to the dean of the College of Literature, Science and the Arts from 1931 to 1934. Vander Velde's staff members were also given official titles: Elizabeth Sparks Adams was named curator and Henry D. Brown was designated as research assistant. Dr. Van, as he was affectionately known, and his growing staff also had a new base for their operation. A suite of rooms in the basement of the Rackham Building were assigned to MHC shortly after Rackham was dedicated in June 1938. This space was sorely needed as the collections had quickly outgrown the single room assigned in the Clements.

It made sense for Vander Velde to want to collect material on the state of Michigan. He was a Michigan native, born October 17, 1890, in Grandville, Michigan, part of a strong Dutch-American community in west Michigan. He graduated from Grand Rapids Central High School in 1908 and enrolled at Hope College in Holland, Michigan. After two years at Hope, Vander Velde

Fig. 75. Lewis G. Vander Velde at the unveiling of his portrait in 1974, with Martha Bigelow, director of Michigan Bureau of History (left) and Elizabeth Sparks Adams (right), the first curator of the Michigan Historical Collection. Bentley Historical Library.

transferred to U-M, earning a bachelor's degree in 1913. Following graduation, he taught history at Culver Military Academy in Culver, Indiana, interrupted by World War I and corresponding military service in 1918–19. After earning an AM degree from U-M in 1921, he served as head of the Department of History at St. Cloud State Teachers College (now St. Cloud State University) in Minnesota from 1921 to 1926. A confirmed bachelor, history was Vander Velde's true passion.

Eager to continue his education, Vander Velde began a doctoral program in history at Harvard University, where he was greatly influenced by historian and scholar Arthur M. Schlessinger Sr., particularly his emphasis on social history. Undoubtedly this emphasis helped spark Vander Velde's enthusiasm for local history. He completed his PhD in 1931, focusing his thesis, "The Presbyterian Church and the Federal Union, 1861–1869," on the crisis of the Civil War years and revealing the rich social history contained within church records. Under Vander Velde, church records would become a major collecting theme for MHC.

While completing his dissertation, Vander Velde returned to U-M in 1928 as an instructor in history. In 1931, he was appointed as an assistant professor of history. He rose through the faculty ranks, becoming an associate professor in 1936 and full professor in 1940. He also served as chair of the Department of History from 1947 to 1953. His teaching specialty was the constitutional history of the United States, but he also developed courses on the political and social history of the state of Michigan.

When Vander Velde began his collecting program in 1935, the National Archives of the United States was less than a year old. In Michigan, the only program actively collecting historical material was the Burton Historical Collection in Detroit. In 1915, businessman and lawyer Clarence Monroe Burton had donated his personal collection of books and manuscripts on Detroit and Michigan history, along with the building it was housed in, to the Detroit Public Library. Given the lack of in-state collecting institutions, naturally the first order of business was soliciting and acquiring material. Vander Velde was clear in stipulating that the major focus was to collect material to make it possible for students to conduct research. Yet he also identified other key objectives: to enhance the university's collection of Michigan newspapers, to preserve historical records, to expand the University Library's collection of secondary works on Michigan history, and to create a historical mindset among the people of Michigan.

Vander Velde worked long and hard to bring about an appreciation of the importance of preserving historical records. He spoke before numerous groups throughout Michigan, espousing the importance of local history. In 1940, in what would be his only article published in *The American Archivist*, Vander Velde wrote on the value of local history, acknowledging that despite having "four years of experience as a collector," he remained "fundamentally a historian."[5] Citing general principles as well as specific policies of MHC, his main points were that local records should be centrally collected by an independent agency "wholly devoted to the care and arrangement of records" to allow historians to make comparisons between similar records from other communities. He argued that records should not be collected for the sake of collecting but for the purpose of making them available. In a key principle that remains in place to this day, Vander Velde affirmed that "acquisitions should be by donation." He acknowledged that "cold financial facts" and limited resources drove that decision but noted that the archives provide an honor to the donor as well as relieving the donor of the preservation responsibilities.

In a talk before a local history conference, Vander Velde cited the earlier American historians at U-M and wondered why historians such as Charles Kendall Adams, Andrew McLaughlin, and Claude Halstead Van Tyne had not concerned themselves with local history. Vander Velde determined that they were all too "preoccupied with rewriting our national and sectional history, to have much concern with local history."[6] By the time he retired in 1960, Vander Velde had done much as a historian to advance the cause of local history (fig. 75). Beyond his work with the MHC, Vander Velde was active in the Historical Society of Michigan, serving on the board from 1944 to 1946 and from 1956 to 1959, and as president during 1958–59. He was appointed to the Michigan Historical Commission in 1950 by Governor G. Mennen Williams and served until 1974.

At the time of Vander Velde's retirement, a bulletin, "The Michigan Historical Collections: The Vander Velde Years: 25 Years of Leadership," written by Ruth Bordin and published in 1961, looked back on Vander Velde's career and the amazing growth of MHC. The modest collection of county histories, city directories, printed reports, and books had grown to more than twenty-five thousand volumes. In addition, when Vander Velde retired, the collections claimed more than one million individual manuscript items, eighty-seven hundred bound manuscript volumes, and fourteen hundred maps. There were also papers of nineteen different Michigan governors, diaries and letters from Civil War soldiers, and papers of prominent businessmen, churches, and organizations.

Nearly as important as the role of a collecting institution, under Vander Velde the collections served as a training ground for a generation of historians and archivists who worked part-time as graduate students or joined the staff. Among them were Philip Mason, who would become the founding director of the Walter P. Reuther Library of Labor and Urban Affairs; authors and historians George May and Eugene Peterson; and Robert M. Warner, who would later serve as director of MHC before going on to serve as the national archivist of the United States from 1980 to 1985.

At the memorial service for Vander Velde, following his death at the age of 85 on October

31, 1975, Warner reflected on the career of his mentor, calling "Dr. Van" a "pioneer in establishing a modern archival agency," an agency where all levels of society were documented, not just the famous and powerful. That was a very new concept when Vander Velde instilled it as a hallmark of MHC. That legacy lives on in today's successor program, the Bentley Historical Library.

Notes

1. The proposal was designated Faculty Research Fund project #327 and may be found in the Horace H. Rackham School of Graduate Studies records, Box 25, Bentley Historical Library.

2. Elizabeth Sparks Adams would go on to become the first curator of the Michigan Historical Collections. Arthur Kooker completed his PhD in 1941 on antislavery activity in Michigan and would later serve as chair of the history department at the University of Southern California.

3. *Proceedings of the Board of Regents*, November 1935, 735.

4. See note in *Mississippi Valley Historical Review* 24 (March 1938), 598–99.

5. Lewis G. Vander Velde, "Local Records," *The American Archivist* III, no. 4 (October 1940), 251–60.

6. Lewis G. Vander Velde, "The University of Michigan Historical Collections after Twenty Years," Michigan in Perspective Conference on Local History Participation, November 7, 1958. Lewis G. Vander Velde Papers, Box 3, Bentley Historical Library.

WALTER N. KOELZ (1895–1989) AND RUP CHAND (1902–94)

Partners in Asian Collecting

Carla M. Sinopoli

An encounter in the Indian Himalayan town of Kulu in July 1930 changed the lives of Michigan zoologist Walter Koelz and Thakur[1] Rup Chand, a Lahuli man from an impoverished noble family (fig. 76). Koelz had traveled to India to serve as the botanist for the Urusvati Institute, a research center directed by Russian expatriates Nicholas and Helena Roerich; Chand was one of a number of young Himalayan men employed by the institute. Koelz and Chand would spend the next thirty years together—as partners, research collaborators, and creators of remarkable collections for University of Michigan museums.

Walter Norman Koelz was born in the small town of Waterloo, Michigan, in 1895, the son of a German immigrant blacksmith. In 1989, he died in the house in which he was born. Koelz earned a doctoral degree in zoology from the University of Michigan in 1920, specializing in ichthyology; the U-M Museum of Zoology houses specimens of Great Lakes fish from his early research. In 1925, he was hired by the United States Department of Fisheries as naturalist for the MacMillan-Byrd expedition to Greenland; ethnographic objects from that expedition are in the University of Michigan's Museum of Anthropological Archaeology (formerly the Museum of Anthropology). From 1927 to 1930, he was variously employed as Michigan's state ichthyologist and as a postdoctoral fellow and assistant curator of fishes in the U-M Museum of Zoology. He eagerly left these positions in 1930 when the opportunity arose to travel to India (facilitated by Harley H. Bartlett, see Michener and Reznicek, this volume). While Koelz subsequently held titular positions at the university, he never again held a salaried appointment.

Thakur Rup Chand was born on April 16, 1902, in the village of Gumrang in Kulu District in what is now the state of Himachal Pradesh, India. He died in Ann Arbor on August 18, 1994. The youngest son in a distinguished but impoverished family, Chand achieved a fourth-grade education before being called home from his boarding school after his father's death to work on the family farms. In 1927, he joined the British Indian Territorial Force, serving until 1930, when he was hired by the Roerichs. Despite his limited formal education, Rup Chand had learned to write in Urdu and was fluent in four regional languages as well as English.[2] When Walter Koelz arrived at the Urusvati Institute to collect botanical specimens and document traditional uses of medicinal plants, Rup Chand accompanied him as an interpreter.

The two soon became intimate partners and collaborators and for much of the next twenty-five years they would travel together in India, Afghanistan, and Iran, acquiring zoological, botanical, and anthropological objects for museums. The University of Michigan museums are the primary beneficiaries of their prodigious collecting efforts. Today, the Museum of Zoology curates more than thirty-two thousand Asian specimens collected by the two men in addition to several thousand Koelz specimens from Greenland and Michigan; the Herbarium estimates that their combined Koelz and Rup Chand collections number between fifteen thousand and twenty thousand specimens; and the Museum of Anthropological Archaeology

Fig. 76. Rup Chand and Walter Koelz (front row) with members of their collecting team, early 1930s. Bentley Historical Library.

Koelz Himalayan Expedition Collection numbers 583 objects, with additional objects from their travels subsequently donated by friends and colleagues Alexander Ruthven, Harley Bartlett, and George Lauff. Walter Koelz/Rup Chand collections are also held at the New York Botanical Garden, Smithsonian Institution, Paris Museum of Natural History, American Museum of Natural History, and Field Museum of Natural History, among others.

It is likely, though never explicitly stated,[3] that the intimate relationship that developed between Koelz and Chand when both worked for the Urusvati Institute contributed to the conflicts that arose between Koelz and the Institute by late 1931. These bitter fights—in which Koelz and the Roerichs challenged each other's ethics, motivations, competency, and actions—came to a head in early 1932 with the termination of Koelz's employment and his return to Michigan in April of that year.[4] The anger and mutual distrust did not end there: in May, the New York-based Roerich Museum sued Koelz for libel in the New York State Supreme Court, requesting $100,000 in damages. The suit was eventually settled out of court when Koelz agreed to write a (grudging)

Fig. 77. Expedition on Himalayan Landscape, early 1930s. Bentley Historical Library.

letter of apology, acknowledging the bona fides of the Urusvati Institute as a legitimate scientific venture.

Koelz did not remain in Michigan long. Even before he left Kulu he had begun writing to Professor Bartlett, seeking funds to remain in India to collect specimens for the university. He redoubled his efforts upon his return to Ann Arbor and within a few months had received funding from the university's Charles Lang Freer Endowment, which had been created in honor of the late Detroit railroad magnate and collector of Asian art to support research in art history. Koelz was awarded an annual stipend of $5,000 from endowment funds, supplemented by $500 from Freer's protégé, art collector Margaret Watson Parker (see McNamara, this volume). With support in hand, Koelz made plans to return to India as leader of the University of Michigan Himalayan Expedition and was granted the title "Honorary Collaborator in Asiatic Research" by the Board of Regents.[5] Funds in hand, in December 1932 Koelz returned to India, and to Rup Chand.

Given the sources of support, in this expedition Koelz was not charged to collect biological specimens. Instead, his primary task was to collect Tibetan[6] sacred art and artifacts for the university's Museum of Anthropology.[7] It was understood, though, that he also would make collections for the Museum of Zoology and the Herbarium.[8] The choice of the anthropology museum rather than the university's Museum of Art and Archaeology as the appropriate home for the collections appears strange today. However, although Japanese and Chinese art was safely ensconced in the canon of American art museums by the 1930s, Tibetan artistic production continued to be seen as too strange and primitive to be art, belonging

Fig. 78. Walter Koelz and vulture. Bentley Historical Library.

Fig. 79. Collecting team, Rup Chand in foreground, early 1930s. Bentley Historical Library.

instead to the more exotic realm of anthropological collecting.[9]

Throughout his life, Walter Koelz was a prodigious diarist, writing an entry nearly every day during his expeditions. The Bentley Historical Library has ten volumes of Koelz's diaries dating from 1933 to 1946; sadly, his diary containing entries from the first six months of the U-M Himalayan Expedition is lost. By 1935, Rup Chand also had begun recording his travels, writing his diaries in Urdu; the Bentley Library Chand collection contains his original diaries from 1935 to 1994. In his later years, while employed by the University Herbarium, Chand transcribed significant portions of his 1935–50 diaries into English.[10] Koelz's extant diaries from the 1932–34

U-M Himalayan Expedition have been published by Sinopoli[11] and portions of his Iranian diaries were published in 1983 by Koelz in collaboration with Henry T. Wright.[12] Koelz and Chand were also avid photographers and the Bentley Library houses hundreds of their photographs and negatives as well as six 8-millimeter black-and-white films made in India in 1936.

During the eighteen-month U-M 1932–34 expedition, Koelz and Rup Chand spent approximately six months traveling and collecting in the mountains of Ladakh, Zanskar, Spiti, and Rupshu (figs. 77 and 78). It was here where the renown of Rup Chand's distinguished family proved particularly helpful, giving them access to local rulers and Buddhist monasteries and to sacred objects that it is unlikely Koelz could have acquired on his own. The men were joined in their travels by several other young men, all childhood friends of Rup Chand,[13] who also had previously worked for the Roerichs (fig. 79). In the winters, when heavy snows prevented mountain travel, the team collected in lowland areas of northern India and Pakistan (then part of British India). In January 1934, Koelz shipped twelve large wooden crates to the university. Contained within them were more than five thousand birds (and a small number of mammals), four thousand dried plants, and nearly six hundred ethnographic and historic objects, including religious (thangka) paintings, bronze and stone images, icon cases, jewelry, textiles, and other objects.[14]

Koelz had hoped to return to the mountains the following summer. Instead, he received word in February 1934 that the Great Depression had affected the university's ability to fund the expedition and therefore he should come back home to attend to his collections.[15] When Koelz returned to Michigan, Chand accompanied him. They settled with Koelz's mother on the family property in Waterloo, where Koelz assisted in cataloging and exhibiting the collections and wrote a manuscript on the Kashmiri shawls.[16]

The two men remained in the United States for less than two years. In January 1936, Koelz and Chand returned to India on a new expedition under the auspices of the United States Department of Agriculture's (USDA) Bureau of Plant Industry. This time their mission was to collect seeds and plant specimens, though they once again collected birds to sell to museums and material culture for Koelz's personal collection. They reprised their Himalayan route, traveling and collecting in peninsular and western India and Afghanistan. Koelz and Chand returned together to Ann Arbor in April 1938.

In July 1939, they began their travels again, in a trip that would last fourteen years. Their work began in Iran and was funded for its first eighteen months by the USDA Bureau of Plant Industry. During this period, Koelz, Chand, and their Tibetan assistants collected "some 3000 samples of seeds, 4000 collections of herbarium specimens, and a comprehensive collection of birdskins."[17] The latter Koelz sold to museums, including the U-M Museum of Zoology and the American Museum of Natural History.[18]

A combination of short-term paid contracts and the sale of as many as fifty thousand bird skins to museums[19] funded Koelz and Chand's Asian travels from 1941 to 1953, a tumultuous period spanning World War II and the independence and division of British India.[20] The team remained in Iran from 1939 to January 1946 when they returned to India, where they remained until October 1953. Assam, in northeastern India, was a particular focus of their ornithological collecting during this period.

Koelz returned to Michigan in 1953 and in 1956 he was awarded the Franklin Meyer Medal by the American Genetic Association for outstanding contributions to agriculture.[21] In many respects a man out of time, this zoologist turned art collector turned botanist exemplified a nineteenth-century model of wide-ranging naturalist that had existed before academic disciplines became hardened. He certainly did not fit, by temperament or interests, the increasingly rigid disciplinary structure of the mid- to late twentieth-century university. Koelz never went overseas again. He spent the rest of his life in his Waterloo home, hunting and tending his garden full of exotic plants from his travels. He rarely went to the university, though counted many U-M botanists, ornithologists, art historians, and anthropologists among the frequent visitors to his treasure-filled home.

In 1956, Harley Bartlett assisted Rup Chand to immigrate to the United States and Koelz's Waterloo home. Chand first volunteered and later was hired by the University Herbarium to process his and Koelz's many plant specimens. In 1960, the two men had an acrimonious falling out. Koelz once again found himself in court—sued for $50,000 by Rup Chand to compensate him for half his share of proceeds from their twenty-five years of collaboration. It was an ugly suit, covered in lascivious detail by local newspapers. Koelz's claims that Chand had been his manservant—and never his partner or collaborator—appalled their many friends, and several testified for the much-loved Rup Chand. In June 1963, the Jackson County Circuit Court awarded Chand a settlement of $15,414, half the funds then in their (formerly) joint bank account. Rup Chand spent the next thirty-one years in a guest house in Ann Arbor provided by generous friends. The two men never reconciled.

Walter Koelz and Rup Chand. Neither could have imagined the lives they would lead. For the University of Michigan museums, their legacy is remarkable: vast and significant collections of birds, plants, and cultural objects, and the rich documentation that gives these collections tremendous ongoing academic value. The sacred art objects collected in the 1932–34 U-M expedition are the ongoing focus of scholarly writing[22] and were exhibited in both the University of Michigan Museum of Art and the Ella Sharp Museum (Jackson, Michigan) in 2013, and in both the Block Museum of Art and New York's Rubin Museum in 2014–15. The zoological and botanical collections include a number of important type specimens and collectively provide insights into historic biodiversity in a part of the world being dramatically affected by climate change. The University Herbarium and Museum of Zoology are collaborating in a project to digitally reunify the Koelz/Chand Asian collections and make them accessible for scholarly research. The lives of these two remarkable men have ended; the research and teaching lives of the collections they made will not.

Notes

1. "Thakur" is an honorific title.
2. Autobiographical note; Rup Chand Papers, Bentley Historical Library.
3. Correspondence between Koelz and Bartlett, and from the Roerichs to U-M President Ruthven, Harley Bartlett, and New York Botanical Garden director Elmer Drew Merrill document the increasing tensions over late 1931 and early 1932 and the mutual suspicion and intense dislike that developed between Koelz and the Roerichs over this period; see Koelz Papers, Bentley Historical Library; Roerich Museum archives; and Merrill Papers, New York Botanical Library.
4. Explicitly stated issues were the Roerichs' unhappiness with Koelz's collection of zoological specimens, his use of the Urusvati Institute's guns, and collecting for other institutions, as well as his insubordination and corrupting influence on Rup Chand and other Indian employees of the Institute.
5. *University of Michigan Regents Proceedings*, Ann Arbor: University of Michigan 1929–1932, 44, 141.
6. Although formerly in British India, the mountainous Himalayan regions where Koelz collected were culturally part of Tibet, and much of Koelz and Rup Chand's cultural collecting focused on objects of Tibetan Buddhist religious practice.
7. He was also expected to serve as advisor for the Smithsonian's Freer Museum.
8. And initially, the Cranbrook Museum of Science, though this relation was short-lived.
9. Clare Harris, *The Museum on the Roof of the World: Art, Politics and the Representation of Tibet*. Chicago: University of Chicago Press, 2014, 21.
10. These were first transcribed by hand, and then later typed. Chand occasionally added additional retrospective commentary on the typescript.
11. Carla M. Sinopoli, *The Himalayan Journey of Walter Norman Koelz*, Anthropological Paper 98. Ann Arbor: Museum of Anthropology, University of Michigan, 2013.
12. Walter Koelz, *Persian Diary 1939–1941*. Anthropological Paper 71. Ann Arbor: Museum of Anthropology, University of Michigan, 1983.
13. Rup Chand Diary, July 25, 1936, entry; Rup Chand Papers, Bentley Historical Library.
14. Letter dated January 13, 1934, on file in the Museum of Anthropology Asian Division, University of Michigan.
15. Noted in Koelz diary entry, February 23, 1934; see Sinopoli *The Himalayan Journey of Walter Norman Koelz*, 220.

16. This previously unpublished manuscript was digitized and published on a CD-ROM included in the book, *Wrapped in Beauty: The Koelz Collection of Kashmiri Shawls*, Grace Beardsley and Carla M. Sinopoli, Anthropological Paper 93. Ann Arbor: Museum of Anthropology, University of Michigan 2005.

17. Koelz, *Persian Diary*, xiii.

18. "Contract between The American Museum of Natural History and Dr. Walter Koelz," May 11, 1939, found in Koelz 1941 diary, Koelz Papers, Bentley Historical Library.

19. The Field Museum of Natural History and U-M Museum of Zoology bird division each purchased twenty-five thousand Koelz/Chand birds for $1 each; see *Jackson Citizen Patriot* newspaper, March 1, 1963.

20. Events that appear to have been noted more by Rup Chand than by Koelz in diary entries.

21. *The Michigan Alumnus*, June 9, 1956, 385–86; among his contributions he was credited for having identified disease resistant melons in India, which helped save the California melon crop.

22. Rob Linrothe, *Collecting Paradise: Buddhist Art of Kashmir and its Legacies*. New Delhi: Serindia Publications, 2015.

Richard Barnes

RICHARD BARNES/TAKING INVENTORY

Amanda Krugliak

When past Institute for the Humanities director Daniel Herwitz and I first met Richard Barnes in 2009, I had the immediate impression he had come from a place much further than New York City . . . other worldly, transported from the nineteenth century into this one.

Barnes's work possesses the meticulousness and formality of an early naturalist in combination with the startling observations of a contemporary poet. The vastness and reach of his work suggest the documentations of a time traveler, the beginnings and the ends of things done and come undone, one undeniably hinged to another. The careful organization of objects alludes also to the inverse—the drawer once opened leads to disorder, the unveiling preludes the complete disarray.

Time falls back on itself like a primordial coil, the ultimate tail chase, an infinite spiral. Barnes's photographs chart the velocity of change, the sense of displacement we experience as the momentum shifts, first meandering, and then speeding up. They measure the recession of our own stories into the distance. Unintentionally, the images become starkly modern, austere, bracing us from the inanimate quiescent, no longer a question of if but when.

Like pictures fallen loose from the box stored on a top attic shelf, we are caught in a scattering of questions, pondering. Who were we then? Who are we now? And how did we ever get to this place? Barnes's photographs lure us to the precipice, to reconsider what we think we know, and offer the possibility of transcendence. The duality of his images suggests both what matters to us and matter-of-factly what nearly is. His photographs capture the reassurance of a more ordered and determinate past as well as the anxiety of a tentative and uncertain future. Barnes alters our perception of space and the passage of time, erasing any clear line between past and present simply by crossing it.

Artistic method and scientific practice overlap, and assumptions of certainty for either would be telling only of an imposter, one version of the truth. Barnes instead is interested specifically in the process, mining the progression from some point of origin or authenticity to what happens along the way. There is the hopefulness of redemption in the notion that regardless of how long the journey takes, there will be some proof, some finding, some effort toward understanding and meaning. His inquiries conjure up sensations of longing for life itself.

Within Barnes's visual ruminations one finds unexpectedly that door in the floor, leading us inward toward reflections on our agency and culpability, our own undoing, and the reality of our inevitable obsolescence. The unaffectedness and effortlessness of Barnes's work mirrors the impermanence of things.

Richard Barnes's relationship with the Institute for the Humanities began in 2009, as our first institute artist in residence. He was commissioned to produce a site-specific installation for the institute gallery, an exhibition that would coincide with LSA's theme year Meaningful Objects, and run concurrently with his exhibition of photographs from *Animal Logic* at UMMA as well as an extensive installation at Cranbrook's Institute of Science.

As the institute arts curator, I had the privilege of accompanying him throughout the U-M museums and collections, forging relationships with wonderful people like Amy Harris, Philip Gingerich, Janet Hinshaw, Carla Sinopoli, Bill Sanders, and Dan Fisher—among countless others—who would become integral to his project and its success. Barnes was given full access to the collections of paleontology, zoology, and anthropology. His days of research uncovered butterflies in envelopes, pressed wildflowers in the Herbarium, birds of extinction with vivid plumage catalogued in drawers.

Richard Barnes's Institute for the Humanities exhibition *Pastperfect/FutureTense* opened in September 2009, the culmination of over a month of residency at U-M. As part of the installation, a prehistoric whale cast still in armature was brought into the light and wheeled from the Ruthven Exhibits Museum down North University like a processional, and hung from the gallery ceiling. The installation also included photographs of molds used in the replication process and prints made from long forgotten lantern slides.

The relationships Barnes forged in that first residency with the institute and U-M museums and collections would continue and their afterlife became the basis for future collaborative projects. In 2010, Barnes accompanied U-M paleontologist Philip Gingerich to Egypt's Valley of the Whales, photographing Gingerich's excavations in the field. This collaboration would lead to a feature story for *National Geographic* in August 2010.

In 2012, with the continued support of the institute, Richard Barnes, U-M anthropologist Jason De León, and I began a five-year collaboration, conceptualizing and creating the nationally recognized *State of Exception*. The installation based upon the research of Jason De León's Undocumented Migration Project curates the traces of migration from Mexico through the deserts of Arizona into the United States. The exhibition includes objects and belongings found in the desert from De León's archive, photographs and video created by Barnes along the border between Mexico and Arizona, and audio interviews with individuals recounting their journeys. The ongoing support of Director Sidonie Smith and the Institute for the Humanities has fostered new installations of *State of Exception* across the United States with thousands of visitors. It opened at Parsons New School in February 2017.

Now, for the bicentennial, and coinciding with the institute's Year of Conversions, Barnes returns as an institute visiting artist to document the process of collections on campus migrating to the new facilities on Varsity Drive.

What is remarkable is the arch of Barnes's relationship to the institute and the university, his affinity and intimate knowledge of our vast collections, inextricable from the process of documenting the migration itself.

The extraordinary work from his original installation in the institute gallery, now followed by the current images in this book, offer documentation of the ongoing process of our museums and collections in real time, a transformation almost impossible to fully grasp because it speaks of a future still unknowable.

The subterranean space in Ruthven where Barnes first immersed himself during his reconnaissance has become part of our shared story that recedes in the distance. New methods and facilities help forge brave new worlds.

Barnes's images as part of this most recent project are in many ways the most affecting, even closer to that point of origin, pared down. Photographs of taxidermy birds beneath plastic sheeting become simple shadows, like ink blots, profound and fleeting, all just a blur. Long-necked fowl relegated to Post-it tagged desk tops look like some surreal moving sale, graceful but a fall from grandeur. The white on white of new storage cabinets and the unsettling edgy gleam of overhead light read like futuristic portals in a science fiction movie built to transport us as well as any trace of our stuff.

A contemplative image of red leaves, taxidermy, and an almost nonchalant human skeleton speaks to Barnes's possible propinquity for once living things and interspecies relationships. The overall emotional tone of the work is reflective and evokes the quiet of acceptance. Barnes as the onlooker, always just outside of the frame, clearly wrestles with the anxiety of complete erasure of our individuality and complexity, and the loneliness of that. Disambiguation becomes personified. Barnes's photographs make no distinction between us and them. We share the melancholic temporality of a common gesture.

Keenly aware and down to earth, Barnes offers a fresh take, looking back while moving forward, on a crazy ride. Students conduct their studies surrounded by preserved animal heads serving witness, holding court. It's hard to tell who is observing who. The tenor of these images is both comical and absurd. And with a healthy dose of levity, Barnes reminds us we are still solidly in the mix, one day at a time. It's just a typical day in the midst of upheaval.

The institute's Year of Conversions has focused on times of change, subtle and extreme, the potential we all have to create new systems and paradigms, to reinvent ourselves and think about things differently . . . the about face, the turnaround, the next page. And what comes after is unpredictable, the consequences of true seismic shifts unforeseeable.

Through catalyst, chasm, or conversion, rather than re-semblance of the routine order of things, there is the potential to be upended, to become something or someone radically different, new again.

Sable Antelope

BEWARE
FINGERPRINTS ARE RECORDED
ALERT

PART III

Museums, Libraries, and Collections

DETROIT OBSERVATORY

1854–2017

The Detroit Observatory, located in Ann Arbor at the corner of Observatory and Ann Streets, was both the first large-scale research instrument on the University of Michigan campus and a major step in the plan of the university's first president, Henry P. Tappan, to transform the campus from a teaching college into a research university. Opened in 1854 and listed on the National Register of Historic Places in 1973, the building and its instruments were completely restored in 1997. Today, the Detroit Observatory—the second oldest building on campus and on its original site—is one of the few extant observatories in America to retain its original telescopes from the 1850s in working condition in their original mounts. Under the supervision of the Bentley Historical Library, the observatory provides a unique experience of architecture, science, and academic life in the earliest days of the American research university.

Fig. 80. Detroit Observatory. Painting by Jasper Francis Cropsey, 1855. Bentley Historical Library.

HISTORICAL OVERVIEW

Some of those gathered in the First Presbyterian Church in Ann Arbor on December 21, 1852, for the inauguration of the university's first "chancellor" must have been a bit crestfallen. Many of them had worked hard in political and financial ways to support the fledgling University of Michigan at least since 1837. But the new chancellor (later called president), Henry P. Tappan, declared in his address that outside Prussia, there were no "true" universities in the world, including in the state of Michigan; those that existed in the United States and Britain, including the "University" of Michigan, were, by Prussian standards, merely "University faculty giving instruction in a College." A true university, Tappan argued, would combine faculty able to conduct original thought and investigation with "books, apparatus, specimens and models . . . sufficient for scientific and learned investigation"[1] (See also Blouin, this volume).

Tappan's address was designed in part to challenge the political leaders in the state to provide the resources for a "true" university. The report in the *Detroit Tribune* (reprinted in the *New York Times*) took the point: "The people of the state must cooperate with him and them [the regents], and their means must be increased. This ought to be done by the Legislature. The funds at the

Fig. 81. Observatory Hill in 1935. The addition to the original observatory was removed in 1954. Bentley Historical Library.

control of the Regents are wholly inadequate to the accomplishment of the grand designs of the founders."[2] Indeed, Tappan had already made the same points in greater detail in his 1851 book, *University Education.*

The years of Tappan's presidency would see substantial progress toward his goals through the hiring of new faculty, the launching of a scientific curriculum supported by collections in natural history, the opening of a chemical laboratory in 1857, the beginning of a fine arts collection in 1856, and the reorganization and expansion of the university's library.[3]

As Patricia S. Whitesell has pointed out in her authoritative history of the Detroit Observatory, Tappan believed that because the great planetary masses composed the universe and therefore helped define time and space, astronomy was a fundamental science and related to all others.[4] The 1817 founders of the University of Michigan had agreed, listing a professorship in astronomy (astronomia) as one of the original thirteen faculty positions.[5] The "Great" Comet of 1843, which blazed across the sky from February through April of that year, had fired the popular imagination regarding astronomy. Further, because railroads and other businesses needed an accurate sense of time—which could be determined only through the use of astronomical instruments—there was a practical demand for astronomical training as well.

In fact, astronomy was first listed as a separate subject in the University of Michigan catalog in 1844–45; it was taught by George Palmer Williams, professor of mathematics and one of only two faculty members at that time, but with no dedicated facilities or equipment (Bentley Historical Library, 2009). So when Detroit businessman Henry N. Walker approached Tappan after his inaugural to offer his help for the university, Tappan proposed working together to open an observatory. Walker was a well-known Michigan politician who had been a supporter of governor and senator Lewis Cass and had served as attorney general and as a member of the state legislature in the 1840s. At the time of Tappan's arrival, he was one of the founders and vice president of the Detroit Savings Bank, and he would later be president of the Detroit and Milwaukee Railroad, postmaster of the city of Detroit, and owner and editor of the *Detroit Free Press* from 1861 until 1875.[6]

On December 29, 1852, Walker hosted a gathering of Detroit notables—including Cass himself—at the Detroit hotel, Michigan Exchange. Tappan and others spoke in favor of the project and the group immediately pledged $7,500 with a promise to go as high as $10,000; twenty-nine of those in attendance pledged more than $500, more than $13,000 today. Because of this beneficence of Detroit residents, the observatory was named after them: Detroit Observatory.[7] To ensure that the observatory would be able ultimately to determine the correct time in Detroit for various business purposes, Walker himself pledged $4,000 (more than $112,000 today) to cover the cost of a meridian circle telescope, the instrument necessary for that work. The total cost of the observatory structure and instruments would be $22,000.

With the funding pledged, Tappan and Walker set out for New York in February 1853 to purchase instruments. There they contracted for a refracting telescope. Tappan continued on to Berlin where he purchased the meridian circle telescope and a sidereal clock, both necessary for the time-telling function. The purchased instruments, fully restored and on display at the observatory today, were:

Fitz Refracting Telescope (installed 1857)

The 12 foot 5/8-inch objective lens of this refracting telescope was one of the largest in the world—perhaps the third largest—when it was completed in 1857. President Tappan ordered the telescope from Henry Fitz of New York, who had become a famous American telescope maker, responsible for 80 percent of all telescopes made in America between 1840 and 1855.[8] The original wooden telescope tube was replaced by steel in 1907. The original objective lens is unique in that it is the largest remaining Fitz lens that has not been reground. The telescope was completely restored in 1997.

Fig. 82. Fitz Refracting Telescope, restored in 1997. Photo by Stephen Graham. Bentley Historical Library.

Meridian Circle Telescope (installed 1854)

Tappan ordered this 6-inch telescope from the firm of Pistor & Martins in Berlin. After it was inspected by Franz Brünnow there, it was shipped to New York, and then on to Ann Arbor by train. The meridian circle's precision technology was used in the nineteenth century to precisely track stars as they crossed the meridian and thereby map the heavens. Its precision was also used for determining time.

Fig. 83. Meridian Circle Telescope, first installed in 1854 to measure sidereal time and to synchronize the State's clocks; restored in 1997. Bentley Historical Library.

Sidereal Clock (installed 1854)

This astronomical clock set to sidereal time was manufactured by Christian Friedrich Tiede of Berlin and was ordered by Tappan on the same trip. Sidereal time is based on the fixed positions of stars, rather than the sun, and is the time at which a star crosses the meridian, thus the pairing of this clock with the meridian circle telescope.

THE OBSERVATORY IS BUILT

While Tappan was traveling in Germany, the location for the observatory was selected on the top of a hill overlooking the Huron River Valley, almost half a mile from the campus—and construction began. Although no original copies of the plans for the observatory exist, they were apparently drawn up by Richard H. Bull, professor of mathematics and sometime architect and astronomer at New York University, who himself provided the correct time to the New York Central Railroad.[9]

Construction of the observatory was an engineering feat. Accurate astronomical viewing required that the telescopes be placed on brick piers reaching to bedrock to prevent vibration. The basement was literally hewn into and buried in the hillside, the brick piers were built up, and the relatively small observatory building was built around them. The basement walls were made of local fieldstone and the structure itself was brick, in the Italianate style with a Greek Revival porch. There were only about 3,000 square feet above grade: on the main floor, the meridian telescope room on the east (with shuttered flat roof above, permitting access to the sky above the instrument), and the director's office on the west, each about 500 square feet. The rotating dome itself at the top of the building contained about 350 square feet. Although there were small gathering areas on the first and second floors, the observatory was designed primarily for the use of the faculty astronomers, as later years would tell.

The Board of Regents toured the completed building in June of 1855 (although the Fitz telescope in the dome was a temporary replacement awaiting the final version). In September, New York artist Jasper Cropsey came to Ann Arbor to paint an image of the observatory and the main campus. His observatory image would become the university's icon.

FOUNDING DIRECTORS

The first two directors of the Detroit Observatory were distinguished scientists, certainly among the pioneers of nineteenth-century astronomy in the world. Franz Brünnow had been the assistant director of the Berlin Observatory when Tappan approached him and the famed director Johan Enke to help supervise the building of the meridian telescope there. Brünnow had received his doctorate from the University of Berlin in 1843 and became the assistant director of the Berlin Observatory under Enke in 1851. In the same year he published (in German) a major work, *Handbook of Spherical Astronomy*. When two prominent American astronomers rejected Tappan's offer to direct the observatory, he turned to Brünnow, who accepted and arrived in Ann Arbor in 1854, in time to supervise the installation of the equipment in the observatory (and to discover problems in the Fitz instrument resulting in the return of two versions of it to their maker and the installation of the permanent telescope in the dome only in 1857). At the time he arrived, Brünnow was the only holder of a PhD degree on the university faculty; in addition to adding new rigor to teaching in the field, he promoted scientific research not only through his own work—for example, publishing important works of his own in 1855 and 1858—but also by establishing the first scientific journal on the campus, *Astronomical Notices*, in 1858. In 1857, Brünnow married President Tappan's daughter, Rebecca, and thus tied his fate to that family. When Tappan himself was unceremoniously fired by the regents in 1863, Brünnow resigned and spent the rest of his successful career in Europe.

Brünnow's most brilliant—and at times only—astronomy student, James Craig Watson, graduated from the university in 1857 and immediately became his assistant at the observatory. Watson received his master's degree and appointment as professor of mathematics and astronomy in 1860. He published *A Popular Treatise on Comets* in 1860 and *Theoretical Astronomy Relating to the Motions of the Heavenly Bodies around the Sun in Accor-*

Fig. 84. Observatory by night. Monthly viewing nights grant visitors the opportunity to observe the night sky through the historical Fitz Refracting Telescope. Bentley Historical Library.

dance with the Law of Universal Gravitation in 1868 while discovering nine asteroids at the same time. He directed the observatory from 1863 until 1878 and was inducted into the National Academy of Science in 1867 at the age of twenty-nine. Working together, Watson and Brünnow used the equipment at the observatory and their connections with other observatories to establish the longitude of its location in 1861. This made accurate timekeeping possible for the first time.[10] Watson left the university in 1878 to direct the Washburn Observatory at the University of Wisconsin in Madison and died shortly thereafter. Today, the James Craig Watson prize in astronomy is awarded every two years by the National Academy of Sciences.

FILLING OUT "OBSERVATORY HILL"

It was under Watson that the observatory itself began to be surrounded by other buildings. In 1868, the regents authorized adding a residence for use of the director and his family. The house was attached to the west side of the building, with an entrance through the director's office. A students' observatory was erected in 1878 to the south of the meridian room, partly in response to student complaints that the observatory's telescopes were seldom available because the faculty used them for research.

By 1908, modern astronomy required a bigger and better telescope, so a large classroom building and a dome were added to the east of the building, and the student observatory was moved to the west of it and its equipment upgraded. The student observatory was demolished in 1922 to make way for the building of Couzens Hall, then a dormitory for nurses. To make up for this loss of teaching space, plans were made to equip Angell Hall with a new, rooftop student observatory and laboratory, with the upper floor of the new building entirely devoted to astronomy. This space opened in 1926–27. But expanded student enrollment in the post-World War Two years kept all astronomy spaces occupied, and alumni from the 1950s and early 1960s will remember using the 1908 student spaces on the observatory hill. The former director's house on the west side was in use for various purposes until 1954, when an expansion of Couzens Hall required its demolition.

PATH TO RESTORATION

In 1963, the Department of Astronomy was moved to the newly constructed Dennison Hall on the main campus and the fates of the department and the observatory hill were disconnected. Increasing smoke from the university's power plant and light pollution from area buildings had made the telescopes less useful, while the new facilities on main campus were able to meet student demand. Faculty astronomers increasingly used off-campus facilities for their research. The 1908 addition was condemned and razed in 1976, leaving the original 1854 observatory building alone on the hill but mostly derelict.

Ann Arbor historical preservationists and U-M faculty in astronomy advocated for saving the observatory and the university ultimately responded to this challenge. A successful fund-raising campaign led to the nearly perfect restoration of this important piece of the university's history. Tappan had written: "I cannot speak of the Observatory without emotion. No one will deny that it was a creation of my own." He was right about that. But it also served as the point of origin for Tappan's vision of a "true" research university. No other place in the university moved so rapidly into the front ranks of research in the nineteenth century and, therefore, set the bar for that work so high. By providing the space, the instruments, and the faculty talent, Tappan had launched the process that would power the university's world-class research mission thereafter.

THE OBSERVATORY TODAY

To visit the Detroit Observatory today, then, is to stand in the place where that mission and process were born. The restoration of the building and its contents, including two historical telescopes,[11] enhance that experience. Regular open houses and viewing nights using the Fitz telescope are scheduled through the cooperation of the Bentley Historical Library and the LSA Department of Astronomy. Tours can also be arranged by appointment. An exhibit showcasing the early history of the observatory, the university campus, and the city of Ann Arbor is on display in the room that once was the director's office; the central room contains a small exhibition of scientific instruments and associated materials. The historical records of the Detroit Observatory and many of the major actors in its history are part of the collections of the Bentley Historical Library.

Entry prepared by the staff of the Bentley Historical Library in memory of Patricia (Sandy) Whitesell.

1398 East Ann Street. The Detroit Observatory offers monthly daylight tours and public viewing nights (hosted in conjunction with the Department of Astronomy https://dept.astro.lsa.umich.edu/detroit.php). Special guided tours can be arranged. Visit http://bentley.umich.edu/about/detroit-observatory for information.

Notes

1. Tappan, *A Discourse*, 20.
2. "The Michigan University – Inauguration of Rev. Henry B. Tappan," *New York Times*, 28 December, 1852. (Reprinted from *Detroit Daily Tribune*, 22 December, 1852.)
3. Six thousand volumes were added by the end of his term in 1863. See Farrand.
4. Whitesell, 40 . See also Tappan, *A Discourse*, 17.
5. Peckham and Steneck, 6.
6. "Henry N. Walker. Death Takes a Well-Known Citizen of Detroit." *Detroit Free Press*, 24 February,1886.
7. Whitesell, 27.
8. Whitesell, 53.
9. "Richard H. Bull," *New York Times*, 3 February, 1892.
10. For an extensive explanation of this process, see Watson in the *Detroit Free Press*, reprinted in the *Michigan Argus*, January 27, 1865. The time recorded on campus was called "Observatory" time throughout Michigan.
11. A third telescope now in the building dates to the 1880s and was used by students. It is not original to the building and was never used in the observatory.

REFERENCES

Deo, Nisha. "Visiting Professor Lectures on Photographer." *Purdue Exponent*, February 13, 2009.

Detroit Daily Tribune,1852.

Detroit Free Press, 1865; 1886.

Farrand, Elizabeth Martha. *History of the University of Michigan*. Ann Arbor: Register Publishing House, 1885.

Michigan Argus, 1865.

New York Times, 1852; 1892.

Peckham, Howard H., and Margaret L. Steneck. *The Making of the University of Michigan, 1817–1992*. Ann Arbor: Bentley Historical Library, 1997.

Stargazing: 150 Years of Astronomy at the University of Michigan." Bentley Historical Library, 2009. http://deepblue.lib.umich.edu/handle/2027.42/111856

Tappan, Henry Philip. "A Discourse, Delivered by Henry P. Tappan, D. D. At Ann Arbor, Mich., On the Occasion of His Inauguration as Chancellor of the University of Michigan, December 21st, 1852." Detroit: Advertiser Power Presses, 1852.

Tappan, Henry Philip. *University Education*. New York: G. P. Putnam, 1851.

Whitesell, Patricia S. *A Creation of His Own: Tappan's Detroit Observatory*. Ann Arbor: Bentley Historical Library, 1998.

MUSEUM OF NATURAL HISTORY

1956–2017

Originally created as the "Exhibit Museum" in the Ruthven Museums Building in 1956, the University of Michigan Museum of Natural History (MNH)[1] presents exhibits and educational programs in natural sciences and anthropology for the benefit of the university community and the public. The Museum of Natural History developed out of an earlier exhibitions department dedicated to displaying the university museums' rich research collections. It became an independent unit in 1956, at a time of reorganization of the university's museums. The invigorated new museum dedicated itself to exhibition design, public programming, and outreach. From its inception until today, MNH displays have primarily focused on evolution and on Michigan's flora and fauna.

Today, MNH is facing dramatic changes as it makes plans to leave the Ruthven Building for a new home in the Biological Sciences Building (BSB) opening in 2019. Like its predecessor, this new museum will build upon the university's rich scientific collections while also emphasizing current faculty research and science education.

THE FOUNDATIONS

From its inception, the University of Michigan has collected, researched, and displayed specimens of natural history. Natural history collections were first displayed in Mason Hall and then in the 1881 University Museum on State Street. Within a few decades, however, the collections had outgrown their space and a new building was needed. In 1925, the state legislature appropriated funds for a structure with one wing devoted to offices and research facilities and the other to exhibits and displays.[2] The vision for the new building entailed, in the words of its director Alexander Ruthven, ideas that continue to inform MNH's work today:

> Museums have two main objects: the increase of knowledge and the diffusion of knowledge. Exhibits should be designed to teach lessons and should not be so loaded with accessories that the lessons are obscured, so greatly stressed that other fields of activity are restricted, or so arranged as to be viewed only with fatigue or confusion.[3]

Completed in 1928, the building housed four museums: Zoology, Herbarium, Anthropology (now Anthropological Archaeology), and Paleontology. Administratively independent, the building's occupants were grouped under the University Museums Division and Ruthven was named director. Ruthven created a "Department of Visual Education" to better serve the educational aspects of the museums, and Museum of Zoology herpetologist Crystal Thompson was named its curator. In this position, she helped to design the Hall of Evolution and exhibitions about biological principles, created to serve undergraduate biology classes. In 1930, an anonymous alumnus donated a collection of wild animals native to Michigan and a popular zoo was erected on the grounds of the new building.

Although Ruthven became president of the University of Michigan in 1929, he retained his position as director of the University Museums until Carl E. Guthe (director, Museum of Anthropology) succeeded him in 1936. The exhibit galleries were a lively, well-attended destination in the early decades and an innovative site for research and teaching. In the early 1930s, Crystal Thompson conducted studies of museum visitors funded by a grant from the Carnegie Corporation.

Fig. 85. Mastodon Display in the Hall of Evolution, Ruthven Museums Building. Museum of Natural History.

When Guthe stepped down in 1943, no one was named to replace him. Instead, a "Museums Operating Committee," comprising the directors of the four museums, ran the University Museums.[4] The exhibits continued to grow. In the 1940s, expeditions of the Museum of Paleontology expanded the dinosaur displays in the Hall of Evolution. The *Anatasaurus* (later renamed *Edmontosaurus*), discovered by Ermine Cowles Case in 1938, was put on display in 1941, and the Owosso Mastodon went up in 1947.[5]

The 1947 appointment of Irving G. Reimann as "Prefect of Exhibits" introduced an emphasis on student and public education. Reimann upgraded many of the existing displays and advocated for children's classes, adult night courses, and a lecture series.[6] In 1949, new exhibits were installed in the Hall of Biological Principles. Innovative backlighting systems and novel color schemes illuminated state-of-the-art glass, wax, and latex models of invertebrate and vertebrate animals, which were produced by eminent artists such as Leopold and Rudolf Blaschka, George Marchand, and the museums' own Carleton W. Angell. Angell and Dr. Alexander Butsch also began to work on the first set of four dioramas portraying the life-worlds of Michigan Native Americans in seemingly timeless settings, which went on display in 1949.[7] During the same year, the Department of Visual Education was renamed the "Exhibit Section."

THE EXHIBIT MUSEUM OF NATURAL HISTORY, 1956–92

On February 14, 1956, the regents approved changes to the university bylaws regarding the administrative relationships of the research museums and the Kelsey Museum, placing them under the College of Literature, Science and the Arts. At the same time, the exhibition section was restructured as a free standing "Exhibit Museum."[8]

Reimann became its first director. Cooperation and collaboration between the research museum curators, faculty, and Exhibit Museum staff were thus important to developing exhibitions, and an advisory committee was created with representatives from the building's other museums, the Departments of Geology and Zoology, and the Exhibit Museum.

At its first meeting, the advisory committee stressed the public function of the museum:

1. The university is a basic community.
2. The instruction should be at the freshman level.
3. Instructive and attractive exhibits would, however, reach the general public at large.
4. Freshman-level exhibits would be likely to reach the general public and attract its interest.[9]

While the research museums emphasized graduate education, MNH's focus on undergraduate education continues to this day. While the creation of an independent exhibit museum formally separated the public and display functions from the research museums' expertise, the public museum has throughout its history partnered with interested research museum curators and departmental faculty to provide scientific content and verify the accuracy of exhibits and programs.

Prior to 1956, the subject material covered in the exhibits was limited to fields represented by the research museums, drawing on their collections and curators' scholarship. Even as this relationship continued after 1956, the Exhibit Museum also realized that many subjects typically seen in natural history museums were not represented in the exhibitions. As soon as the museum gained independence, Reimann began to plan exhibits in astronomy, geology, and human physiology.

The planetarium and astronomy exhibits opened in 1959,[10] as did the georama—a three-dimensional model landscape depicting a range of geological formations.[11] Reimann began a drive to raise money to display an *Allosaurus* skeleton from the Cleveland-Lloyd quarry in Utah. The bones arrived in the fall of 1960 and the display was completed in 1965.[12] A partial *Stegosaurus* skeleton was added in 1967.

Docents and Exhibit Preparators[13]

Two significant aspects of the museum span the time before and after it became autonomous: a longstanding tradition of docents and the professionalization of exhibit preparators, which led to pioneering diorama displays.

Elmer Berry was appointed docent in 1937 and Gail Bailey wrote the "Suggested Outline for Developing Docent Program" in 1949. Also in 1949, Heather Thorpe came to the museum from the Buffalo Museum of Science and became the first senior docent. She significantly expanded the docent program and initiated the first Docent Manual. Although it is not clear when the docent program began to recruit exclusively from university students, the student docent program was in full swing by 1965.[14] In the 1970s, Ruth Haag succeeded Gail Bailey, creating, among other things, a free service guide to assist the public on the weekends.

Two key exhibit preparators and exhibition designers joined the staff in 1947: Dr. Robert S. Butsch and Dr. William A. Lunk. Both contributed greatly to new exhibits and both would later serve as museum director. By 1965, the museum had four full-time preparators along with a full-time cabinetmaker, Joachim (Joe) Knueppelholz.

Around 1957, Butsch began work on a set of eight additional Native American dioramas depicting groups from across North America, which were completed in 1962. During the same years, preparator and diorama artist George Marchand began work on a series of seven dioramas called *Life through the Age*, which he completed in 1969. These became the prototype for the dioramas that Marchand and Reimann sold to museums across the country.[15] Approximately seventy-five didactic exhibits, mostly created by Lunk, became the "backbone" of the Hall of Evolution. In 1963, the zoo was disbanded and torn down, amid public protest, to make room for

Fig. 86. Presentation of Exhibit on Biological Principles, 1940s. Bentley Historical Library.

a new $1.2 million addition to the Ruthven Museums Building.[16] The *Dunkleosteus* head was added to the displays in 1966,[17] and two years later the museum purchased the Transparent Anatomical Woman ("TAM") from Rush Studios in Chicago in 1968. Butsch finished the bog-to-forest diorama the same year, after approximately three years' work.

After twenty-six years of service, Reimann retired in 1973. He was replaced in 1974 when Butsch was given a half-time appointment as director. The early 1980s were fraught with staff cuts and loss of rooms to the Museum of Anthropology and University Library. In 1984, exhibit preparator John Klausmeyer was hired in preparation for Butsch's upcoming retirement, to produce exhibits and to take over Butsch's Museum Methods class. Exhibit preparator William A. Lunk became interim director in 1985. During the post-Reimann period, a schedule of producing three new temporary exhibits per year emerged, at the cost of maintaining and developing the permanent exhibits.

In 1988, UMMZ curator of Insects and professor of biology Thomas E. Moore was given a half-time, five-year appointment as director. The focus on changing temporary exhibits continued, and few changes were made to the permanent exhibits.

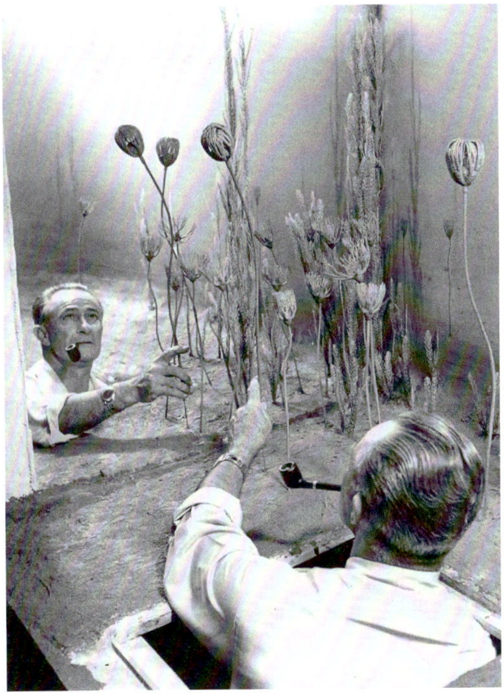

Fig. 87. Exhibit Preparator and Artist, George Marchand, c. 1963. Museum of Natural History.

Moore's directorship was marked by staff turnover and the replacement of the *Allosaurus* skull with a new, accurate cast. The museum's membership program was instituted during this period.

A MUSEUM FOR THE TWENTY-FIRST CENTURY (1993–2017)

William R. Farrand, professor of geology and curator in the Museum of Anthropology, was appointed half-time director in 1993. Under his leadership, the museum's public engagement, which had declined after Reimann's retirement, increased dramatically through the development of new public programs and events, and new permanent exhibits were added with public support.

In 1994, a Traveling Discovery Kit on the topic of *Endangered Species* carried the museum's education efforts into area schools. A Members Night was organized, and more than one thousand people attended the museum's first Family Day. In a nod to tight budgets, a donation box was added to the Rotunda lobby.

The Exhibit Museum also reached out for community support and involvement with a Buy-A-Bone fundraising campaign in 1994–95 to obtain a cast skeleton of *Deinonychus*. The skeleton marked the first substantial addition to the Hall of Evolution in thirty years. More than thirty-five hundred visitors attended the *Deinonychus* Family Weekend. The successful Buy-A-Bone campaign was repeated for the *Back to the Sea: The Evolution of Whales (Dorudon)* display in 1997 and again in 1999 for *Pterosaurs: When Reptiles Ruled the Sky (Anhanguera)*.

At the end of the 1990s, the museum held its first summer camp program and Family Halloween Party. It also created a Board of Advisors (1998), drawn from the faculty and the external community, to provide expertise and counsel to the director and staff. Later, two elected student docent representatives were added to the board.

After Farrand's retirement in 2000, Associate Director Amy S. Harris became interim director and then director in 2002, a role she continues today. Harris has had a succession of three- and five-year terms as a full-time, professional museum director, and is not a member of the faculty, unlike Moore and Farrand before her. In 2005, as part of a larger administrative reorganization plan, the museum was placed in the LSA Undergraduate Education Division. Harris now reports to the associate dean for Undergraduate Education instead of directly to the dean.

Under Harris's direction, public engagement and activity in exhibits continued to expand. Special emphasis was placed on providing support for faculty teaching and research, on undergraduate engagement, and on connecting the public with university research. A vision of the museum as a forum for public engagement around current science issues led to programs such as the Science Café series, launched in 2007.

Harris led the museum's community partnership with the Ann Arbor District Library, funded by a federal grant from the Institute for Museum and Library Services. Debuting in 2000, the "Family Reading and Science" program combined hands-on science activities for children with encouragement to read library books on the topic. Multiyear grants from the Community Foundation for Southeast Michigan and the Toyota USA Foundation made it possible to expand the program to more than forty libraries in southeast Michigan. The museum developed a middle school STEM careers program, *Science for Tomorrow*.

Beginning in 2003, the museum organized its winter term exhibit and program activities around an annual theme. In 2006, the museum led the LSA Winter Theme Semester, "Explore Evolution," prompted by the museum's participation in a NSF-funded project led by the Nebraska State Museum. Each of the six partner museums received a copy of the *Explore Evolution* exhibit to keep permanently. In anticipation of this new exhibit, LSA funded the renovation of space in the back of the fourth floor to create a 2,000-square-foot temporary exhibit gallery.

Theme semesters expanded the museum's relationships with faculty and collaborations with U-M and community organizations, and provided an efficient and rich means to serve the museum's multiple audiences. The museum continued to lead or jointly lead LSA theme semesters, including "The Universe: Yours to Discover" (2009),

Fig. 88. Biological Science Building, the new home for the Museum of Natural History, digital projection. The building will open for academic units in 2018 and for the public museum in 2019. Design Team SmithGroupJJR/Ennead Architects.

"Meaningful Objects: Museums in the Academy" (2010–11), "Water" (2011), and "Understanding Race" (2013). Major traveling exhibits hosted in association with theme semesters included *Hubble Space Telescope: New Views of the Universe* (2011) and *Race: Are We So Different?* (2013).

In 2004, donors supported the creation of a butterfly garden in front of the museum to make better use of the grounds for education. The garden was expanded in 2010 when the building entrance was renovated, the interior entryway restored, and the Museum Store moved and expanded.

The museum successfully raised funds to renovate the 1959 planetarium in 2007, converting from analog to a full-dome digital video projection system. School and public attendance grew following the renovation, and the planetarium increasingly participated in supporting faculty research outreach and national grant-funded programs.

Numerous annual public programs took root, including Discovery Days, Behind-the-Scenes Day (offered in cooperation with the research museums), ID Day, the Family Halloween Party (peak attendance exceeded four thousand), and a Butterfly Festival. Free weekend Dinosaur Tours and Hands-On Demonstrations also encouraged public visitation and engagement.

Permanent exhibits were updated. Major exhibit projects in the Hall of Evolution funded by "Buy-A-Bone" fundraising campaigns included updates to *Edmontosaurus* (2003), *Mastodons and the Ice Age (Buesching Mastodon)* (2005), and *Back to the Sea: The Evolution of Whales (Basilosaurus)* (2011). In 2011, woodworking and metalworking activities were terminated because of concerns about fire. Two exhibit preparators were laid off and the museum experimented with an outsourcing model for exhibit fabrication.

The student docent program grew and thrived, providing fifty to sixty undergraduate students each year with an active, and often transformational, learning experience. Up to two thousand undergraduates visit the museum each year with their classes or for assignments.

The Museum of Natual History remains a popular destination for school fieldtrips. The early 2000s were, however, marked by severe cuts to public school funding, increased transportation costs, and greater emphasis on standardized curriculum and testing, which led teachers to reduce field trip activity in favor of classroom time. Education staff responded by tying programs to mandated curriculum standards and expanded marketing efforts, which helped slow the drop in attendance. Even as school attendance has dropped, public attendance has expanded, now exceeding 150,000 annually.

By the late 1990s, it became apparent to

Fig. 89. Opossum diorama with blooming Trillium in the Michigan Wildlife exhibit. Museum of Natural History.

Fig. 90. Set of dioramas showing Native American people. The dioramas were completed in 1962. In 2011, the dioramas were removed from the exhibition floor. Museum of Natural History.

museums around the world that evolving cultural norms as well as advances in museum studies and anthropology required changes in exhibition practices. What was deemed an acceptable visualization of Native American history in previous decades was no longer considered appropriate. The museum's fourteen miniature dioramas of Native American cultures came under increased scrutiny from both Native and non-Native visitors, criticized for their portrayals of Native peoples in miniaturized, ahistorical scenes and for their presence in a museum of "natural" history. In 2011, the dioramas were moved to storage where they are still available for study by university classes.

Shortly after the museum became autonomous in 1956, Director Reimann expressed the need for a new and separate building. He reiterated this statement several times during his tenure and in 1962 wrote to the dean that the facilities were taxed "very nearly to their limit" and that Ruthven's building was "badly planned for public attendance and exhibit purposes."[18] Today, the lack of climate control and other limitations of a nearly ninety-year-old building present concerns about the safety of specimens and comfort of visitors.

In 2014, the regents approved a new Biological Sciences Building, which will include a new Museum of Natural History,[19] expected to open by early 2019. The structure and design of the museum will enhance its ability to support the university's and LSA's research and teaching missions, as well as to attract the general public. Academic research labs will be visible to visitors, enabling faculty and students to share their research with the public. The facility will also include a presentation space, two public "investigate labs," a student-curated exhibit area, and an enhanced digital dome theater. An all-new exhibit program is in development, making use of existing specimens and artifacts and updating both the science and the exhibit experiences for students and the general public.

Perhaps Alexander Grant Ruthven best described the University of Michigan Museum of Natural History's philosophy when he said:

> The Museum is not finished. It never can be if it is to be a functioning unit of the University. My earnest hope is that it will grow and change with the times and serve science and the University with increasing effectiveness.[20]

Entry prepared by Amy Harris and Lori Ann Dick.

Open Monday-Friday 9 am–5 pm; Sunday 12 pm–5 pm; 1109 Geddes Avenue. http://www.lsa.umich.edu/ummnh

Notes

1. The name change from Exhibit Museum to Museum of Natural History became official in 2011.
2. Guthe, 474; Rogers, 1140.
3. Ruthven, 7.
4. Rogers, 1441.
5. Ibid.
6. Museums Records, Bentley Historical Library.
7. Kellum, 277–79.
8. University of Michigan Regents Proceedings, University Bylaws, Section 10.06 edited and section 10.09 added.
9. "Advisory Committee of the Exhibits Museum Minutes, March 29, 1956" (Museums Papers, Bentley Historical Library).
10. U-M News Service Bulletin, April 15, 1959, on file in Museum of Natural History.
11. *Ann Arbor News*, 1959, on file in Museum of Natural History.
12. *Ann Arbor News*, October 14, 1960; University Record, July 1965, articles on file in Museum of Natural History.
13. Exhibit preparators are specialists who prepare specimens for museum displays.
14. Interview with Elizabeth Cardinal, student docent in 1965 and daughter of Dr. Robert S. Butsch, exhibitor preparator and later director of the museum.
15. These institutions include the Field Museum of Natural History, Carnegie Museum of Natural History, and American Museum of Natural History.
16. *Ann Arbor News*, April 3, 1963; *Ann Arbor News*, August–September 1962; articles on file in the Museum of Natural History; The "Mort Map," Bentley Historical Library.
17. *Ann Arbor News*, June 29, 1966; article on file in Museum of Natural History.
18. Letter from Reimann to Dean Burton D. Thuma, November 23, 1963. Museums Papers, Bentley Historic Library.
19. The museum is still deciding whether to change the name of the museum to better match its twenty-first century focus.
20. Ruthven, ii.

REFERENCES

Angell, Carleton Watson. *Catalog of Works of Art by Carleton Watson Angell*. Ann Arbor: University Museums, 1955.

"Beauty and Utility Combine in New Museum." *Michigan Alumnus* 34, no. 31 (1928): 679–82.

Guthe, Carl E. "The Museums at Michigan." *Michigan Alumnus* 43, no. 11 (1937): 469–77.

Kellum, Lewis B. "The University Museums." In *The President's Report for 1948–1949*, 275–81. Ann Arbor: University of Michigan Official Publication, 1950.

Maxted, Jean. "Eventful History Accompanies University Museums Growth." *Michigan Daily* (Ann Arbor, MI), April 2, 1939.

Rogers, J. Speed. "The University Museums." In *The University of Michigan: An Encyclopedic Survey Part VIII*, edited by Walter Arthur Donnely, Wilfred Byron Shaw, and Ruth W. Gjelsness, 1431–42. Ann Arbor: University of Michigan Press, 1956.

Ruthven, Alexander G. "Description of the Building." In *The University Museums Building of the University of Michigan*. Ann Arbor: University of Michigan, 1929.

Ruthven, Alexander G. *A Naturalist in a University Museum*. Ann Arbor: University of Michigan Alumni Press, 1931.

MUSEUM OF ZOOLOGY

1913–2017

The University of Michigan Museum of Zoology (UMMZ) is a center for the study of animal diversity focusing on the evolutionary origins of the planet's animal species, the genetic information they contain, and the communities and ecosystems they help form. Now part of the Department of Ecology and Evolutionary Biology (EEB), UMMZ houses world-class collections—totaling over thirteen million specimens—resulting from almost two hundred years of regional and global biodiversity studies.

HISTORICAL OVERVIEW

1837–1913 Foundations

UMMZ's collections date to the foundation of the state of Michigan and the establishment of the State Geological Survey, which was designed to explore the geological features of the region and to inventory its flora, fauna, and fossils (see Caple and Williams, this volume). The resulting specimens were housed in the university's Cabinet of Natural History. Dr. Abram Sager headed the botanical and zoological departments of the Geological Survey and his specimens formed the initial nucleus of what would become UMMZ's collections. In addition to documenting the new state's resources, the collections also served an instructional purpose, as the first zoology classes consisted of lectures supplemented with demonstrations of cabinet specimens.

In 1855, Alexander Winchell was appointed professor of geology, zoology, and botany. He expanded the zoological curriculum and its associated teaching collection, extending it well beyond Michigan fauna and transforming the "cabinet" into the Museum of Natural History. Winchell hired Carl Rominger, M.D., as assistant curator, paleontologist, and taxidermist in 1863. By 1867, the zoological collection had up to sixteen thousand specimens.

Prior to the 1870s, the university's natural history collections were primarily national in scope. This changed dramatically when the university hired Joseph Beal Steere to lead his five-year circumglobal expedition to collect materials for the museum. The Steere Zoological Collection alone totaled sixty thousand specimens, including twenty-five thousand insects, fifteen hundred mollusks and eight thousand birds. This flood of new collection material resulted in a space crunch that spurred the funding and construction of the 1881 University Museum Building.

In 1879, zoology became a separate division of the University Museum (from botany and paleontology) and Steere was appointed professor of zoology and curator of the museum. Following his resignation in 1894, Jacob Reighard was appointed professor of zoology and director of the zoological laboratory and museum. He continued as head of the Department of Zoology until 1925. Reighard made three museum hires: Herbert Sargent (1898), Charles C. Adams (1903), and Alexander G. Ruthven (1906).

1913–56 FORMALIZING THE MUSEUM

In 1913, at Ruthven's request, the regents formally changed the name of the University Museum to the Museum of Zoology, with Ruthven as director. In addition to Ruthven, the full-time scientific staff of the newly recognized UMMZ included Asa Wood, curator of birds; Crystal Thompson, scientific assistant in charge of fish and invertebrates;

Fig. 91. Loggerhead Shrike, Lanius ludovicianus, collected in 1837 by Abram Sager.

and Helen Thompson, scientific assistant in charge of amphibians and reptiles.

Over the next eight years, Ruthven supervised the transformation of the museum. The Section of Botany was transferred to the newly formed University Herbarium and a dynamic and far-reaching recruitment process formed the UMMZ into six major divisions, each with at least one full-time curator: Mammals (Lee Dice), Birds (Norman Wood), Reptiles and Amphibians (Ruthven, Helen Thompson Gaige), Fishes (Carl Hubbs), Insects (Frederick Gaige), and Mollusks (Mina Winslow). These divisions still exist today, and from a contemporary perspective it is notable that more than one hundred years ago, two of the seven curators were women.

Ruthven was adept at outreach. Under his leadership, the Museum of Zoology attracted a large and active group of influential amateur naturalists. Informally known as the Detroit Naturalists Club, they participated in many aspects of the museum's program. Some members became nationally recognized authorities on specific taxonomic groups as well as honorary curators of the museum. Others were significant benefactors, supporting the establishment of UMMZ publications series, financing numerous national and international expeditions, and purchasing or donating important collections. Indeed, a number of UMMZ's most important extant endowments stem from the generosity of club members.

The net effect of all this new activity was the rapid expansion of the museum's collections and its curators' academic output. Space in the Museum Building was quickly outgrown. In 1925, after many years of lobbying by Ruthven, the legislature appropriated funds for a University Museums Building, later renamed the Ruthven Museums Building.

When the new building opened in 1928, it held four separate museums and a section of exhibits. Ruthven was appointed to the newly created post of director of University Museums and Frederick Gaige was made assistant director of the Museum of Zoology. Ruthven's appointment was short-lived. He was recruited to the presidency of the university in 1929. Gaige became UMMZ's director, a position he held until 1945.

Although Gaige's directorship spanned the Great Depression and World War II, UMMZ's collections and academic programs continued to grow and develop. The Fish Division in particular expanded dramatically. Its dynamic curator, Carl Hubbs, was interested in systematics, zoogeography, development, and the importance of hybridization in fish evolution. By the time he left in 1944, UMMZ had become the leading ichthyological center in North America, training more than forty PhDs and growing the collection from five thousand to almost 2,100,000 specimens. Hubbs was succeeded by his former student Reeve Bailey in 1944.

Josselyn Van Tyne was appointed curator of the Birds Division in 1931 and initiated an era of tremendous growth in collections and the ornithology program—successfully acquiring major private collections and establishing an outstanding ornithological library. He and his students concentrated on life history and anatomical studies of birds.

In the Mammal Division, Lee Dice specialized in mammalian genetics. In 1934, he became director of the separate Laboratory of Vertebrate Genetics, relinquishing his curatorship in 1938. William Burt, a behavioral ecologist, succeeded him and was joined by Assistant Curator Emmett Hooper, a specialist in Mexican mammal diversity.

Once Ruthven became president, Helen Gaige assumed responsibility for the Division of Herpetology. The research and training program in reptiles and amphibians had become highly regarded, attracting leading students, including some who would subsequently become divisional curators.

As curator of the Mollusks Division, Mina Winslow engaged in several regional and international collecting trips prior to her resignation in 1929. She was replaced by one of the honorary mollusk curators, Calvin Goodrich, a specialist in North American pleurocerid snails who sampled extensively in southern US drainages prior to their 1930s' industrialization. In 1934, UMMZ graduate Henry van der Schalie was made assistant curator, becoming curator in 1944, and established a prominent research and teaching program in snail-borne diseases and mollusk biodiversity.

Fig. 92. Illustration by E. N. Fischer of a new species of frog collected by Ruthven in 1914 in British Guiana. Ruthven named the species, Hyla helenae, for his colleague, Helen Thompson Gaige. Ruthven, 15.

Fig. 93. Image of a male paratype of Appalachia arcana, collected by Theodore H. Hubbell in Michigan near Oscoda, Iosco Co. on Aug. 29, 1934. UMMZ I-00079769. Hubbell and Irving 1938.

Fig. 94. Holotype of Cyprinodon alvarezi, a Mexican desert pupfish, collected by Miller and Huddle in 1961. The UMMZ has 3,195 holotypes, specimens used when formally describing a new species.

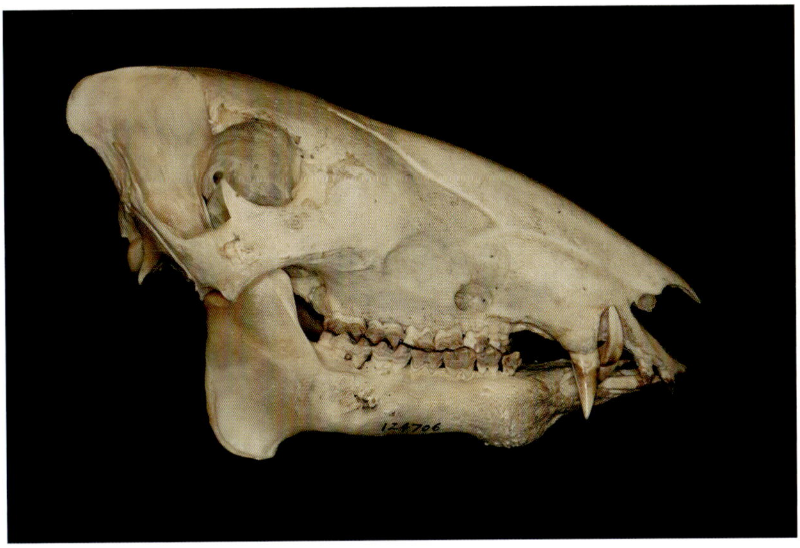

Fig. 95. A skull of Catagonus wagneri, a species of peccary from Paraguay, collected in the late 1970s by Phil Myers. Photo by Phil Myers.

Frederick Gaige's specialization was neotropical ants, but his research output was constrained by directorial demands. Nonetheless, Insect Division functioning was enhanced by the activities of multiple honorary curators who took on responsibility for major insect groups and added significantly to the collection.

At the cessation of World War II, UMMZ experienced an abrupt and significant loss when both Frederick and Helen Gaige retired. These two individuals had played significant roles in the program for decades and their loss was keenly felt. The museum turned toward its former students to fill these and other gaps: J. Speed Rodgers was appointed director and T. H. Hubbell curator of Insects. This pattern of hiring former UMMZ students continued during the ten years of Rodgers's directorship (e.g., Charles Walker as curator of Herpetology and Robert Miller as curator of Fishes). Exceptions were the hiring of William Gosline, an ichthyologist from Stanford University, and Robert Storer, an ornithologist from UC Berkeley.

1956–82 Change and Continuity

Rodgers died unexpectedly in 1955 and was succeeded by T. H. Hubbell, who served as director until his retirement in 1968. This was a time of significant reorganization and development. Teaching and research programs were broadened, ties with the zoology department were strengthened, systematic biology came to prominence in the curriculum, and a research wing was added to the museum with monies from the National Science Foundation.

A longstanding issue for the museums concerned the curators' role in teaching. Since 1928, curators had ad-hoc voluntary associations with teaching departments—the Department of Zoology in the case of UMMZ. In 1956, the "research museums" were placed under the dean of LSA and UMMZ curators entered into an expanded role in undergraduate and graduate teaching as half-time faculty, with tenure based in the Department of Zoology. While the museum retained curatorial appointments as well as a separate budget and director, its long-term fate and direction were now inextricably tied to its academic department.

Six curatorial appointments were made during Hubbell's leadership, including Insect curator Thomas Moore (1956), a specialist in cicadas, and Bird curator and former UMMZ student Harrison Tordoff (1957), who studied Fringillidae systematics and crossbill genetics. Richard Alexander was recruited as curator of Insects in 1957. In 1963, UMMZ graduate John B. Burch was appointed curator of Mollusks. Two new herpetologists were hired in the mid-1960s: Donald Tinkle (1965) as a curator of Amphibians and Reptiles and Arnold Kluge (1965) as a professor in the Department of Zoology and research associate in UMMZ (he became curator of Amphibians and Reptiles in 1967).

Nelson Hairston served as UMMZ director from 1968 to 1975. He differed from his predecessors in that he was a faculty member in the Department of Zoology rather than a UMMZ curator. Hairston was a well-regarded ecologist who performed much of his fieldwork at the Edwin S. George Reserve, which was administered by UMMZ. Burt retired in 1969; Douglas Lay served as curator of Mammals for three years, followed by Robert Farentinos, who had a one-year term. Gerald Smith, a former UMMZ student, was appointed in 1969 as curator of Fishes. Robert Payne was appointed curator of Birds in 1970, replacing Tordoff, and Ronald Nussbaum was hired as curator of Amphibians and Reptiles in 1974, replacing Walker.

Donald Tinkle was UMMZ director from 1975 until his premature death in 1980 at age fifty. Three hires were made under Tinkle: Philip Myers (initially a visiting curator of Mammals) as assistant curator (1976); Alex Tompa as curator in the Mollusk Division (1977); and William Hamilton as curator of Insects (1978). Hamilton was arguably the most prominent evolutionary theorist of his generation and was recruited back to his native United Kingdom in 1984 to accept a position as Royal Society Research Professor at Oxford University. Upon Tinkle's death, Robert Storer became the interim director from 1980 to 1982. Two curators were appointed during this period: Barry O'Connor as a curator of Insects (1980) and William Fink as curator of Fishes (1982), replacing Reeve Bailey.

1982–2011 Challenges and Contraction

In 1982, William Dawson was made director of UMMZ, serving in that role until 1993. This was a period when molecular biology was developing and Dawson presciently foresaw the value of incorporating a molecular perspective into UMMZ research, making the provision of a Laboratory of Molecular Systematics (overseen by biology professor Wesley Brown) a prerequisite of his appointment. During Dawson's directorship, three assistant curators were hired: Mary McKitrick, curator of Birds (1986–93), Doug Eernisse, curator of Mollusks (1986–94), and Priscilla Tucker, curator of Mammals (1988–present).

Richard Alexander assumed the UMMZ directorship from 1993 to 1998. One of his first duties concerned searches for new Bird and Mollusk curators. This led to the appointment of Bird curator David Mindell in 1994 and Mollusk curator Diarmaid Ó Foighil in 1995.

Gerald Smith was UMMZ director from 1998 to 2002, the last three years also holding directorship of the Herbarium. This was a momentous period for the Department of Biology,[1] the tenure home for UMMZ curators. In 2001, following a 1999 external review, LSA dean, Shirley Neuman, implemented the recommendation that the department be split into two distinct entities: the Department of Molecular, Cellular, and Developmental Biology (MCDB) and the Department of Ecology & Evolutionary Biology (EEB). UMMZ and the Herbarium were linked with the latter. Three curators retired during Smith's directorship: Alexander and Moore from the Insect Division, and Burch from Mollusks.

Smith completed his directorial service in 2002. An external search for a new director proved unsuccessful. David Mindell was appointed interim director for one year (2002–2003) and director from 2003 to 2005. Searches led to the appointment of L. Lacey Knowles as curator of Insects in 2003 and Thomas Duda as curator of Mollusks in 2004.

Smith and Kluge retired in 2003, leaving the Fish and Bird Divisions with a single curator each. In 2004, LSA reduced the number of curators in UMMZ from twelve to eight. Mindell resigned from the UMMZ directorship in 2005 and was succeeded by William Fink (2005–11). Attrition of the curatorial ranks continued with the retirement of curator of Birds, Robert Payne, in 2007. This was exacerbated by the departure of the remaining bird curator, David Mindell, in 2008, leaving the division that held representatives of two-thirds of the planet's bird species without a curator or research and teaching program.

In 2009, the university decided to address a chronic and long-ignored safety issue regarding UMMZ's enormous ethanol-preserved collections by relocating them to a new facility that would meet current safety standards. This initiative represented a substantial investment in the design and construction of a new state-of-the-art ethanol collection facility located in an off-campus building on Varsity Drive that was already home to the Herbarium. Construction of the facility and associated research and preparatory space was completed in December 2011 and the "wet" collections, comprising approximately five million specimens, were successfully relocated by August 2012.

In 2010, LSA decided to relocate the remaining collections from the three research museums (UMMZ, UMMAA, and UMMP) occupying Ruthven to the Varsity Drive "Research Museums Center" (RMC) facility. This required the design and construction of environmentally controlled collection spaces, preparatory laboratories, research space, and offices. Following the installation of archival grade cabinetry, UMMZ collections comprising over eight million specimens will be relocated beginning in 2016.

In 2011, UMMZ was formally subsumed in the Department of Ecology and Evolutionary Biology. The museum directorship became an associate chair position, reporting to the EEB chair, and charged with responsibility for oversight of faculty curation, collection staff, and relevant operating budget and endowment funds. That same year UMMZ's collection capabilities were enhanced by an NSF-funded project to install a liquid nitrogen freezer facility, allowing for high-quality preservation of specimen biomolecular structure.

2011–The Present

Diarmaid Ó Foighil served as associate chair for UMMZ Collections from 2011 to 2013. Daniel Rabosky was recruited in 2012 as a faculty member and UMMZ curator of Amphibians and Reptiles.

Priscilla Tucker was appointed associate chair for UMMZ Collections from 2014 to 2017. She is overseeing the migration of UMMZ collection databases to a single platform, Specify, a project supported by LSA, and the move of the extensive UMMZ dry collections to the new Research Museums Center.

The museum today retains the six divisions envisioned by Ruthven, but under a very different organizational structure. Curatorial numbers have been reduced from a maximum of twelve to, with the retirement of Nussbaum in 2015, eight. Nonetheless, UMMZ remains a vital center of research and teaching, and its collections, some nearly two hundred years old, provide a remarkable resource for the study of global biodiversity and evolution.

CURRENT SCOPE OF THE COLLECTIONS AND RESEARCH PRESENCE

UMMZ houses approximately thirteen million vertebrate and invertebrate specimens, including over three thousand type specimens (table 1). The collections are global in scope, though with certain areas better represented than others due to the research interests of individual curators and their students. They continue to grow by approximately 6,700 specimens per year.[2] They remain of vital importance to biodiversity researchers at U-M and at other institutions, both nationally and internationally, as exemplified by activities such as loans, research visits, and online database searches. Indeed, as interests in global biodiversity, climate change, and long-term evolutionary and ecological processes continue to grow, so too does the research importance of UMMZ's collections, built over nearly two centuries of scientific research. And, as in the university's first zoology offerings in the mid-nineteenth century, the collections continue to serve an important educational function to U-M classes, visitors from K–12 classes, and the local community.

All divisions of the museum are participating in ongoing digitization efforts to make collection information available to the public and to

Table 1. Size, composition, and areas of taxonomic and geographic concentration of the UMMZ collections (data from 2015).

Collection	Size	Types	Composition/Strengths
Birds	210,000 specimens	320	All geographic regions represented, with holdings of 2/3 of the species in the world
Fishes	3.5 million specimens	624	Rich in material from U.S., Mexico, Asia, Japan, Neotropics, and Africa
Insects	4 million insects and 2 million Acari	1,249	Worldwide in scope; outstanding collections of Acari, Orthoptera, and Odonata
Mammals	150,000 specimens	130	Representation is broad and deep, including largest Peromyscus collection
Mollusks	4 million specimens	400	One of the largest collections of freshwater mollusks and land snails in North America
Reptiles and Amphibians	433,354 specimens	472	Second largest collection in world; geographic strengths in North America and Indian Ocean area

research. These efforts began in 1995 with the creation of the Animal Diversity Web, an online database of animal natural history, distribution, and classification (http://animaldiversity.org/about) led by UMMZ curator Phil Myers. UMMZ research and collections information are also being disseminated on a number of national and international databases,[3] opening possibilities for new kinds of "big data" driven comparative research unimagined when Alexander Ruthven was envisioning the work of the Museum of Zoology more than a century ago.

CURRENT RESEARCH

The research foci of UMMZ's eight curators vary both taxonomically and geographically and serve as snapshots of the kinds of engaged biodiversity research that have taken place since UMMZ's inception.

Reptiles and Amphibians curator Rabosky studies fundamental aspects of among-lineage differences in species richness and phenotypic diversity, especially in vertebrates. He has developed both empirical and theoretical lines of research. His main taxonomic focus is Australian lizard and snake fauna and he has developed extensive collections from his fieldwork.

In the Insect Division, O'Connor specializes in Acari (mites) evolution, systematics, and taxonomy. Over the past three decades, he has built one of the world's great mite research collections, describing many new species and training a new generation of acarologists. Curator Knowles is a global leader in the development of probabilistic methodologies for phylogenetic and phylogeographic analyses; her main taxonomic focus is Orthoptera (grasshoppers and crickets). She cohosted a NSF/UMMZ-funded Species Tree Workshop in January 2009 that attracted 170 participants interested in new approaches for estimating species trees and that resulted in the 2010 book, *Estimating Species Trees: Practical and Theoretical Aspects.*

Mammals curator Tucker studies mammalian evolutionary pattern and process. She contributed to early studies of mammalian Y chromosome evolution and pioneered the use of Y-linked DNA sequences to trace paternal ancestry in phylogenetic reconstructions. More recently, she has studied the genetic underpinnings of the species boundary, conducting a genome-wide investigation of admixture in a European hybrid zone between two closely related species of house mice.

In the Mollusk Division, curator Ó Foighil's background is in marine mollusk evolution and systematics and his research has continued in this vein, and also branched out to work on freshwater and terrestrial taxa, in association with Mollusk

Fig. 96. Dwarf bearded dragon, Pogona minor, collected using a pit trap at Lorna Glen in Western Australia in April of 2015 by Dan Rabosky, Alsion Davis-Rabosky, Pascal Title and Greg Schneider. Photo by Pascal Title.

curator John Burch and his students. In particular, Ó Foighil used Burch's samples of now-extirpated partulid tree snail populations to study the evolutionary history, systematics, and conservation biology of this endangered Pacific Island radiation. Curator Duda is an innovative marine evolutionary biologist studying predatory cone snails in the genus *Conus*. He studies the *Conus* radiation on many levels, from a macroevolutionary perspective across the radiation to within-species analyses that link evolution at specific genes (paralyzing conotoxins) to organismal ecological performance; Duda has also assembled a world-class collection of *Conus* species.

In the Bird Division, Ben Winger, fellow in the Michigan Society of Fellows (2015–18), has recently been appointed as assistant curator of Birds. He studies seasonal migration in songbirds and the biogeographic history of speciation in Andean birds. Alison Davis Rabosky joined the Division of Reptiles and Amphibians as Assistant Curator in 2016. She studies color polymorphism and phenotype evolution of sociality, and population structure in island populations of reptiles.

Entry prepared by Priscilla Tucker.

Research collections can be visited by appointment. Contact information available at http://www.lsa.umich.edu/ummz

Notes

1. The Department of Zoology was placed under the "Division of Biological Sciences" in June 1975; the division was renamed Department of Biology in February 1986 (*Proceedings of the Regents of the University of Michigan, 1972–75*, 1315–16; *Proceedings of the Regents of the University of Michigan, 1884–87*, 537–38).
2. Data averaged over the past 4.5 years.
3. For example, IDigBio (https://www.idigbio.org), gBif (Global Biodiversity Information Facility, http://www.gbif.org), VertNet (http://vertnet.org/).

REFERENCES

Hubbell, T. H., and Irving J. Cantrall. "A New Species of Appalachia from Michigan (Orthoptera, Acrididae, Cyrtacanthracridinae)." *Occasional Papers of the Museum of Zoology*, No. 389. Ann Arbor: University of Michigan (1938).

Ruthven, Alexander. "The Amphibians of the University of Michigan-Walker Expedition to British Guiana." *Occasional Papers of the Museum of Zoology*, No. 69. Ann Arbor: University of Michigan (1919).

MUSEUM OF PALEONTOLOGY

1928–2017

The University of Michigan Museum of Paleontology (UMMP) is a research museum committed to the study of life's history and the factors that have interacted to influence its broad patterns. The museum's research often begins with the study of fossils (the remains or traces of ancient organisms), but it inevitably broadens to consider their environmental context and their relations to other organisms, both those of their own time and those that lived before and after. Paleontology is one of the most interdisciplinary of the natural sciences, not only because its subject matter is at once biological and geological, but also because its methods of investigation are frequently a complex mix of both of these fields as well as physics, chemistry, and other natural sciences, augmented by methods drawn from statistics, mathematics, and computer science. Life is surely the most complex of natural phenomena to attract the attention of science, and life's history poses the additional challenge of probing "deep time" and integrating information about biotas and habitats that are in some cases profoundly different from those we observe today.

The history of UMMP can be traced back to the earliest decades of the University of Michigan. Its early collections grew through a combination of systematic exploration by Michigan's first state geologists and more opportunistic acquisitions of notable specimens from individuals with various ties to the state or university. The impetus behind the development of collections was originally pedagogical, serving to enrich the teaching program and public profile of the university. As both the paleontology collections and the research emphasis of the university grew, so too did expectations that faculty would contribute to their fields through scholarly research and related activities. The collections thus began to serve as an archive

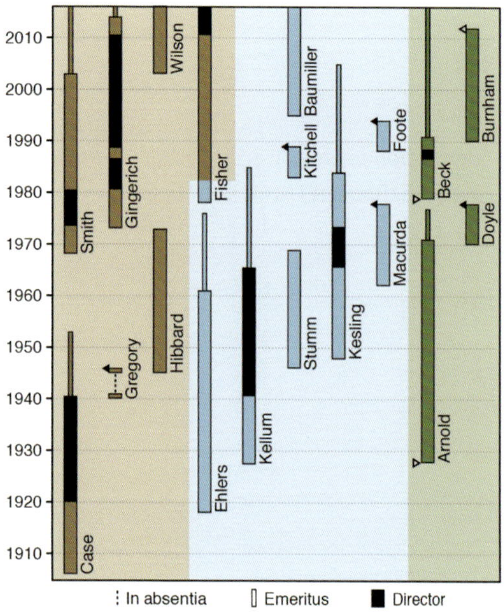

Fig. 97. Tenure-track faculty of the Museum of Paleontology, from E. C. Case onward. Faculty were added (left to right on diagram) in vertebrate paleontology (brown region on left), invertebrate paleontology (blue region in center), and paleobotany (green region at right). Vertical succession does not imply continuity of research area. Open arrows mark moves into or out of UMMP from or to other U-M departments. Closed arrows mark departure from U-M.

of research results and materials that would seed the development of new scholarly projects.

By the end of the twentieth century, UMMP had one of the foremost (and largest) university-related paleontological collections in the United States. In recent decades, its program of graduate education in paleontology has been ranked consistently in the nation's top four, and UMMP is especially known for offering abundant opportunities for field work and for maintaining a scholarly tradition that balances innovations in theory with

Fig. 98. *Dorudon atrox*, a primitive whale from the Eocene of Egypt, collected by Prof. Gingerich and his team. Composite of UMMP 101215 and 101222.

a dedication to empirical studies of organismal form, function, and evolutionary pattern. With university teaching and research in paleontology still our primary goals, we leave to our sister unit, the Museum of Natural History, all exhibit programs open to the public, although we often collaborate with them in developing these exhibits. To develop greater synergies between our teaching and research programs, we are working constantly to improve management of our collections and are now breaking new ground in providing digital access to high-resolution, photorealistic 3D models of paleontological specimens.

HISTORICAL OVERVIEW

Foundations: 1837–1928

The University of Michigan's engagement in paleontology can be traced back to the first meeting of its Board of Regents in 1837 and the creation of a Cabinet of Natural History. The collections were initially included with other natural history collections, including mineralogy and geology, as part of the University Museum. As recounted by Lewis B. Kellum, the first fossils were collected by Douglass Houghton on Isle Royale.[1] Houghton had been named the first state geologist and was one of the first faculty members appointed by the university. When the state legislature created the Michigan Geological Survey in 1838, it explicitly decreed that specimens not required by the state should be deposited with the university (see Caple and Williams, this volume).

Following Houghton's death in 1845, Abram Sager, professor of zoology and botany, took over control of the paleontological collection until 1855, when Alexander Winchell was appointed as professor of geology, the first to occupy this position since Houghton. As curator of the University Museum, Winchell oversaw all collections and all teaching in paleontology. Picking up the work that Houghton had begun, Winchell was also named state geologist and directed the Michigan Geological Survey from 1859 to 1861 and, following the Civil War, from 1869 to 1871. In addition to more technical research contributions, he became well known for many popular books such as *Sketches of Creation* (1870). In 1873, Winchell left Ann Arbor to become the first chancellor of Syracuse University, but his tenure there was short-lived due to economic difficulties unrelated to his leadership.[2] After teaching briefly at Vanderbilt, he returned to the University of Michigan in 1878, taking up his former position as professor of geology and paleontology. He also continued his writing, producing works such as *World-Life* (1883), described by Kellum as "at that time probably the only book in the English language covering in a systematic manner the entire field of earth history."[3] Winchell initiated the practice of specimen catalogs, which became critical as the collection grew. On the national level, following his return to Michigan, Winchell chaired the committee that set up the Geological Society of America and served as its president in 1891, the same year in which he died.

During the interval between Winchell's two

Fig. 99. *Placenticeras*, a Cretaceous ammonite with tooth-marks of a mosasaur (marine reptile), showing interaction between an ancient predator and its prey. UM 35484.

appointments at U-M, the paleontology collection was briefly overseen by professors of geology Eugene W. Hilgard (1873–75) and William H. Pettee (1875–76) and assistant professor of paleontology and curator of the University Museum, Joseph B. Steere (1876–79). With Winchell's return to his position as professor of geology and paleontology, Steere was promoted to professor of zoology.[4] After Winchell's death, teaching duties in paleontology were taken over by Israel Russell (1892–1906), professor of geology, followed by Alexander Ruthven (1906–13) (see Miller, this volume).

Another significant influence on the early history of paleontology at U-M was Carl Ludwig Rominger. His appointment was only as assistant curator and only for two years (1864–66), during Winchell's first term of service. However, during this time, he made a series of systematic collections sampling the fossils of Paleozoic units throughout the Great Lakes region, setting a standard for comprehensive representation in the paleontological collection. Later, Rominger made available (and eventually sold) to the university a large collection of European fossils that again boosted the profile and value of the collection. Rominger served as state geologist from 1870 to 1884, and until his death in 1907 deposited specimens in the paleontological collection.

Both Winchell and Rominger had studied the full diversity of fossil organisms available to them, but their special expertise had been in studies of marine invertebrates, groups well represented in Michigan's Paleozoic strata. A new direction was thus set with the appointment of assistant professor of historical geology and paleontology, Ermine Cowles Case, in 1907. Case was a vertebrate paleontologist and refocused the museum's fieldwork on rocks of late Paleozoic and Mesozoic age, located in western and southwestern parts of the country. Case was in the field virtually every summer, greatly expanding both the research collection and the diversity of paleontological specimens exhibited in the University Museum.

The paleontological and geological collections were granted their own space (though even then it was called the Museum of Geology) for the first time in 1915 when the Department of Geology became one of the original occupants of the newly opened Kraus Natural Science Building.

Together with specimen preparator William Buettner, Case (promoted to professor of historical geology and paleontology in 1912) (see Gingerich, this volume) oversaw the move of the collections into Kraus. In 1919, invertebrate paleontologist George M. Ehlers was hired, reestablishing representation in that subfield. The Museum of Geology was given its own budget, separate from that of the Department of Geology, in 1926–27, by which time the move of its paleontological collections into what would become the Ruthven Museums Building was already planned.[5] In 1928, when the transfer of the collections into the new Museums Building was complete, the name of the unit was changed to Museum of Paleontology, bringing the museum to essentially its present identity, configuration, and location.

Fig. 100. *Quercus columbiana*, leaf of a Miocene oak from Idaho. UMMP 73739.

The First Decades: 1928–46

In conjunction with the move into the new Museums Building, Lewis B. Kellum joined the Museum of Paleontology as curator of Mesozoic and Cenozoic invertebrates, and just a year later, in 1929, Chester A. Arnold was added as curator of paleobotany. Museum collections now grew through fieldwork in all three of these subdisciplines, as well as from transfers within the university (e.g., the Ford-Mitchell collection of dental and cranial specimens from the School of Dentistry, 1928)[6] and outside acquisitions.

Largely through Case's initiative, UMMP established itself as a source of published research with the 1924 launch of the series *Contributions from the Museum of Geology*. In 1928, following the change in the museum's name, this series became *Contributions from the Museum of Paleontology*. It is still published today along with the separate series, *Papers on Paleontology*, for monographic works.

Case retired in 1941 (though he remained active in research for an additional decade as professor emeritus), and Kellum became director. At the same time, the Board of Regents converted the Museum of Paleontology from an entity within the Department of Geology to an independent unit reporting directly to the president of the university (as was already the case for other museums).[7] In conjunction with this shift, UMMP curators who were contributing to the teaching mission of the university were reappointed half-time in a department and half-time in the museum, reflecting their mix of instructional and curatorial duties. Ehlers and Kellum's departmental appointments were in the Department of Geology, but Arnold's departmental half was in the Department of Botany. To rebuild vertebrate paleontology after Case's retirement, Joseph T. Gregory was hired with an appointment split between the Museum of Paleontology and the Department of Geology. Gregory's interests paralleled those of Case in many respects, and he had great promise, but within a year, the United States entered World War II, and Gregory joined the Air Force. When the war was over, he briefly returned to Michigan but then continued his career at Yale University and the University of California, Berkeley.

Growth and Return to the College: 1946–56

Following Gregory's unexpected departure, Claude W. Hibbard, a Michigan PhD (1941, zoology) who had quickly established an active program at the University of Kansas, was hired as assistant professor of geology and associate curator of vertebrate paleontology. Hibbard focused most of his research effort on late Cenozoic mammalian faunas of the Great Plains but regularly dealt as well with Pleistocene vertebrates that turned up in the Great Lakes region. Also in 1946,

Erwin C. Stumm was added as associate curator of invertebrates. In a break from the previous pattern of split appointments, Stumm was hired full-time in the Museum of Paleontology. By 1949, the petroleum industry was developing and with it the prominence of micropaleontology, leading to Robert V. Kesling's appointment, which was split between the Museum of Paleontology and the Department of Geology.

During Case's career and Hibbard's first year, the Hall of Evolution in the Ruthven Museums Building had been overseen by the Museum of Paleontology. In 1947, this responsibility was transferred to Irving G. Reimann, who oversaw the exhibits department in the University Museums;[8] as noted above, this arrangement continues today with the Museum of Natural History as the direct descendant of the earlier exhibits department. Another change came almost a decade later, in 1956, when Dean Charles E. Odegaard successfully pressed for the budget of the Museum of Paleontology (along with other natural science museums) to revert from control of the president to the College of Literature, Science and the Arts (see Sinopoli, "Museum Building/s" this volume).

Finding Strength beyond Our Numbers: 1956–2017

In the late 1950s, the number of curators in the Museum of Paleontology grew to six (see fig. 97 for details) under the directorship of Kellum, when Stumm and Kesling were added. Ehlers's retirement was balanced by Donald B. Macurda's arrival, but within a few years Kellum retired and Stumm died, reducing curatorial numbers to four. Although Stumm's line was not reauthorized, Gerald R. Smith, who studied both fossil and extant fish, was added in 1969. Smith's appointment was split between the Museum of Paleontology and the Museum of Zoology, but he nonetheless offered courses in both the geology and biology departments.

Meanwhile, institutional strength in paleontology also developed in sister units. A vertebrate paleontologist, John A. Dorr Jr. (1952–86), was appointed full-time in what had become the Department of Geology and Mineralogy (later, Geological Sciences). Likewise, Frank H. T. Rhodes, a specialist on conodonts (now recognized as primitive chordates), was added to the Department of Geology and Mineralogy in 1967. He quickly rose into administrative roles as dean of the college in 1971 and vice president for academic affairs in 1974 before leaving to assume the presidency of Cornell University in 1977. Since 1956, Charles B. Beck, working on early seed plants, had been in the Department of Botany (later, Biology) prior to joining the Museum of Paleontology following the departure of James A. Doyle. In 1982, Catherine E. Badgley, working on vertebrate taphonomy and community paleoecology, was appointed as visiting assistant professor in geological sciences through the Michigan Society of Fellows. At the conclusion of this three-year term, she became a research scientist in the Museum of Paleontology and subsequently rose to faculty positions in the Residential College and the Department of Ecology and Evolutionary Biology. Following in her footsteps, a number of additional young faculty members have enriched our program via the Michigan Society of Fellows.

Today, the museum has four curatorial lines. Philip Gingerich retired in 2014 after four decades of accomplishment in elucidating the origins and early history of several mammalian groups, most notably primates and whales, and contributing even more broadly to studies of evolutionary rates and patterns of morphologic change. In his place, UMMP has been joined by Matt Friedman, who is making fundamental contributions to our understanding of the early radiation of vertebrates and the relationships of major groups of boney fishes. Jeffrey Wilson is known for unique insights into the anatomy, functional morphology, and biogeography of many reptilian groups, and especially the largest of terrestrial vertebrates, the sauropods. Tomasz Baumiller is a pioneering investigator of biotic interactions and biomechanics, dealing especially with echinoderms, but equally conversant with quantitative models of taphonomic histories and macroevolutionary patterns. Daniel Fisher committed his early efforts to elucidating

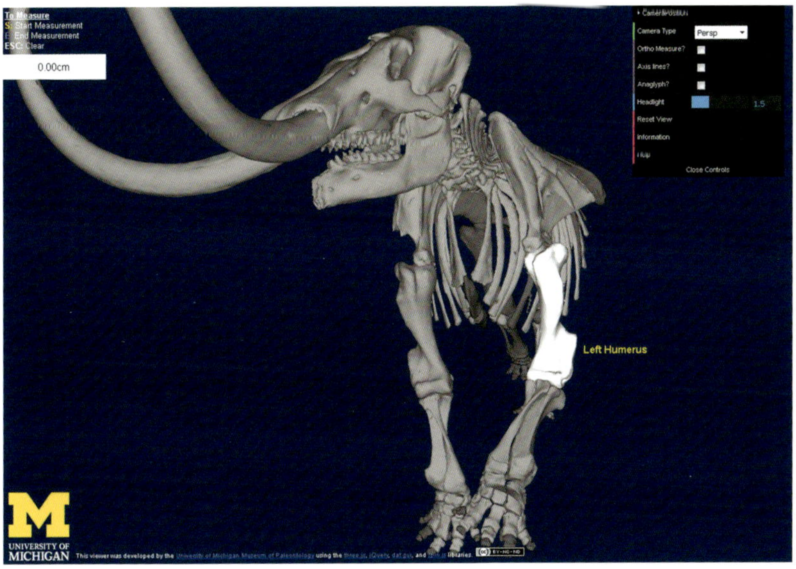

Fig. 101. BonePicker module of the UMORF website, showing the Buesching mastodon skeleton. Mousing over a bone identifies it; double-clicking opens a 3D model of that bone.

the paleobiology of primitive arthropods and echinoderms, but he soon started to explore the role of temporal data in phylogenetic analysis, the paleobiology of Pleistocene proboscideans, and the causes of their extinction. Both individually and as a group, our research portfolio has broadened since the first decades of our unit's history—we now address a wider range of problems with a more diverse repertoire of methods.[9] We also maintain an active outreach program through an organization of avocational paleontologists (Friends of the UMMP), started by Kesling over forty years ago and still growing. For all this, our greatest pride is the many students we have trained who are now serving our profession in diverse ways.

ORGANIZATION OF THE COLLECTION

The collection of the Museum of Paleontology has grown substantially throughout its history. In the early years, the pedagogical priority drove a conscious effort to broaden its scope, even though no single museum could realistically aspire to assemble a comprehensive archive of the history of life. Over time, emphasis shifted to focusing serially on areas of investigation that excited the research interests of individual curators. This shift coincided with a major transformation in the field of paleontology, diminishing the relative emphasis on descriptive studies serving taxonomic and biostratigraphic goals and promoting instead paleoecology, functional morphology, biogeochemistry, phylogeny, and investigation of micro- and macroevolutionary patterns. These kinds of studies require more intensive collecting of larger samples from well-controlled stratigraphic and environmental contexts. The best of these efforts still yield new fruit, as subsequent generations of paleontologists return to refine or extend the work of their forebears.

UMMP has tried to maintain expertise in vertebrates, invertebrates, and plants, with the exact balance shifting according to the activities and identities of our staff and students. Within each of these three broad divisions is a "type collection" (specimens used in descriptions of new taxa), a "systematic collection" (specimens organized according to the taxonomic hierarchy), a "stratigraphic collection" (assemblages of co-occurring material preserving associations of ecological significance), and a "teaching collection" (specimens for routine student use). Summaries of specific strengths of the collection can be found on our website.

Winchell's early handwritten catalogs survive today only in part, having been long ago superseded by voluminous card files. Now those too have mostly given way to various database applications. In concert with the U-M Museum of Zoology and the Herbarium, UMMP is currently engaged in a major effort to import all our collection data into an open-source collection management database system known as Specify. This software is supported by the US National Science Foundation and a broad community of client institutions, and it should handle our needs well. Dissemination of collection information through the World Wide Web and global data networks will allow our collections to be accessed by scholars and the public worldwide.

In 2014, UMMP launched another initiative to make the collections more accessible and useful to students, researchers, and the public—the University of Michigan Online Repository of Fossils.[10] Paleontologists depend on an intimate understanding of complex three-dimensional forms, and yet most of the media available for portraying these configurations are decidedly two-dimensional. To capture data on 3D shape, we use point digitizers, scanning digitizers, and photogrammetry, but to make use of these data, we developed a web-based interface that allows 3D models to be displayed and interactively manipulated on almost any desktop or mobile platform. Starting with specimens that would attract broad interest, we have expanded coverage to represent more of the ongoing work of the museum. We hope to extend this to our teaching collections, our type collections, and representative material from all new research. For those interested in our history, we expect this to be an exciting new chapter.

Another new development for our collection is its move to new quarters in the Research Museums Center. This will provide better climate control and updated laboratory space in conjunction with similar facilities for other natural science collections. This move precedes the shift of our primary office and laboratory space into the new Biological Sciences Building now taking shape just north of the Ruthven Museums Building. Collectively, these events will launch a new era for paleontology at the University of Michigan, but one that we trust will continue to recognize the deep fascination we all share for the remains of life's past and the insight into processes that will also control its future.

Entry prepared by Daniel C. Fisher.

Research collections can be visited by appointment. Email: Paleo-Museum@umich.edu

http://www.lsa.umich.edu/paleontology

Notes

1. Kellum, 1487.
2. Winchell Papers.
3. Kellum, 1488.
4. Ibid., 1489.
5. Alexander Ruthven, April 17, 1925, letter to Professor E. C. Case, Museums Papers, Bentley Historical Library. While the paleontological collections moved into the new museum building, the geological collections remained in the Kraus Building under the auspices of the Department of Geology.
6. The human crania in this collection were subsequently transferred to the Museum of Anthropology in 1996.
7. Kesling, 1.
8. Kellum, 1501. See also entry on Museum of Natural History, this volume.
9. Summaries of current research projects are available at http://www.lsa.umich.edu/paleontology
10. https://umorf.ummp.lsa.umich.edu/wp

References

Kellum, Lewis B. "The Museum of Paleontology." In *The University of Michigan: An Encyclopedic Survey, Part VIII, Libraries, etc.*, edited by Walter A. Donnely, Wilfred B. Shaw, and Ruth W. Gjelsness, 1487–1502. Ann Arbor: University of Michigan Press, 1956.

Kesling, Robert V. "History of the Museum of Paleontology, 1940–1975." Ann Arbor: University of Michigan, 1975. (PDF available at http://www.lsa.umich.edu/paleontology/about/history)

Winchell, Alexander. *Sketches of Creation: A Popular View of Some of the Grand Conclusions of the Sciences in Reference to the History of Matter and of Life*. New York: Harper & Brothers, 1870.

Winchell, Alexander. *World-Life: Or, Comparative Geology*. Chicago: S. C. Griggs, 1883.

Winchell Papers. Syracuse University Archives. Accessed January 2016 (available at http://archives.syr.edu/collections/chancellors/sua_winchell_a.htm)

HERBARIUM

1921–2017

The University of Michigan Herbarium (UMH) is the third-largest university herbarium in North America. It provides stewardship of 1.75 million plant and fungal specimens, including twenty-three thousand type specimens (the designated specimens used in describing new taxa).[1] UMH has globally important collections of lichens, bryophytes (mosses), ferns, marine algae, fungi, and seed plants. It contains complete representation of the Michigan flora and unparalleled collections of taxa from tropical America and elsewhere.

BUILDING OF THE COLLECTIONS, 1837–81

Provisions for the first "collection of specimens" of the University of Michigan were made in 1837, months after the state of Michigan and the U-M Ann Arbor campus were established. Following Regent Henry Schoolcraft's request to build museum cabinets on the new campus, the university received specimens from about nine hundred plant species collected on the Geological Survey led by Douglass Houghton, beginning in 1837 (see Caple and Williams, this volume).[2]

The regents signaled their commitment to botany by hiring Asa Gray as the first U-M professor in 1838, although ultimately Gray's contributions were only symbolic. Because he was hired before the campus had buildings or students, Gray was granted a paid leave of absence in 1839 and $5,000 to acquire books in Europe for the U-M library. In May 1842, with the campus still under construction, Gray moved to Harvard, where he became the most renowned botanist in American history.

Abram Sager, botanist on the Geological Survey, was appointed Gray's successor as professor of botany and zoology in 1842, a position he accepted without a regular salary until he began teaching the university's first courses in botany in 1845.[3] Although Sager resigned in 1855 to become professor of obstetrics and physiology, his interest in plants continued. In 1866, he donated his private collection of twelve hundred species to the university.[4]

Sager's replacement, Alexander Winchell (1855–73), had transferred from physics and civil engineering. As professor of geology, zoology, and botany and curator of the University Museum, Winchell improved the administration of the museum's collections and incorporated specimen-based laboratory sessions into his botany courses.[5]

Mark W. Harrington and Joseph Beal Steere were promising undergraduate students who worked in the museum under Winchell in 1868. Following his graduation, Steere traveled to the Amazon basin, Peruvian Andes, and South East Asia (1870–75) (see Barndt, this volume) where he collected 1,156 botanical specimens (mainly mosses and ferns). Harrington, appointed assistant curator of plants in 1869, was named assistant professor of geology, zoology, and botany in 1873—and charged with expanding the Herbarium. In 1876, he visited the Royal Botanical Gardens in Kew to complete the identification of Steere's fern collections, resulting in the university's first botanical publication.[6] Volney M. Spalding, another Winchell student, assumed Harrington's duties while Harrington was on leave in Europe and China, and succeeded him in 1877.[7]

INSTITUTIONALIZING THE COLLECTIONS, 1881–1920

In 1881, the regents decided that the professor of a subject should be the curator of the associated collections in the University Museum. The first faculty-curators in the Department of Botany were plant ecologist Volney M. Spalding (1881–1904); Charles Albert Davis (1905–1908), an instructor in forestry with an interest in Michigan flora; and Forest Buffen H. Brown (1908–12), who studied floodplain vegetation in Ypsilanti, Michigan.

By 1912, the growth of the collections required some divisions of space and labor along taxonomic lines. The first division separated cryptogams (spore-producing ferns, fungi, mosses, lichens, and algae) from phanerogams (seed plants). Calvin Henry Kauffman (1912–21) became the first curator of the Cryptogamic Herbarium and Henry Allan Gleason (1912–19) the first curator of the Phanerogamic Herbarium. Both men made lasting contributions. Kauffman's 1918 publication, *Agaricaceae of Michigan*, was considered the most comprehensive treatment of North American Agaricaceae species to date. Gleason's alternative to Fredric Clements's 1916 climax theory of succession—later known as the individualist theory of plant succession—remains a fundamental idea in plant ecology.[8]

The Museum of Zoology also curated botanical specimens, creating its own herbarium in 1912. Its specimens came from the Michigan Biological Survey that had been initiated in 1904 and was led by Alexander Ruthven (see Miller, this volume). Charles Keene Dodge, who had been active in the botanical phase of the survey, contributed five thousand plant specimens and was given the nonsalaried title of associate curator of botany in the Museum of Zoology.[9] After his death in 1918, the university received an additional thirty-five thousand specimens from Dodge's estate.

Fig. 102. One of 80,000 UMH marine algae specimen that was digitized as part of an NSF initiative (www.macroalgae.org).

1913 FIRE

In 1913, the main plant collections were housed in the Botanical Laboratory on the fourth floor of the south wing of historic University Hall, with some collections stored in bundles in the attic. A building fire—at a time when water lines on campus had been closed off for repair—destroyed all the bundled specimens and damaged many of the fourth floor specimens. Because there were no specimen catalogs at that time, it is not known exactly how many specimens were lost. Historical specimens in UMH that exhibit water or smoke stains serve as a reminder of this disastrous event.

UNIVERSITY OF MICHIGAN HERBARIUM, 1921–2017

In 1921, the Phanerogam, Cryptogam, and Museum of Zoology herbaria were united into the newly established University of Michigan Herbarium. UMH had its own budget and its director reported to the president. Thus it had the administrative rank of a school or college within the university. This changed in 1956 when UMH was incorporated into the College of Literature, Science and the Arts (LSA), taking on a rank akin to a college department. In 2011, UMH became a unit within the Department of Ecology and Evolutionary Biology.

Since their beginnings, the Herbarium's collections have had several homes. Following their initial placement in a professor's house in the 1830s, the collections were moved to Mason Hall (1841), the south wing of University Hall (1891), and the Natural Science (Kraus) Building (1915). In 1928, the Herbarium relocated to the University Museums Building. Due to its burgeoning growth, substantial parts of the collections were moved to the Museum Annex in 1947, but space was still insufficient. With support from the National Science Foundation, the Herbarium moved to the North University Building in 1960, where it remained until the building was razed in 2002.

As a temporary provision in 2001, LSA moved the Herbarium to a warehouse on Varsity Drive, about six miles south of campus. By 2010, it became clear that this location would be permanent and the other natural history collections—the Museums of Zoology, Paleontology, and Anthropological Archaeology—are joining UMH in the renovated facility, renamed the Research Museums Center (RMC). Wih this move, the natural history research museums will be reunited for the first time in seventy years, though in an off-campus location isolated from daily interactions on central campus.

THE BARTLETT COLLECTIONS

An important contributor to the explosive growth of the collections was Harley Bartlett, an unsalaried curator and director of the Matthaei Botanical Garden (see Michener and Reznicek, this volume). In 1957, Bartlett donated his personal collection of nearly five hundred thousand specimens to UMH, almost doubling the size of the collection and changing "the Herbarium from a medium-sized and mostly locally oriented institution to a large and more diversified one, with interests in many parts of the world."[10] Included in the Bartlett donation were the important Asian collections of Walter Koelz and Rup Chand (see Sinopoli, this volume). The full assimilation of the Bartlett material was completed only in 1990.

Transfer to LSA Administration

The Herbarium's transfer to LSA in 1956 provided greater accessibility to graduate students.[11] Between 1950 and 1970, approximately one-third of all PhD graduates in botany had a UMH faculty-curator as a major advisor. The Herbarium's curators (who had joint faculty appointments in the Department of Botany) provided first-rate graduate training in plant systematics. In 1974, UMH was ranked one of the top three herbaria in the United States, alongside the New York and Missouri Botanical Gardens.[12]

RESEARCH AND COLLECTIONS

The Herbarium's collections are organized into five collection types: fungi, algae, bryophytes and lichens, ferns, and vascular plants. Of these, vascular plants count for by far the largest portion of the collection (1.1 million of 1.7 million specimens). At different periods, each of these groups has had associated faculty-curators.[13] All areas of the collections are used in ongoing research projects and are of tremendous scientific importance.

Fungi (280,000 Specimens)

The world-class fungal collections of UMH are due in large part to U-M's leadership in the field of mycology. Notably, mycologists directed the Herbarium for its first seventy-five years.[14] Calvin Kauffman's work (director from 1921–30) built up collections of mushrooms of the Great Lakes region, but he also collected across North America and internationally. Kauffman's student Edwin Mains (director from 1930–59) studied North American rust fungi (Pucciniomycetes) and earth tongues (Geoglossaceae). In collaboration with C. L. Lundell, Mains also documented fungi from rain forest in British Honduras (now Belize). Bessie Benice Kanouse was appointed as curator of Fungi in 1926 and worked in close association

Fig. 103. Students work on backlog of 30,000 unmounted plant specimens and learn botany fundamentals.

with Mains. Alexander Smith (director from 1959–72), a student of Kauffman and Mains, was a leading expert in mushrooms, puffballs, and boletes. Smith contributed an astounding ninety-two thousand specimens to UMH and wrote the popular *Mushroom Hunter's Field Guide*, which has sold over one hundred thousand copies since its first publication in 1958.

Robert Lynn Shaffer joined Smith as curator of Fungi (1960–93) and directed UMH from 1975–86. He was a leading expert in the systematics of *Russula*, a large and difficult genus of fleshy fungi. Robert Fogel (curator of Fungi, 1978–2003) studied truffles and initiated an online catalog of UMH fungal specimens. Presently, Curator Timothy James (since 2009) holds the Wehmeyer Chair in the Taxonomy of Fungi. In addition to his research on parasitic fungi, fungal genomics, and the phylogeny of basal fungal lineages, James oversees specimen digitization, teaching, and public outreach regarding fungal biology. Collection manager Patricia Rogers, who also manages lichens and bryophyte collections, assists James with fungal specimen curation.

Algae (96,000 Specimens)

The first curator of Algae (nonsalaried) was William Randolph Taylor (1930–66). Taylor studied marine algae from the Pacific and Atlantic basins. He was part of the "Crossroads" atomic project of the US Navy, which performed biological surveys before and following nuclear bomb testing. UMH maintains these collections, including a type specimen collected in 1946 on Elugelab before the island was bombed back into the sea. Following his death, a colleague wrote, "Taylor's death marks the end of an era when a single individual could be an authority of entire phyla of organisms from all parts of the world."[15] Michael Wynne was appointed curator of Algae in 1978 (until 2007). During his career, Wynne described a total of twenty-nine new genera and ninety-two new species of marine algae from around the world. Wynne recently participated in an NSF digitization project to publish eighty thousand UMH macroalgal specimen images and data online.[16]

Bryophytes (163,000 Specimens) and Lichens (57,000 Specimens)

A major boost to the lichen collection was the 1929 purchase, with financial help from Dr. Howard A. Kelley, of the lichen library and herbarium of Professor Bruce Fink of Miami University. Fink's herbarium was rich in types and historical collections. William Campbell Steere, the grandson of Joseph Beal Steere, became first curator of Bryophytes (1945–50) and is credited with transforming bryology from an amateur hobby to a biological profession in North America. Howard Alvin Crum, a doctoral and postdoctoral student with Steere, was appointed curator of Bryophytes and Lichens from 1965 to 1995; his crowning achievement was the two-volume *Mosses of Eastern North America* (1981) written with Lewis E. Anderson.

Ferns (90,000 Specimens)

UMH has an exceptional worldwide collection of ferns dating to the 1870s Steere collections. Warren Herbert Wagner Jr. was appointed curator of Pteridophytes (nonsalaried) in 1961. Although his academic home was the Department of Botany, Wagner was an important ally of UMH, bringing it attention, graduate students, and collections. Wagner's greatest contribution to the Herbarium was his role in acquiring the fern collection of Professor E. B. Copeland, which contained an inordinate number of type specimens; he was also instrumental in obtaining the collections of the American Fern Society. His wife, Florence Wagner, a research scientist in the Herbarium, studies fern evolution through the lens of cytology.

Vascular Plants (1,100,000 Specimens)

The vascular plant collection is the largest in the Herbarium, and the focus of most of our curators, research scientists, and collection managers. John Henry Ehlers, a botany instructor with interest in Michigan flora, was named curator of Phanerogams from 1916 to 1939. In 1934, U-M established an agreement with the Carnegie Institute to perform a biological survey of the Mayan area of Central America. As a result, UMH hired Cyrus Longworth Lundell in 1934. In addition to botany, Lundell had interests in Mayan culture and archeology. During an aerial survey in 1931, he had discovered a lost Mayan city, which he named Calakmul; it was the first of sixteen Mayan sites he would discover. Lundell named 450 plant species from Guatemala, including ancestors of squash, cacao, pinto beans, and other crop plants. Lundell succeeded Ehlers in 1939 (until 1944) as curator of Phanerogams and Ferns.

Elzada Clover was another colorful historical figure associated with UMH. Having received her PhD from U-M Department of Botany, Clover was appointed assistant curator at the Botanical Gardens and assistant professor in botany in 1935. She was among the first botanists to catalog plants in the Grand Canyon along the Colorado River—and she and U-M teaching assistant Lois Jotter were the first women to raft the entire length of the Grand Canyon. Clover's cacti collections, which include type specimens, were an important contribution to UMH.

Rogers McVaugh, appointed curator of Vascular Plants in 1946 (until 1979), was renowned for his work with neotropical plant groups, especially the flora of western Mexico. His collaboration with former student William R. Anderson culminated in the eight-volume series *Flora Novo-Galiciana* covering a large portion of western Mexico. Largely because of McVaugh's efforts, UMH has arguably the best collection of Mexican vascular plants outside Mexico.

Edward Groesbeck Voss, a McVaugh student, was hired in 1955 to work on the *Flora of Michigan*, on which he spent the next forty years. He was appointed as curator in Vascular Plants in 1960. In 1972, the Michigan senate passed a resolution to honor Voss for the publication of the first volume of *Michigan Flora*. The second volume was honored with the Henry Gleason Award from the New York Botanical Garden. The third (and final) volume was published in 1996.

William R. Anderson was appointed as associate curator of Vascular Plants in 1974 (until 2002) and served as UMH director from 1986 to 1999. Anderson specialized in the tropical family Malpighiaceae. His wife, Christiane Anderson, was appointed assistant research scientist in 1979 and worked on the same family. The Andersons' research and collections extended the strengths of UMH more firmly into tropical South America.

Anton A. Reznicek joined the Herbarium as an assistant curator in 1978 and is currently the sole full-time curatorial appointment in the unit. He is a sedge specialist and U-M's foremost expert on Michigan flora. Richard K. Rabeler joined the staff as an adjunct research investigator and collection manager in 1987. His research focuses on systematics of the Caryophyllaceae.

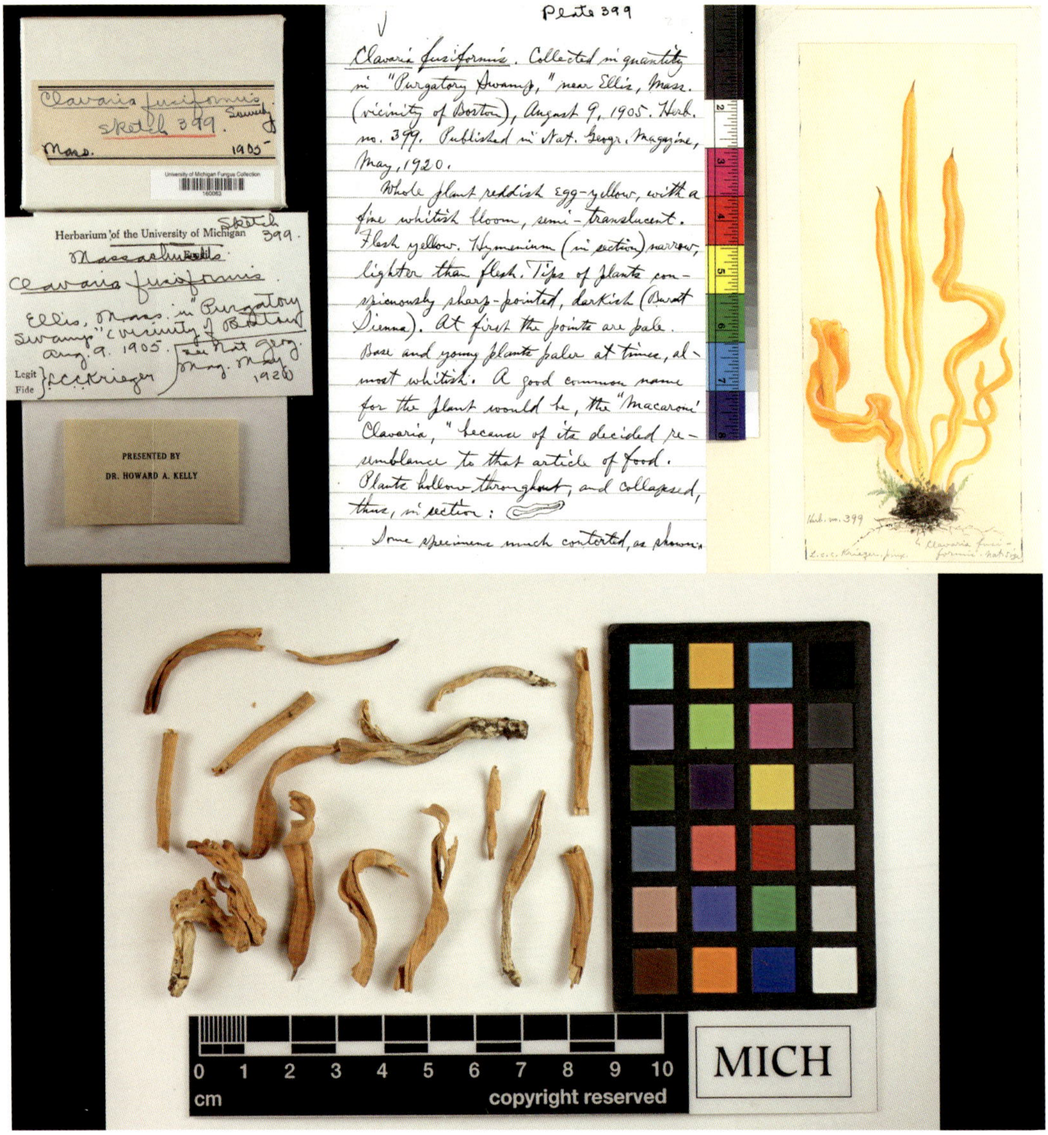

Fig. 104. This 1905 specimen of *Clavaris fusiformis* is one of over 200,000 digitized UMH fungal collections. This specimen was photographed beside its field notebook entry and an H. V. Kriegel painting of the fresh specimen.

In 2005, Christopher W. Dick was appointed assistant curator of Vascular Plants; he advances biogeographic research in Central and South American forests. Paul E. Berry was appointed as curator in 2006 and served as director from 2006 to 2015. Berry edited the eight-volume *Flora of the Venezuelan Guayana*, and his current research focuses on the hyper-diverse genera *Croton* and *Euphorbia*. Dick took over the role of director in 2015 and will likely be the last UMH director, given that UMH and UMMZ will merge under a single directorship in 2017.

MICHIGAN FLORA OUTREACH

Botanical clubs and volunteers from Ann Arbor have been closely associated with the U-M collections since their inception. Alexander Winchell, in his report on Michigan flora in 1861, acknowledged the contributions of Mary H. Clark, who had developed an herbarium associated with a seminary for young ladies. Mark Harrington recognized volunteer work of Ann Arbor citizens in the 1870s. The tradition of engaging the public in education and botanical exploration continues to this day. Anton Reznicek, while writing the *Manual of the Michigan Flora* with Ed Voss, established the Michigan Flora website,[17] which features continuously updated information on Michigan's nearly three thousand native and introduced vascular plant species. Reznicek and Collections Manager Beverly Walters have vetted thousands of plant photos for the site, and they have solicited thousands of specimens from citizen botanists that have extended range limit data for many Michigan species.

The current faculty and staff look forward to continuing the tradition of public engagement, stewardship, and primary research into the plants and fungi of Michigan and the Great Lakes region.

Fig. 105. 1914 specimen of balsam willow (*Salix pyrifolia*) from the herbarium of C. K. Dodge.

UMH IN THE TWENTY-FIRST CENTURY

By the year 2000, UMH had accumulated an enormous wealth of specimens numbering approximately 1.75 million. However, basic data associated with these specimens were largely hidden from the outside world. Few specimens were data based, and even this scant information (with the exception of some fungal collections) was not available online.

UMH began specimen digitization efforts (imaging, transcribing label information, assigning geographic coordinates) on its extensive (>23,000) type collections in 2011 with funding from the Mellon Foundation. Since then, the National Science Foundation has supported UMH with seven awards for specimen digitization. These projects have focused on macroalgae, lichens and bryophytes, micro- and macrofungi, and select vascular plant groups. UMH has completely digitized and published data online for more than five hundred thousand specimens. These specimen data are the foundation of research in systematics and macroecology as well as for conservation, education, and public outreach. Our institutional goal is to database the entire collection. This will allow the public, and future curators, to better understand and explore our collections, their usage and scientific value, and target collections for future research.

The unification of the research museums at the RMC will create many opportunities for collaboration. For example, with the establishment of a molecular genetics lab, students and postdocs will be able to study genomes of any of the millions of plant, fungal, and animal specimens under our stewardship. A micro-CT scan facility will enable researchers to examine the internal structures of fossils, woody fruits, and dried flowers. A newly established liquid nitrogen facility will enable us to curate tissue samples and DNA. This will provide U-M researchers with opportunities to study biological diversity using genomic or transcriptomic (expressed gene) data sets.

The Herbarium will continue to train the next generation of biodiversity scientists by offering undergraduate internships and graduate student curatorial assistant (GSCA) scholarships. These opportunities introduce students to the fascinating history, diversity, and research potential of the U-M natural history collections.

Entry prepared by Christopher W. Dick.

3600 Varsity Drive. The Herbarium can be visited by appointment.

https://www.lsa.umich.edu/herb

Notes

1. The following account does not include all of the many individuals who have contributed to the growth of the Herbarium, nor does it mention many important specimen acquisitions. For more detailed period histories, see Mains for 1837–1954, McVaugh and Shaffer for 1940–74, and Berry for 1974–2013.
2. Mains, 1442. Schoolcraft and Houghton had taken part in other expeditions in the Great Lakes region. Houghton, Abram Sager, and John Wright who accompanied Houghton on the First Geological Survey of Michigan as zoologist and botanist had all studied under botanist Amos Eaton at the Rensselaer Institute.
3. Mains, 494.
4. See Winchell, 7. No specimens from the Sager collection seem to have survived. Sager's herbarium may have been destroyed in the 1913 fire of the first University Museum, otherwise lost, or deaccessioned.
5. Mains, 495–96.
6. See Harrington.
7. Harrington's expertise and research interests included astronomy, and after his resignation he became a professor of mathematics and astronomy in China. When he returned to U-M in 1879, the university named him professor of astronomy and director of the Detroit Observatory. See Mains, 497.
8. Gleason never did enjoy fame as an ecologist. He left Michigan in 1919 to focus on his taxonomic interests at the New York Botanical Garden.
9. Mains 1447–48. After Dodge's death in 1918, Ruthven appointed Cecil Billington honorary curator of this collection until 1921.
10. McVaugh and Shaffer.
11. Ibid.
12. Ibid.
13. In 2016, the Herbarium had two faculty-curators (Berry and Dick), two research scientists (Rabeler and Anderson), and one full curator (Reznicek) working on vascular plants; one faculty curator (James) works on fungi.
14. The mycologists who directed UMH were Calvin Henry Kauffman, 1921–30; Edwin Butterworth Mains, 1930–59; Alexander Hanchett Smith, 1959–72; and Robert Lynn Shaffer, 1975–86.
15. Hillis.
16. See www.macroalgae.org
17. See http://michiganflora.net

REFERENCES

Berry, Paul E. *Recent History of the University of Michigan Herbarium (1976–present)*. Ann Arbor: University of Michigan Press, 2017.

Clements, Frederic E. *Plant Succession: An Analysis of the Development of Vegetation*. Washington, DC: Carnegie Institution of Washington, 1916.

Gleason, Henry A. "The Individualistic Concept of the Plant Association." *Bulletin of the Torrey Botanical Club* 53 (1926): 7–26.

Gleason, Henry A. "The Structure and Development of the Plant Association." *Bulletin of the Torrey Botanical Club* 44, no. 10 (1917): 463–81.

Harrington, Mark W. "The Tropical Ferns Collected by Professor Steere in the Years 1870–1875." *Journal of the Linnean Society of London: Botany* 16, no. 89 (1877): 25–37.

Hillis, Llewellya. "William Randolph Taylor, 1895–1990." *British Phycological Journal* 27, no. 1 (1992): 1–2.

Mains, Edwin B. "The Department of Botany." In *The University of Michigan: An Encyclopedic Survey Vol. II*, edited by Wilfred B. Shaw, 493–505. Ann Arbor: University of Michigan Press, 1951.

Mains, Edwin B. "The Herbarium." In *The University of Michigan: An Encyclopedic Survey Vol. IV*, edited by Walter A. Donnelly, Wilfred B. Shaw, and Ruth W. Gjelsness, 1442–55. Ann Arbor: University of Michigan Press, 1958.

McVaugh, Rogers, and Robert L. Shaffer. "History of the University of Michigan Herbarium 1940–1970" [*sic* 1975] (available at http://um2017.org/2017_Website/History_of_Herbarium_2.html)

Winchell, Alexander. "Statement of Operations in the Museum, in the Department of Geology, Zoology and Botany and the Department of Ethnology and Relics, for the year ending 20th September 1866." Ann Arbor: University of Michigan, 1866.

MUSEUM OF ANTHROPOLOGICAL ARCHAEOLOGY

(Museum of Anthropology)

1922–2017

The Museum of Anthropological Archaeology (originally named Museum of Anthropology) is a research and collection museum in the University of Michigan College of Literature, Science and the Arts (LSA) that is focused on the study of humanity's past. Although formally established in 1922, the earliest of the museum's collections came to the university in the mid-nineteenth century through the first University Museum. Today, its collections number over three million archaeological and ethnographic artifacts, comparative specimens, and associated documentation derived from more than a century of anthropological research across six continents on periods spanning from a half million years ago to the present. As a nonexhibiting "research museum," the museum's mission integrates research and teaching with collection care and documentation.

HISTORICAL OVERVIEW

Foundations: 1837–1922

The beginnings of anthropological collecting at the University of Michigan can be traced to its founding decades. In 1840, Douglass Houghton donated a Chippewa birch bark canoe; by 1862, the Museum of Natural History included a division of "Ethnology and Relics," containing objects from Michigan's native communities and eclectic donations from students and alumni.

The university acquired its first well-documented anthropological collections in the late nineteenth century. The 1871–75 expedition of Joseph B. Steere brought ethnographic and archaeological artifacts from the Amazon, Peru, Taiwan, and the Philippines. In 1885, the Chinese Imperial Government donated the contents of their display at the 1884–85 New Orleans World Cotton Centennial Exposition, in recognition of President James B. Angell's 1880–81 service as US minister to China.[1] In 1893, the Smithsonian donated "duplicate" ceramic vessels that had been collected by the Bureau of American Ethnology from Native potters in the US Southwest.[2]

Less systematic acquisitions also came to the University Museum, including archaeological artifacts and human remains donated from across Michigan. Objects from the collections began to be used in teaching in 1892 when Harlan I. Smith, a freshman from Saginaw, arranged for Professor Francis Kelsey to offer the first archaeology course taught with museum artifacts. Smith also organized the museum's first permanent exhibition of anthropological objects and conducted an inventory of Michigan's archaeological sites and collections.

In 1913, the Natural History Museum was renamed the Museum of Zoology, with Alexander Ruthven its director. Responsibility for anthropological collections now rested uneasily under zoology and Ruthven, who sought to restrict his reconfigured museum to only zoological materials and rejected a number of offered archaeological and ethnographic collections.

The First Decades: 1922–44

Steps toward creating a separate anthropology museum began in 1917 when President Hutchins and Kelsey were approached by alumnus and colonial administrator Dean Worcester about sponsoring an archaeological expedition in the US Philippines. After some reluctance (eventually overcome by funding from Horace Rackham), the university approached Carl E. Guthe, a U-M alumnus and recent anthropology PhD from Harvard, to direct

Fig. 106. Chimú-Inka ceramic vessel from the Joseph Beal Steere Collection, Pacasmayo, Peru, collected 1871–75. UMMAA 6265.

it. Guthe agreed, on the condition that the university would create an anthropology museum and department. The Museum of Anthropology was formally established in 1922 as a division of the University Museum, with Guthe given the title of associate director of anthropology.[3] The Department of Anthropology was created six years later. Guthe spent three years directing excavations in the Philippines, amassing a large collection for his new museum. He returned to Ann Arbor in 1925 and was formally installed as the now-independent museum's first director. In 1928, the museum found its home on the fourth floor of the new Museums Building (later renamed the Ruthven Museums Building).

At its creation, the Museum of Anthropology had two divisions: the Archaeology Division supervised by Guthe and the Great Lakes Division overseen by the recently retired dean of the Homeopathic Hospital, Wilbert B. Hinsdale. By 1926, the museum's annual report listed three more divisions: the Orient, Ethnology/Ethnobotany, and Human Biology.[4]

Building on Ruthven's vision, Guthe sought to create a research museum—with collections generated through carefully documented scientific excavations led by university researchers. And even as anthropology as a discipline had by the 1920s largely moved away from its early home in museums,[5] Guthe remained convinced that students should learn about human cultures through encounters with objects in museum collections.

The collections grew explosively in the museum's first decades. In 1927, the museum became the site of a national Ceramic Repository of prehistoric eastern US ceramics as well as the main repository for the collections of the Michigan Archaeological Society. In 1929, the museum created the nation's first Ethnobotanical Laboratory; it began its first publication series in 1932. By 1942, only twenty years after its founding, the museum had accessioned some sixteen hundred separate collections (including those transferred from the earlier University Museum). Many came from the kinds of scientific expeditions that Guthe had envisioned. Others were donated by individuals or came from purchases and exchanges with established museums, including the British Museum, Smithsonian, Phillips Academy, American School of Prehistoric Study in France, and Academy of Natural Sciences of Philadelphia.

The Griffin Years: 1944–75

Guthe left Michigan in 1944, leaving behind a staff of four: James B. Griffin, curator of Archaeology; Volney H. Jones, director of the Ethnobotanical Laboratory and Ethnology Division; Kamer Aga Oglu, curator of the Division of the Orient; and Emerson Greenman, curator of Great Lakes Archaeology. Griffin was appointed the museum's interim director in 1944 and director in 1946, following a failed effort by the administration to unify the anthropology department and museum under a single director.[6] Griffin led the museum until his retirement in 1975, continuing its commitment to scientific fieldwork; in 1950, he established the Radiocarbon Laboratory, the second of its kind in the country.

By the mid-1940s, the museum's curators held joint faculty appointments in the Department of Anthropology. In 1956, administrative authority over the research museums and the newly autonomous Exhibit Museum of Natural History was shifted from the central administration to the College of Literature, Science, and the Arts (LSA). This shift further strengthened the museum's educational mission, closely linking the joint faculty-curators to student training and cementing its status as a national leader in education and research in anthropological archaeology. Curators' priorities increasingly came to focus on conducting and publishing new research (in many cases, generating new collections) and away from the care of existing collections.

The curatorial staff doubled under Griffin, growing explosively in the mid-1960s with the addition of curators for newly created divisions of European, Mesopotamian, Latin American, and Environmental Archaeology. The long-dormant division of Biological Anthropology was revived. The 1960s and 1970s were a period of intellectual fervor in archaeology and the museum's new curators were leaders in the theoretical and methodological revolutions of what came to be called "New" or "Processual" archaeology. By the time Griffin retired in 1975, the museum's reputation as *the* leading archaeological program in the United States was firmly established and the doctoral program in archaeology (formally overseen by the Department of Anthropology but physically based in and integrally associated with the museum) was the top-ranked program in the country.

1975–2016

Griffin's retirement marked a shift in the trajectory of museum leadership. While the first two directors had led the museum for twenty-three and thirty-one years respectively, over the subsequent forty years leadership has rotated among eight individuals (see below) and the curators as a whole constitute the museum's decision-making body. These changes were initiated under Ethnology curator Richard Ford, who served as director from 1975 to 1983 and again in 2002–2005. Ford also sought to refocus attention on the care and documentation of collections and expanded the museum's commitment to undergraduate teaching.

Over this forty-plus-year period, the museum has continued to grow and change, fueled by departures and retirements, opportunistic hires, and changing research priorities. In 2014, the museum's curators voted to change its name to Museum of Anthropological Archaeology. Today, the museum's collections are organized into eleven divisions based on geography and/or analytical focus: Africa, Asia, Europe, Great Lakes, Latin America, and North America, Archaeology, Ethnology, Archaeobiology (including Ethnobotany and Zooarchaeology), Human Biology, and Analytical Collections. Not all divisions have curators, and one curatorial appointment, "circumpolar archaeology," is not associated with permanent collections.

Director	Years
Richard Ford	1975–83
Jeffrey Parsons	1983–86
John Speth	1986–89
Henry Wright	1989–92
John O'Shea	1992–97
Robert Whallon	1997–2002
Richard Ford	2002–2005
Carla Sinopoli	2005–11
Joyce Marcus	2011–16

The museum's collections also have continued to grow. They quickly outgrew their space in the Ruthven Building, expanding to additional storage spaces across campus. The greatest growth has been in assemblages from archaeological fieldwork in North America. While archaeological collections have also been added from international research, today archaeological materials are rarely allowed to leave their source countries and the museum adheres to the UNESCO 1970 Convention on the export of cultural property and does not accept undocumented archaeological collections. Nonetheless, between 2002 and 2014, 256

Fig. 107. Ceramic vessel, Zuni Pueblo (New Mexico), collected in 1881 by the Colonel James Stevenson Expedition of the Smithsonian Institution and transferred to University of Michigan in 1893. UMMAA 7318.

collections of archaeological, ethnographic, and comparative specimens were accessioned.

The changing ethical and legal landscape of archaeological research and collecting has affected both what and how the museum adds to its collections and the status and disposal of existing collections. With the passage of the Native American Grave Protection and Repatriation Act (NAGPRA) in 1990, the museum began a process of inventorying, reporting, consultation, and repatriation of Native American human remains, funerary objects, sacred objects, and objects of cultural patrimony. This process, still ongoing, has been controversial and challenging, as tensions between archaeological interests in researching these materials came into conflict with tribal priorities to recover ancestral remains and objects. In 2009, the U-M Office of the Vice President for Research undertook a more active role in coordinating NAGPRA activities and the museum hired its first permanent NAGPRA staff.

Throughout its history, the museum has hosted researchers from North America and abroad to study the collections. Although renowned in the scholarly world, without a dedicated exhibition space, the museum's remarkable collections are not widely known on the U-M campus. Nonetheless, hundreds of undergraduate and graduate students have analyzed collections for thesis and other research projects and many thousands of students from anthropology and many other departments have encountered objects from the collections in classrooms and museum visits. In addition, the museum collaborates with the Museum of Natural History and other campus and public exhibiting museums in developing archaeology exhibitions and programming. The museum's website now features virtual exhibitions of portions of the collection.

2017 and Forward

The Ruthven Museums Building was a high-quality facility when it opened in 1928. However, lack of climate control, insufficient space, and other problems of an aging structure in a time of increasingly technologically demanding science made it problematic as a twenty-first-century museum building. In 2012, LSA decided to move the collections of the research museums to the Research Museums Center, a newly refurbished storage and research facility south of campus. While this move separates the collections from the academic heart of the university, it will ultimately reunite the greater part of the museum's collections in a single location that adheres to contemporary standards of curation.

SCOPE OF THE COLLECTIONS

Although the museum has changed and grown over the last ninety-plus years, to a significant extent it continues to exemplify the vision of its first two directors. Its collections—built through curatorial research, opportunistic and systematic collecting, and donations—are vast, numbering among the largest and most important anthropological research collections in the country.

The Great Lakes division curates archaeological materials from Michigan, surrounding Great Lakes states, and Ontario. The division's early role as a repository of the Michigan Archaeological Society coupled with nearly a century of fieldwork throughout Michigan have resulted in a comprehensive systematic collection of one million-plus objects from over twenty-three hundred archaeological sites, most dating before European contact.

Collections in the North America division also number more than a million objects. The Ceramic Repository consists of type collections of prehistoric ceramics and associated documentation from across eastern North America. Important today for its role in the history of North American archaeology, the collection also exposes students and researchers to a remarkable breadth of archaeological materials. Museum field projects in the east and southwest United States have contributed important, well-documented, scientific collections to the division.

The Asian collections include archaeological and ethnographic objects and nearly five thousand glass photograph negatives of Dean Worcester. While not as extensive as the North American division, the Asia division contains some of the museum's founding collections, from Steere's 1871–75 expedition, Guthe's 1922–25 Philippines Expedition, and the 1885 Chinese Government Collection as well as more than eighty subsequent archaeological and ethnographic collections.

Most important of the Near Eastern Archaeology collections are the unique fieldwork-based, comprehensive archaeological collections derived

Fig. 108. Tara, painting collected from Likir Monastery in 1933 by Walter N. Koelz UMMAA 17458.

from curator Henry Wright's surveys and excavations in Iran and Syria.

The Latin American Archaeology collections consist of pre-Columbian ceramics from Mexico and Peru (including those collected by Steere in the 1870s) and some Central American and Amazonian materials.

The European Archaeology collections consist primarily of Lower and Middle Paleolithic artifacts acquired through exchanges and donations early in the museum's history. Recent additions come from research at neolithic to Bronze Age sites in southeastern Europe.

Archaeological collections in the Africa division include prehistoric stone tools from

Egypt and a small assemblage of Stone Age artifacts.

Ethnobotanical research has been a focus of the museum from its beginnings, with research by curators Gilmore, Jones, and Ford generating sizeable collections of plant materials from archaeological sites, ethnographic collecting, and comparative specimens. From the 1930s through the 1970s, researchers regularly sent specimens for analysis and the division produced more than seven hundred reports. The collections, numbering more than thirty-three thousand samples, are especially strong in materials from Michigan and the US Southwest.

In contrast to the botanical collections, most animal bones from archaeological collections are curated with associated materials in their respective geographical division. The Zooarchaeological Laboratory contains comparative specimens and is especially strong in New and Old World domesticates.

The Human Biology division curates human skeletal remains from archaeological sites. Remains from the United States are subject to NAGPRA and have been or are in the process of being transferred to tribes. Other remains come from museum excavations in the Philippines and Iran.

The Analytical Collection consists of nonartifact materials, mostly sediments and geological specimens, used to study ancient environments and identify raw material sources.

The Ethnographic collections, divided into Asian and global collections, number more than sixteen thousand objects. They span from the nineteenth-century beginnings of the University Museum to recent collections of traditional earthenware ceramics from the US Southwest, the Philippines, and Myanmar. The collections are especially strong in Native American (especially Anishinaabe) and Southeast Asian baskets, Southwestern ceramics, Polynesian tapa clothes, and South and East Asian textiles, paintings, and ornaments.

In addition to the objects it curates, the museum holds more than one hundred thousand photographs and slides, as well as maps, field notes, and other collection and site documentation. With global archaeological heritage threatened by war, climate change, and development, the objects in the collections and the documentation associated with them constitute an invaluable resource, essential for continued research on the human past.

Fig. 109. Neolithic flint points, desert north of the Fayoum, Egypt, donated by Enoch E. Peterson in 1935. UMMAA 6645.

CURRENT RESEARCH

The museum's curators, emeritus curators, affiliated researchers, and students continue to be leading researchers: in the field and in the study of collections. Archaeological research today involves sophisticated technology and large collaborative teams. While fewer museum field projects generate large new collections, the collections continue to grow from work in North America and smaller accessions, as well as from short- and long-term loans of materials brought back for study from students' and curators' fieldwork around the world.

Today's curators continue to conduct research in Michigan (John O'Shea's research beneath Lake Huron), the southeastern US (Robin Beck's work at the Berry site in North Carolina), Asia (Carla Sinopoli in Iron Age to medieval southern India; Henry Wright on state formation in China), Latin

America (Kent Flannery and Joyce Marcus in Peru and Mesoamerica; Raven Garvey in Chile and Argentina), Europe (John O'Shea on Bronze and Copper Age southeastern Europe; Robert Whallon on Paleolithic Montenegro), and Africa (Brian Stewart on Stone Age periods in Lesotho and South Africa; Henry Wright on Madagascar state formation). Archaeology doctoral students work in all these countries and many more. And each year, undergraduate and graduate students from across the university conduct new research on the museum's collections and dozens of scholars from around the world come to Ann Arbor to study the collections of the Museum of Anthropological Archaeology.

New analytic techniques from the material and biological sciences—such as instrumental neutron activation, inductively coupled plasma mass spectrometry, AMS radiocarbon dating, phytolith and starch analysis, stable isotope studies, and DNA analysis, among many others—are allowing old collections to be studied in new ways to address questions that the first generation of the museum's curators could not have imagined.

As the Museum of Anthropological Archaeology enters its second century, curators and staff are working to digitize the collections and to upgrade from the museum's old and incomplete database to a new, more effective and web-accessible system that will enhance access to the collection for scholars and the public, assuring that the irreplaceable record of the human past resulting from nearly two centuries of anthropological collecting will continue to contribute to knowledge for centuries to come.

Entry prepared by Carla M. Sinopoli.

Research collections can be visited by appointment. Email: Anthro-Museum@umich.edu.

https://www.lsa.umich.edu/ummaa

Notes

1. *Proceedings of the Board of Regents*, June 1886, 560–61.
2. In its early decades, the Smithsonian donated "duplicate" objects from its collections to museums and libraries across the United States. The University of Michigan was among the recipients of such donations (see discussion in Walsh, 2002; Nichols, 2014).
3. *Proceedings of the Board of Regents*, June 1922, 507.
4. Report of the Director, 6.
5. Conn, 32.
6. Peace, 19.

REFERENCES

Conn, Steven. *Do Museums Still Need Objects?* Philadelphia: University of Pennsylvania Press, 2010.

Nichols, Catherine A. "A Century of Circulation: The Return of the Smithsonian Institution's Duplicate Anthropological Specimens." *Museum Anthropology* 37, no. 2 (2014): 144–59.

Peace, William. "Introduction: The University of Michigan's Department of Anthropology: Leslie White and the Politics of Departmental Expansion." *Michigan Discussions in Anthropology* 16, no. 1 (2006): 1–32.

"Report of the Director of the Museum of Anthropology to the Board of Regents, July 1, 1925 to June 30, 1926." *University of Michigan Bulletin* 28, no. 26: 1–14.

University of Michigan Board of Regents. *Proceedings of the Board of Regents.* Ann Arbor: University of Michigan (1837–present). (Available at http://quod.lib.umich.edu/u/umregproc)

Walsh, Jane M. "Collections as Currency." In *Anthropology, History and American Indians: Essays in Honor of William Curtis Sturtevant*, edited by William Merrill and Ives Goddard, 201–10. Smithsonian Contributions to Anthropology 44. Washington, DC: Smithsonian Institution Press, 2002.

KELSEY MUSEUM OF ARCHAEOLOGY

1928–2017

The Kelsey Museum of Archaeology supports teaching and research on Classical, Egyptian, and Near Eastern archaeology through stewardship of its rich collections, an active exhibitions program, and sponsorship of ongoing field research. As an exhibit museum at a public university, the Kelsey also embraces an active public outreach mission, engaging with local residents and visitors of all ages, from kindergarten through retirement.

The museum is housed in two connected buildings. Newberry Hall on State Street is a Richardsonian Romanesque structure originally built in 1891 for the Student Christian Association; it has served as the home of the Museum of Classical Archaeology (subsequently renamed the Kelsey Museum) since 1929. Since 1981, it has also housed the university's Interdepartmental Program in Classical Art and Archaeology (IPCAA). In 2009, the Upjohn Exhibition Wing was opened off the west side of Newberry Hall, providing secure and climate-controlled storage facilities for the collections and 6,800 square feet of exhibition space.

The contents of the Kelsey Museum number approximately one hundred thousand objects and have two main components: (1) a representative collection of Classical, Egyptian, and Near Eastern artifacts, acquired by gift and purchase, mostly between the 1890s and the 1970s; and (2) finds from a number of U-M excavations in Egypt and the Near East, largely carried out in the 1920s and 1930s.

Approximately fifteen hundred objects are on permanent view. In addition, the museum mounts two to three special exhibitions per year, providing opportunities for displaying materials usually held in storage as well as for hosting occasional traveling shows. The special exhibitions are teaching and research-driven projects, undertaken by the museum's curators and occasional guest curators. Curators who hold joint faculty appointments often organize classes around their exhibitions and work with students in every stage of the exhibition process.

In addition to maintaining and exhibiting its collections, the museum is an active field research unit, sponsoring archaeological projects in Italy, Greece, Turkey, and Egypt. These provide research and training opportunities to numerous undergraduate and graduate students.

HISTORICAL OVERVIEW

Foundations: 1855–1928

The Kelsey Museum has its roots in classical studies, one of the first subjects taught at the University of Michigan. In 1855, Henry Frieze, then-chairman of the Department of Latin, began to acquire plaster and terracotta casts of ancient sculptures and other artifacts for use in teaching (see Choberka, this volume). They were displayed in various buildings (University Hall, the original Mason Hall, and the University Library—all no longer extant) until they were moved to Alumni Hall in 1910. Most casts were disposed of or destroyed over the course of the twentieth century; some notable surviving specimens include casts of several sections of the sculpted frieze of the Parthenon, now on view in the Hatcher Graduate Library, and a replica of the Venus de Milo, displayed in the Martha Cooke House (a women's residence hall on South University Avenue).

After Frieze died in 1889, he was succeeded by Francis W. Kelsey, who continued as professor of Latin and departmental chair until his death

Fig. 110. Coffin of Djehutymose. Saite period, c. 685—525 BC, Egypt. KM 1989.3.1.

in 1927. Kelsey was a remarkable individual (see Pedley, this volume). He served as president of both the American Philological Association and the Archaeological Institute of America, and his influence on the development of classical studies and archaeology at U-M—and throughout North America—is still being felt today. Today's Kelsey Museum is largely his brainchild in two respects: his emphasis on the collection and use of actual artifacts rather than casts for teaching purposes, and his sponsorship of archaeological field projects.

Kelsey began acquiring archaeological artifacts in 1893. He continued to add to the collections over the next three decades. Most of the objects he purchased came from Italy, North Africa, and Egypt, and most dated to the Roman period. In addition to archaeological artifacts, he was particularly interested in acquiring papyri (ancient texts written on papyrus)—and the hope of finding new papyri in archaeological contexts motivated the first university-sponsored field projects in the Mediterranean region.

In 1924, Kelsey secured funding to excavate at various sites around the Mediterranean, including Pisidian Antioch in Turkey, Carthage in Tunisia, and Karanis in Egypt. Interestingly, he also investigated (but eventually abandoned) the possibility of an excavation at Olynthos in Greece—the site of one of the Kelsey Museum's newest archaeological field projects. The finds from Karanis ultimately came to constitute nearly half the museum's holdings.

Kelsey died in 1927, and in 1928 the "Museum of Classical Archaeology"[1] was established as a separate administrative unit under the University Museums. In 1929, the museum was given a permanent home in Newberry Hall (which had been leased to the university for classroom space since 1921). In 1937, Newberry Hall was purchased by the university, and in 1953 the museum was renamed the Francis W. Kelsey Museum of Archaeology.[2]

Establishing the Museum: 1928–73

The first forty-five years of the museum saw the completion within a decade of the excavations at Karanis and other sites in Egypt and the Near East (Dime [ancient Soknopaiou Nesos] and Terenouthis in Egypt, Seleucia on the Tigris in Iraq, and Sepphoris in Israel); study and publication of the results of these projects; limited further acquisitions by gift and purchase; and the resumption of fieldwork in Egypt in 1958 (documentation of the monastery of St. Catherine), followed by excavations in Syria (at the Roman and mediaeval sites of Dibsi Faraj and Qasr al-Hayr) and Libya (at the classical sites of Apollonia and Cyrene). The muse-

Fig. 111. Glass vessels, Roman period, 1st to 4th centuries CE, Egypt. Flask, KM 5936, goblets, KM 5965 and KM 5966.

Fig. 112. Alabaster statuette. Seleucia, Parthian period, 1st to 3rd centuries CE, Iraq. KM 1.6118.

um's first director, John G. Winter, served from 1929 to 1950. Subsequent directors included E. E. Peterson (1950–61), George Forsyth (1961–69), Theodore Buttrey (1969–71), and Louise Shier (1971–73). For most of these years, the museum had a very small staff consisting only of the faculty director, one or two full-time curators, a "technician," and a secretary. In 1956, the Kelsey Museum joined five other campus museums in being placed under the dean of the College of Literature, Science and the Arts. By the end of this period, the museum was no longer the dynamic research and teaching unit it had been in its early years, and its future was in serious doubt.

Growth and Transformation: 1973–2017

A new phase began in 1973 with the appointment of John G. Pedley as director, charged with developing a new sense of mission and integrating the museum more fully into the life of the university. During his years as director (1973–86), Pedley charted a path that has been pursued and developed by his successors, Elaine Gazda (1986–96), Sharon Herbert (1996–2013), and Christopher Ratté (2013–present). Hallmarks of Pedley's legacy include the development of the professional staff of the museum, the appointment of academic curators with joint appointments in the museum and in academic departments, the institution of a thematic exhibition program, new publications, new fieldwork, a new acquisitions program, and a close relation with the Interdepartmental Program in Classical Art and Archaeology. A major achievement of Elaine Gazda's tenure as director was the accreditation of the museum by the American Alliance of Museums in 1990. Sharon Herbert oversaw the construction of the Upjohn Exhibit Wing, largely funded by a gift of eight million dollars from Ed and Mary Meader of Kalamazoo in 2003—at the time the largest gift in the history of the College of Literature, Science and the Arts. The new addition made it possible to triple the number of objects on permanent display and provided space for a regular program of special exhibitions. The staff of the museum has now grown to over twenty persons.[3]

SCOPE OF THE COLLECTIONS

The finds from excavations sponsored by the university in Egypt and the Near East constitute the majority of the objects in the Kelsey Museum, most dating to the Roman period. In most cases, the original excavation records are also stored at the museum. The museum also contains a representative collection of Classical, Egyptian, and Near Eastern objects, dating from the Neolithic to the Middle Ages, and an important collection of archival photographs. Table 2 below presents a rough breakdown of inventoried artifacts:[4]

Table 1

Period of artifact	*Number of artifacts*
Old World prehistory	140
Pre-dynastic and Dynastic Egypt	630
Hellenistic and Roman Egypt[1]	50,250
Ancient Near East[2]	15,280
Greece and Cyprus	1280
Italy and Roman provinces (excluding Egypt)	10,700
Late Antique and Byzantine	2,870
Islamic	7,000

[1] Mostly from the excavations at Karanis.

[2] Mostly from Seleucia on the Tigris and dating to the Hellenistic and Parthian periods (third century BC to second

Kelsey's first acquisition in 1893 was a purchase of a lot of 109 items (lamps, vases, and building materials) from the Musée Lavigerie in Carthage. In the same year, he bought another 1,096 objects from dealers in Tunis, Rome, Capri, and Sicily. These purchases joined a group of several thousand coins donated to the university in the 1880s.

Kelsey continued to add to the collections until his death. His acquisition, through gifts and purchases, ranged from pottery and terracotta figurines to painted stucco and inscribed tombstones to Roman brick stamps, objects of daily life, Roman glass vessels, Egyptian tomb sets, and Greek papyri from Egypt. As a result of Kelsey's collecting, the museum now has the largest collection of Latin inscriptions in the Western Hemisphere. In 1925, Kelsey contracted with the well-known Italian archaeological artist Maria Barosso to paint a set of almost full-scale replicas of the wall paintings in the Villa of the Mysteries at Pompeii.

After Kelsey secured funding to excavate at sites around the Mediterranean in 1924, the nature of the collections began to change, thanks to traditional partage (division) arrangements that gave the university title to significant portions of the finds from the excavations.

Two collections are of particular importance: from Karanis in Egypt and Seleucia on the Tigris in Iraq. Nearly half the objects in the collections (seventeen thousand objects plus twenty-nine thousand coins) come from the Graeco-Roman farming town of Karanis in the fertile Fayum region of Egypt, about forty miles southwest of Cairo. The excavations at Karanis, carried out from 1924 to 1935, were exemplary for their focus on the residential and industrial areas of the settlement as well as on the temples and other monumental buildings, and for their careful documentation of the artifacts of daily life. According to the Egyptian government practice of the time, the university received two-thirds of the finds. As a result, the Kelsey has the world's largest collection of Graeco-Roman objects from Egypt outside Cairo, all from controlled scientific excavations. Of special interest are the coins and coin hoards (still incompletely studied), the well-preserved organic materials (including baskets, brooms, rope, textiles, and a wide variety of wooden objects and furniture), and an important collection of glass vessels, as well as bronze and stone sculptures and other works of art. The finds from Karanis and other sites excavated in Egypt,

Fig. 113. Textile. Coptic period, 3rd to 7th C AD, Egypt. KM 22710.

together with the excavation records, also stored at the Kelsey, constitute an invaluable corpus of evidence for study and research on all aspects of life in an ancient town in Roman Egypt. They continue to provide research opportunities to students and scholars at U-M and around the world.

After Karanis, the next largest collection of excavated finds in the museum are those from Seleucia on the Tigris in Iraq (a major ancient metropolis, founded by one of Alexander the Great's generals just south of modern Baghdad), comprising thirteen thousand objects mainly of the Hellenistic and Parthian periods (third century BC to second century AD). The excavations at Seleucia were carried out from 1928 to 1932 and again from 1935 to 1936. Among the finds are pottery (including one of the largest collections of Parthian pottery outside Iraq), lamps, seals and sealings, coins, jewelry, architectural stucco, and a very important collection of terracotta, bone, and alabaster figurines. The finds from Seleucia are of particular importance in light of the recent devastation of the Baghdad Museum and considerable recent damage to the site of Seleucia itself.

Subsequent Michigan-sponsored fieldwork at other sites further augmented the Kelsey collections. By the 1960s, however, antiquity laws had changed significantly, and most finds from foreign excavations could no longer be legally exported to the United States.

While excavations account for the majority of the museum's holdings, the remainder comes from purchases, bequests, and gifts, which continued well after Kelsey's death. The largest donation was the collection of 3,336 artifacts amassed by Peter Ruthven, mostly during the 1930s. This far-ranging collection encompasses the late Roman period through the twelfth century in Egypt, and includes glass, bone, glazed ceramics, wood, textiles, magical amulets, beads, bracelets, papyrus, stone, and bronze.

Among the most popular objects from the collection currently on display are the Dynastic Egyptian holdings, including the painted coffin of the Saite-period (seventh–early sixth century BC) Egyptian priest Djehutymose, donated in 1931 by Albert M. Todd, a businessman from Kalamazoo known as the "Peppermint King," and a number of other objects acquired from the Egyptian Antiquities Service in the 1930s and again in the 1950s.

The last major period of acquisitions by purchase took place in the mid-1970s, in an effort by then-director John Pedley to improve the museum's collections in, in his words, "Greek and Roman fine art." Especially notable were the purchases of several Roman portraits—including a very fine head of Polydeukion, the foster son of the Graeco-Roman man of letters, Herodes Atticus (acquired jointly with the University of Michigan Museum of Art)—and a red-figure amphora attributed to the Berlin Painter.

Since the 1980s, the museum has not pursued a program of active acquisitions. The museum does, however, continue to collect "passively," accepting donations when they are in accord with the 1970 UNESCO convention on cultural property (which establishes strict guidelines for the acceptable exportation of cultural materials from their countries of origin to other countries) and when they complement the collections, either by adding to existing strengths or by filling in gaps. A notable such acquisition was the gift in 1991 of 158 provenanced stamp seals from Tepe Giyan, Iran (circa 4000 BC), making the Kelsey's holdings of seals in that category comparable with those of the British Museum and the Louvre.

CURRENT RESEARCH

Current museum-sponsored research falls into two categories: research on the collections and ongoing field research. Collections research is often undertaken in connection with special exhibitions, resulting in published exhibition catalogs and collections of essays. It frequently includes materials analysis of objects in the collections, carried out by the conservators, students, and curators.

In addition to exhibition catalogs, recent Kelsey publications include a monograph on the Djehutymose coffin by curator Terry Wilfong and a book on collectors whose gifts have enriched the museum, and the collections they donated, by former curators Lauren Talalay and Margaret Root. Kelsey materials are also regularly studied by external scholars.

The field projects sponsored by the museum include both excavations and surveys and bring a wide variety of different theoretical and methodological approaches to the study of ancient societies. This pluralistic vision has helped make U-M a national center for teaching and research in Classical, Egyptian, and Near Eastern archaeology.

A major focus of current Kelsey-sponsored field research is the development of the Greek and Roman city. The excavations at Gabii near Rome by Nicola Terrenato (Kelsey research associate and professor of classical studies) are shedding new light on the early history of Rome itself. A project begun in 2014 by Lisa Nevett (also a Kelsey research associate and professor of classical studies) focuses on domestic architecture and the evidence for daily life in the short-lived town of Olynthus in northern Greece. An archaeological survey of the port city of Notion in Turkey, begun in 2014 under the direction of Kelsey director Christopher Ratté, is studying the history of urbanism in western Asia Minor following the conquests of Alexander the Great.

In addition to these projects at classical sites, Janet Richards (Kelsey research associate and professor of Near Eastern studies, and former curator) is directing a large-scale investigation of a late third- and early second-millennium BC mortuary landscape at Abydos in southern Egypt. Former Kelsey director Sharon Herbert supervised a long-term program of excavations at Tel Kedesh, a Phoenician settlement in northern Israel.

CONCLUSION

The Kelsey Museum is one of the university's most effective public faces. Archaeological exploration has broad popular appeal, and an archaeology museum provides a forum for discussion of a wide range of contemporary as well as historical concerns—such as imperialism, past and present; changing religious attitudes; or the long-term history of the Middle East. Together with the Museum of Anthropological Archaeology, the Kelsey Museum is uniquely suited to support and promote the University of Michigan's commitment to dialogue with the past, to students, scholars, and the public at large.

Entry prepared by Christopher Ratté.

434 South State Street. Open Tuesday–Friday 9 am–5 pm; Saturday–Sunday 1 pm–4 pm. Email: kelseymuse@umich.edu. https://www.lsa.umich.edu/kelsey

Notes

1. *Proceedings of the Board of Regents*, November 1928, 804.

2. From 1941 through 1946, the museum was renamed Museum of Art and Archaeology, which also included responsibility for the art collections in Alumni Memorial Hall. Regental action in the summer of 1946 divided the collections again into two independent museums. *Proceedings of the Board of Regents*, November 1945, 165.

3. The director and four academic curators; an associate director; two collections managers; two conservators; two exhibition preparators; two educators in charge of outreach to both the university and the general public, especially schoolchildren; an editor; the coordinator of the graduate program; a graphic artist; an administrative support staff of two persons; and a security staff of four persons.

4. Because of gaps in the current computer records, now being revised, these numbers are approximately ten thousand short of the actual numbers of objects in the collections, but they give a general sense of the proportions of objects in major categories.

REFERENCES

Menegazzi, Roberta. *Seleucia al Tigri: Le terrecotte figurate dagli scavi italiani e americani.* Monografie di Mesopotamia 16. Florence: Le Lettere, 2014.

Pedley, John G. *Life and Work of Francis Willey Kelsey: Archaeology, Antiquity, and the Arts.* Ann Arbor: University of Michigan Press, 2011.

Talalay, Lauren E., and Margaret Cool Root. *Passionate Curiosities: Tales of Collectors and Collections from the Kelsey Museum.* Ann Arbor: Kelsey Museum Publications, 2015.

University of Michigan Board of Regents. *Proceedings of the Board of Regents.* Ann Arbor: University of Michigan (1837–present). (Available at http://quod.lib.umich.edu/u/umregproc)

Wilfong, Terry G. *Life, Death, and Afterlife in Ancient Egypt: The Djehutymose Coffin in the Kelsey Museum of Archaeology.* Ann Arbor: Kelsey Museum Publications, 2013.

MUSEUM OF ART

1856–2017

The collection held by the University of Michigan Museum of Art (UMMA) traces its origins to an assemblage of casts and copies called the Art and Antiquities Collection that was exhibited on the Ann Arbor campus in the mid-nineteenth century. In the intervening years, UMMA has become one of the leading university art museums in the country, with a collection of over twenty thousand objects in all media—from paintings and sculptures to screens, ceramics, works on paper, photographs, and digital media.

The museum facility includes the historic beaux arts-style Alumni Memorial Hall, completed in 1910, and the modern Maxine and Stuart Frankel and the Frankel Family Wing, which opened in 2009. The addition was built for the twenty-first century, in line with the museum's mission "to transform individual and civic life by promoting the discovery, contemplation, and enjoyment of the art of the world." The museum offers a dynamic schedule of special exhibitions, while its numerous galleries, auditorium, and mixed-use spaces host performances, readings, and talks. UMMA serves more than two hundred thousand visitors on site each year and a global audience via its website, social media, exhibition tours, and publications.

HISTORICAL OVERVIEW

First Years: 1852–66

Interest in the usefulness of a collection of visual art began as early as 1852 when Detroit portrait painter Alvah Braddish was appointed professor of the Theory and Practice of Fine Arts and was provided with a room for the "reception of specimens of Art" (although what he received there is unclear). Three years later, Latin professor Henry Simmons Frieze began acquiring maps, photographs, and plaster and terra cotta copies of antiquities to illustrate his lectures in the Department of Classics. In 1856, this fledgling collection of several hundred pieces, then called the Art and Antiquities Collection, was put on display in a Gallery of Fine Arts in the South College Building (Haven Hall), which also housed the university's library. This was only the fourth university art collection in the nation, founded alongside those at Bowdoin College, Harvard, and Yale, and predated major civic art museums in the area, including the Art Institute of Chicago (1879), the Detroit Institute of Arts (1885), and the Toledo Museum of Art (1901) (see Choberka, this volume).

Frieze published the first catalog of what he called the "Museum of Art and Antiquities"[1] in 1857, shortly before the collection was installed in the old North College (Mason Hall). In 1862, a group of Ann Arbor citizens helped the university acquire local sculptor Randolph Rogers's marble masterpiece *Nydia, the Blind Flower Girl of Pompeii*, a character in Sir Edward Bulwer-Lytton's popular 1834 novel *The Last Days of Pompeii*. The sculpture remains on view at Alumni Memorial Hall and stands as the first important original work of art in the collection of the University of Michigan Museum of Art.

By the mid-1860s, the art collection included full-size casts of ancient statues and busts, two hundred terra cotta models, engravings and photographic views, historical medallions, and copies of modern sculptures. By the time Frieze was named curator of the fine arts collection in 1866,[2] the University of Michigan boasted the largest university art museum in the country.

Fig. 114. *Begin the Beguine,* Max Beckmann (Germany, 1884–1950), 1946, oil on canvas. Museum purchase, 1948/1.103.

Collection Expansion: 1866–1910

As the Art and Antiquities Collection continued to grow, university personnel paid increasing attention to its preservation and role within the university community. The revised 1876 catalog[3] listed 1,789 objects. Over the next twenty years, the collection was augmented by several large gifts, including the contents of Randolph Rogers's sculpture studio in Rome and a substantial portion of the collection of Michigan banker Henry Clay Lewis—then the most valuable donation the university had ever received—including paintings that remain bedrocks of the American collection: Charles Wimar's *Attack on the Emigrant Train* (1856) and Eastman Johnson's *Boyhood of Lincoln* (1865–67).

Accommodating the expanding collection became a matter of recurring concern. The collection was moved to the new library building in 1883. After Frieze's death in 1889, the new curator, professor of Greek Martin Luther D'Ooge, published another revised catalog,[4] which referred to the collection as the Gallery of Art and Archaeology and listed 175 paintings, 205 sculptures, 36 architectural ornaments, and 1,789 gems and medallions. By the end of D'Ooge's tenure, the museum had outgrown its space once again. The collection required a new, permanent home.

Alumni Memorial Hall and Emergence of UMMA: 1910–49

Fundraising for a dedicated structure to house this art and archaeology collection began in 1894 and gained momentum when it was combined with the Alumni Association's memorial committee's efforts to commemorate university graduates who had died in the three great wars of the nineteenth century (the Civil, Spanish-American, and Mexican-American wars). Alumni Memorial Hall, a 41,000-square-foot building designed by the Detroit firm Donaldson and Meier, opened in 1910 to accommodate the offices of the Alumni Association and the Museum of Art and Archaeology. The following year, the Department of Fine Arts was established there as well, consolidating the collecting and teaching of art.

New strides in professionalization brought the formation of a permanent acquisitions committee (consisting of the university president, the collection curator, and other institutional leaders) in 1910 and the initiation of a program of rotating exhibitions. The inaugural exhibition at Alumni Memorial Hall, *American and Oriental Art*, consisted of loans from the esteemed collection of Detroit industrialist Charles Lang Freer. The exhibition established a commitment to the global history of art that remains an important part of UMMA's mission to this day.

In 1928, a Museum of Classical Archaeology was created and installed in Newberry Hall, formally dividing the Art and Archaeology collections. This process of separation was completed in 1946 with the establishment of two separate institutions: the Museum of Archaeology (renamed the Kelsey Museum of Archaeology in 1953) and the University of Michigan Museum of Art.

Fig. 115. Ewer with silver fittings, China, Jiangsu province, Jingdezhen, Ming dynasty (1368–1644), Yongle mark and period (1403–25), porcelain with underglaze cobalt decoration, silver spout and lid. Gift of Mr. and Mrs. John A. Pope for the James Marshall Plumer Memorial Collection, 1968/1.50.

Even as the classical archaeology collections left Alumni Memorial Hall, new art donations came in. Non-Western collections were given a substantial boost in 1936 when Freer's friend and fellow collector Margaret Watson Parker left more than six hundred works of art to the university, including Japanese wood block prints, nineteenth-century French paintings, Asian ceramics and a world-class collection of pieces by James McNeill Whistler. The Parker gift remains the most important collection acquired by the museum to date.

In conjunction with UMMA's independent

status, Jean Paul Slusser, professor of drawing and painting, was named the first director. His astute collecting efforts, focused on acquiring modern and contemporary art, brought German painter Max Beckmann's *Begin the Beguine* (1946) to the museum, as well as works of art by European and American cubists and surrealists, such as Fernand Leger, Stuart Davis, Roberto Matta, and Gerome Kamrowski, and German expressionist artists Käthe Kollwitz, Max Pechstein, and others.

Modernization and University Collaborations: 1949–79

The second half of the twentieth century witnessed the museum strengthening its ties with its academic environment. Charles H. Sawyer, UMMA's second director, collected to support the university's core curriculum, which at that time meant acquiring Old Master paintings from the Western tradition. Some of the objects acquired during this time include Guercino's *Esther before Ahasuerus* (1639), Jacopo del Casentino's panel *Enthroned Madonna and Child with Saints and Angels* (c. 1325), and Joos van Cleve's *St. John on Patmos* (c. 1525). Asian collections were further strengthened by donations from U-M art historian James Marshall Plumer, made between 1961 and 1983. The collection of the surrealist artist Chet LaMore, given in 1972, contained a small but significant group of African objects, the kernel of UMMA's collection from that continent. In the 1960s, Sawyer began to offer seminars at the museum. These developed into the Museum Practice Program, a predecessor of the current Museum Studies Program. Thus, Sawyer's vision of the museum as a training ground for future museum professionals had a lasting impact on its programming and mission going forward.

When Alumni Memorial Hall closed for renovations in 1966, the Alumni Association offices moved to the Michigan Union, providing UMMA with its own space for the first time in more than one hundred years of university art collecting. In 1968, Sawyer formed the Friends of the Museum of Art, a membership organization, in the newly renovated space; in 1975, during Director Bret Waller's tenure, community volunteers instituted the Docent Program. Both groups strengthened UMMA's ties to its community beyond the university walls and broadened opportunities for support and outreach. During Waller's term, the museum acquired several significant modernist works of art, including Charles Ginnever's *Daedalus*, a monumental public sculpture, and paintings by Helen Frankenthaler, Richard Diebenkorn, and Joan Mitchell. Photography also became an important collecting area during the 1970s, with the initial gift of Margaret and Howard Bond in 1979 marking the beginning of their enduring contributions to the museum.

Fig. 116. Power Figure (*nkisi nduda*), Yombe peoples, Democratic Republic of Congo, late nineteenth century, wood, cloth, fiber, animal hides, feathers, mirror, glass, metal, resin, medicinal substances, pigment. Gift of Candis and Helmut Stern, 2005/1.191.

Exhibitions and Interpretations: 1979–2004

In the last two decades of the twentieth century, the UMMA exhibitions program became increasingly ambitious, using key collection objects as the point of origin for scholarly inquiry. History of art professor Joel Isaacson curated the critically acclaimed *The Crisis of Impressionism 1878–1882* (1979–80), while curators Carole McNamara and Annette Dixon organized *Monet at Vétheuil: The Turning Point* (1998), an exhibition that still holds the museum's exhibition attendance record. This focus on "thoughtful blockbusters" helped to double UMMA's attendance in just a few short years, raising the institution's profile both nationally and internationally.

During the 1980s, under the direction of Evan Maurer, the museum began to seriously build African art into one of its key collecting areas. Notable gifts came from docents and collectors Agnes Miner and Margaret Coudron and their husbands, and from Candis and Helmut Stern, whose donations between 1987 and 2005 significantly expanded the collection. The Sterns later donated to the museum's renovation project (the auditorium is named in Helmut's honor) and, in 2015, endowed UMMA's first curator of African art.

Director Maurer also introduced a small but representative collection of Native American art to the collection. The establishment of a fund for acquisitions—initiated in 1989 by collector and philanthropist W. Hawkins Ferry—allowed museum staff to target gaps in the collection and purchase significant works of art previously out of reach.

Reimagining the Museum: 2004–2009

Between 2004 and 2009, UMMA underwent a $41.9 million expansion and restoration project that more than doubled the size of the museum with the addition of the 53,000-square-foot Maxine and Stuart Frankel and the Frankel Family Wing. This project created total gallery space of 94,000 square feet and restored much of the original architectural detail of Alumni Memorial Hall. The building project, spearheaded by Director James Steward and designed by architect Brad Cloepfil and Allied Works Architecture, sought to harmonize the architectural addition with the university's surroundings, while creating an inviting space. Transparent glass walls and a lantern-like façade project the lively museum complex into public space. The museum's award-winning architecture created a multidisciplinary meeting place for the arts, allowing an increasingly broad series of public programs that operate in parallel with its collections and displays. Examples include experimental music and dance performances by the School of Music, Theatre and Dance, and the Creative Writing program's distinguished Zell Visiting Writers Series.

Advancing a University Art Museum for the World: 2010–present

Joseph Rosa, director from 2010 to 2016, has fostered exhibitions and programs that reflect the intermedia approach of much contemporary art today, allowing the museum to function as a laboratory for artistic experience both on and off the walls. An institution-changing Andrew W. Mellon Foundation grant awarded in 2011 created new staff positions to increase the use of the collections by students and faculty. UMMA staff guide students through the galleries and arrange visits to the print and object study rooms to view selections of art from storage. Recent visitorship numbers illustrate the success of this program. During the 2011–12 school year, 2,232 students in 67 U-M classes visited UMMA, while three years later, 7,697 students in 300 classes visited—an increase of nearly 350 percent in the number of students and nearly 450 percent in the number of classes.

SCOPE OF COLLECTIONS

Asian

Comprising more than five thousand objects made over the past five millennia, the Asian art collection is the largest in Michigan, with particular strengths in Chinese and Japanese painting, Chinese and Korean ceramics, Indian miniature

Fig. 117. Alumni Memorial Hall museum entrance during *Artscapade* 2014. Photo by Leisa Thomson.

painting, and Buddhist sculpture. The collection originated in Margaret Watson Parker's bequest and has been sustained through an acquisition fund established by her husband, U-M professor of ophthalmology Dr. Walter Parker, and donations. The Parker fund has allowed for the purchase of many important Chinese scrolls.[5] A collection of paintings by twentieth-century Chinese painter Chang Ku-nien, given by Dr. Chang-Yang and Mrs. Shirley Chang in 2006, complements the earlier paintings.

Other noteworthy Asian collections include two hundred prints of Kabuki actors, given by James Hayes, and a substantial number of Indian prints and artifacts from Walter and Nesta Spink, who also donated an important collection of Western prints. More recently, UMMA received a significant collection of Japanese kimono from Howard and Patricia Yamaguchi, including stunning examples of textile designs by Minagawa Gekka. The Korean collection was greatly expanded with a major acquisition of 250 ceramics, paintings, and other objects through the generosity of Bruce and Inta Hasenkamp along with the support of Elder and Mrs. Sang-Yong Nam. Support from the Korea Foundation and U-M alumnus Woon-Hyung Lee helped to build a new gallery for this material.

African

Initiated in 1972 with the acquisition of the LaMore Collection, the African art collection now numbers more than one thousand sculptures, masks, architectural and decorative elements, textiles, metalwork, and ceramics, made by peoples throughout the continent. Director Mauer, a specialist in art of Central Africa and the Congo region, particularly encouraged collectors to turn their attention to African objects. Maurer advised Professor and Mrs. Horace M. Miner[6] and businessman and collector Helmut Stern, helping the latter to amass one of the most significant holdings of Congolese art in the world, ninety pieces of which came to UMMA in 2005. U-M alumni James and Vivian Curtis donated more than two hundred works of African and African American origin to the museum, including a significant collection of goldweights. With the opening of the Frankel Wing in 2009, the newly named Robert and Lillian Montalto Bohlen Gallery of African Art provides a dedicated space for the display of African art and material culture.

European and American

Western art has been a central focus of collecting since Randolph Rogers's sculpture *Nydia* entered the permanent collection in 1862. The foundational bequests by Henry Clay Lewis and Margaret Watson Parker have been augmented by carved ivories and enamels, paintings on panel and canvas, sculpture, metalwork, ceramics, and tex-

Fig. 118. Mark di Suvero's *Orion* steel sculpture in front of UMMA's Stuart Frankel and the Frankel Family Wing. Photo by Leisa Thomson.

tiles, dating from the late medieval period to the early twentieth century. Under Director Sawyer, European art, including Old Master drawings and prints, became a special area of focus. Director James Steward developed the eighteenth-century collection with the addition of the grand manner portraits of General Maximilien Foy by François Gérard and of Lord Charles Spencer by Joshua Reynolds, and the dramatic historical painting *The Dead Soldier* by Joseph Wright of Derby. The collection also includes rich holdings in eighteenth- and nineteenth-century French art by artists such as William-Adolphe Bouguereau, Jean-Baptiste Perronneau, Eugène Delacroix, and Claude Monet. American art works of the nineteenth and twentieth centuries—including portraits by Rembrandt Peale and Ammi Phillips, Hudson River School landscapes, and an exceptional early Whistler seascape—allow the museum to tell a broad story about the history of art in the United States.

Modern and Contemporary

Beginning with Slusser, UMMA has collected contemporary art including the major stylistic movements of the early twentieth century—cubism, surrealism, German expressionism, and abstract expressionism—as well as key works by later twentieth-century artists such as Franz Kline, Josef Albers, and David Smith. In the 1960s, a bequest by Florence L. Stol introduced modern masters such as Picasso, Léger, and Henry Moore. In 1993, UMMA received one of the most significant donations in its history when five paintings from the traveling exhibition *Picasso and Gris: Paintings from the Carey Walker Foundation* were given to the museum in recognition of its innovative educational programming during the exhibition's tour in Ann Arbor. More recently, the Lannan Foundation gift contributed works by Judy Chicago, Marcel Duchamp, Keith Haring, Robert Morris, Louise Nevelson, Man Ray, and Dorothea Rockburne, while Jack A. and Noreen Rounick gifted 167 works from the 1970s and 1980s, including paintings by Kenny Scharf and the monumental sculpture *Ternary Marker* by Beverly Pepper. In 2007, U-M professors Ernst Pulgram and Frances McSparran donated a magnificent collection of German and Austrian expressionist art, including works on paper by Gustave Klimt, Egon Schiele, and George Grosz. Across a range of media, the modern and contemporary collections at UMMA are able to demonstrate the breadth of artistic output in the twentieth and twenty-first centuries. The museum continues to actively collect today, and Director Rosa established dedicated galleries for design, photography, and new media.

Prints, Drawings, and Photographs

The museum's substantial collection of works on paper ranges from medieval illuminated manuscript leaves and Old Master drawings and prints to contemporary, experimental techniques. Highlights include outstanding print impressions by Marc Chagall, Albrecht Dürer, Robert Motherwell, Rembrandt van Rijn, and Francisco Goya, and the world-renowned collection of more than 180 etchings and lithographs by James McNeill Whistler from Parker's 1936 gift. The photography collection, which began in earnest in the 1970s, now comprises nearly five thousand images that cover the history of the medium—from early daguerreotypes and salted paper prints by William Henry Fox Talbot to examples by some of the most iconic practitioners of the medium, including Edward Steichen, Alfred Stieglitz, Ansel Adams, Richard Avedon, and Walker Evans, as well as contemporary artists such as Sally Mann and Candida Höfer.

THE UNIVERSITY'S CULTURAL HEART

Since the 2009 expansion, UMMA's visitorship has doubled—the museum now welcomes more than two hundred thousand people a year, of which over thirty thousand participate in educational programming, including classes, workshops, concerts, plays, lectures, and symposia. The architectural design of the new Frankel Wing seeks to capture the interest of the accidental visitor or passerby, encouraging access to the galleries and to the museum's diverse range of programming. As it looks toward the university's bicentennial and beyond, UMMA is transforming virtual as well as physical access to its collections, enabling its vision to be a university art museum for the world.

UMMA Leadership

Henry Simmons Frieze	1866–89
Martin D'Ooge	1889–1911
Herbert Richard Cross	1911–22
Bruce M. Donaldson	1922–40
Jean Paul Slusser (first director)	1946–56
Charles H. Sawyer	1957–72
Bret Waller	1973–80
Clifton Olds (interim director)	1980–81
Evan Mauer	1981–88
Graham Smith (interim director)	1988–90
William Hennessey	1990–97
Carole McNamara (interim director)	1997–98
James Christen Steward	1998–2009
Kathy Huss, Ray Silverman, Ruth Slavin (interim directors)	2009–10
Joseph Rosa	2010–16[7]

Entry prepared by Emily Talbot and Pamela Reister.

525 South State. Open Tuesday through Saturday, 11 am to 5 pm, Sunday 12 to 5 pm. http://www.umma.umich.edu

Notes

1. *Descriptive Catalogue, 1857.*
2. Regents records refer to this appointment as "curator of so much of the Museum as relates to the fine arts," Regents Proceedings, 155.
3. Frieze, 1876.
4. D'Ooge, 1892.
5. Wu, 2000.
6. African objects donated by the Miners can be viewed online at http://quod.lib.umich.edu/m/musart?type=boolean;view=thumbnail;rgn1=musart_cl;q1=-Gift%20of%20Prof.%20and%20Mrs.%20Horace%20M.%20Miner
7. As this volume goes to press, UMMA is searching for its next director.

REFERENCES

Descriptive Catalogue of the Museum of Art and Antiquities, University of Michigan. Ann Arbor: University of Michigan (1857).

D'Ooge, Martin L., *Catalogue of the Gallery of Art and Archaeology in the University of Michigan.* Ann Arbor: University of Michigan (1892).

Frieze, Henry, *Descriptive Catalogue of the Museum of Art and History in the University of Michigan.* Ann Arbor: University of Michigan (1876).

Proceedings of the Board of Regents of the University of Michigan, 1864–1870. Ann Arbor: University of Michigan.

UMMA: A Beacon for Art. Ann Arbor: University of Michigan Museum of Art. (2009).

Wu, Marshall P. S., *The Orchid Pavilion Gathering.* Ann Arbor: Regents of the University of Michigan (2000).

NICHOLS ARBORETUM

1906–2017

Nichols Arboretum was created as a unique collaboration between the city of Ann Arbor and the University of Michigan to provide a botanical garden for studying both natural flora and planted gardens in close proximity to campus. The current 123-acre site is cherished as a community place of refuge as well as a site for research and teaching.

In his testimony before the regents in 1906, botany professor Frederick Newcombe explained the academic need for a botanic garden:

> In this University we have been able to work without a botanic garden, because we have been able to use the whole surrounding country for our field of study. But one can foresee that in twenty-five years the farmer's woodlots will be put under cultivation for timber, the bogs will be drained, and no land will be left where vegetation can be studied in natural conditions. The land of the proposed donation is ideal for the purposes of bringing together in a small area all the vegetation of this climate, giving to each plant society its characteristic conditions. The tract is made up of hills, ravines, level low land, a bog, and a rivulet.[1]

The efforts to secure a botanical garden and arboretum for the campus had been a long-fought battle, arising repeatedly through the 1800s. By 1899, Professor Julius Schlotterbeck of the Department of Pharmacy and V. M. Spalding of the Department of Botany took matters into their own hands, creating some sixty planting beds on campus behind the library. As Schlotterbeck wrote, their plan "was to bring in from the woods and acquire by purchase as many medicinal and economic plants as would flourish in this latitude and in the conditions which could be provided on the campus." Plants were arranged by family to showcase plant relationships. The thirty members of the graduating class of the Department of Pharmacy in 1899 further embellished the evolving botanical garden/arboretum by each contributing "a tree that would be of interest from a medicinal or economic point of view." In addition, a large collection of cannas and a variety of hardy shrubs were added to enhance the campus area. Gradually, with trampling of plants and crowding from campus development, the site became less suitable for the active botanical garden and arboretum that faculty sought, and efforts continued to find a more suitable location.[2]

In 1906, a breakthrough occurred when Walter and Esther Nichols offered to donate thirty acres of land near to campus. Adjacent to the Nichols land was an additional twenty-five acres owned by the city of Ann Arbor, known as the Mummery tract. Another gift of land from the Detroit Edison Company brought the total acreage to approximately eighty acres. Professor Frederick Newcombe, head of the Department of Botany, and George P. Burns, also associated with the Department of Botany and a park commissioner of the city, were instrumental in forging a cooperative arrangement with the city to create a joint botanical garden and arboretum for use by the university and by the Ann Arbor city schools. In 1907, Burns was appointed director of the new botanical garden.

The landscape design for the property was created in 1906 by acclaimed Chicago landscape architect Ossian Cole Simonds, a U-M alumnus (engineering, 1878). Upon visiting the site, Simonds noted that "Ann Arbor is to be congratulated on acquiring such a beautiful park as can be made out of the Arboretum and Mummery tracts. . . . The land is quite varied and this will

Fig. 119. Main Valley facing north towards Huron River. Matthaei Botanical Gardens and Nichols Arboretum.

mean an exceptional opportunity to grow plants of all kinds for there will be abundant light and shade."[3] Simonds developed a design that capitalized on the site's scenic beauty while providing settings for the study of plants and natural systems. He proposed keeping the valleys open and planting the hills and boundaries to increase the overall sense of enclosure. The major road through the property, eventually known as Nichols Drive, dropped 185 feet from Geddes Road to the Huron River and provided scenic overlooks.

Between 1907 and 1915, the Botanical Gardens and Arboretum experienced multiple changes in leadership. Burns left the university in 1910 and was succeeded by Dr. Charles Herbert Otis, who completed his book *Michigan Trees* while serving as director. In 1915, Dr. Henry A. Gleason was appointed director; he began constructing greenhouses at a new botanical gardens site on Iroquois (thought to be more conducive to the types of botanical research and teaching sought by the botany faculty). In 1916, Professor Aubrey Tealdi, chair of the Department of Landscape Design, took charge of the old botanical gardens property. In January 1923, changes were made in the deed from the Nichols family and the name of the property formally became Nichols Arboretum.[4]

In the early years, there was much experimentation with plant collections, particularly recently introduced Asiatic species, many of which did not prove to be particularly hardy or were not adapted to the calcareous soils of the region. A few, such as shrub honeysuckle (*Lonicera maackii* and *L. tatarica*) and common buckthorn (*Rhamnus cathartica*), were heavily planted in portions of the Arboretum. Eventually they escaped from cultivated areas and invaded many of the woodland areas where they are still a problem.

Tealdi deserves credit for creating many of the early plantings in the Arboretum, shaping key collections such as the Peony Garden, Heathdale, lilacs and crabapples, and various trees and shrubs throughout. He received occasional input from Simonds, who continued to teach intermittently in the landscape design program, commuting from his office in Chicago. As Tealdi noted in an article in the *Michigan Alumnus* in 1922, "the aim of an Arboretum is to grow a collection of native and foreign plants in such a way as to show their habit when grown either as specimens or in groups and masses, and to make them readily available for study both by students and by the general public."[5] He further noted that seventy-five species of native trees and shrubs had been found on the site prior to its development. Many of the introduced plants were acquired from the Arnold Arboretum of Harvard University.

Tealdi's method for shaping the spaces and plantings combined the traditional phylogenetic approach of grouping plants by families with aesthetic groupings that took advantage of the unique beauty and topography of the site. He noted that "our aim is to develop both a scientific collection and a landscape park in which beauty of the several parts and of the whole is the main consideration." He also placed great emphasis on the native flora, predicting that the Arboretum would "become a haven of quiet one hundred years from now when our rich native flora will have become a thing of the past in most places," pointing out its value as a refuge for birdlife, and linking it to "the widespread movement for the conservation of national scenery, flora and fauna."[6]

Fig. 120. Peony Garden in the late 1920s, photo by George Swain. Bentley Historical Library.

One of the key gardens shaped by Tealdi was the Peony Garden, created through a gift of peonies from William E. Upjohn in 1922 and an allocation of $2,500 from the Board of Regents to construct and care for the garden.[7] Work began in 1923 and the garden opened to the public in 1927.[8] Plants were arranged in twenty-seven display beds; inventory maps, created in 1924, 1929, 1937, 1949, and 1972, were used as guides for visitors to find specific peonies in the garden. While there had been several previous efforts to renew the garden, the current Peony Garden Initiative began in 2007 under Curator David Michener's direction, supported by an advisory board made up of American Peony Society members and other experts from the United States and Canada. A recent project has been the addition of tree peonies to the garden, which had been originally planned but never implemented. The Peony Garden, widely recognized for its outstanding collection of historic herbaceous peony cultivars, became a founding member of the Peony Collaborative in 2008 as part of the Plant Collections Network of the American Public Gardens Association.

Among Tealdi's great challenges, however, was the pressure for the Arboretum to be used more for recreation purposes than for educational and other passive uses. Tealdi noted that the "scientific development of the arboretum and its use for sports are incompatible. You can't keep a lion and a lamb in the same cage."[9] The conflict became even more pronounced in 1921 when engineering professor F. N. Menefee proposed that the university convert portions of the property to a winter sports complex. President Burton was sympathetic to this idea as a way of combating the "evils of student life" during Michigan's long winters. To make the complex, toboggan runs would be constructed on some of the steepest hills and the Main Valley would be dammed and flooded to create an ice rink. Tealdi vehemently fought the idea, noting the sensitivity of the Arboretum landscape and plantings; in a letter to President Little in 1927, he pointed out the damage to plant collections by skiers and tobogganers. President Ruthven convened a committee in 1934 to consider the property's future. Ultimately, the decision made was to safeguard the Arboretum and to refrain from developing it with winter sport facilities.[10]

Shortly after, Tealdi retired and Professor

Fig. 121. Prairie at Alex Dow Field. Matthaei Botanical Gardens and Nichols Arboretum.

Fig. 122. Arboretum in the Fall. Matthaei Botanical Gardens and Nichols Arboretum.

Harlow Whittemore took over. Whittemore shared much of Tealdi's vision, suggesting that it would be a mistake to try to rival other arboretums. Rather, Nichols Arboretum should demonstrate its value to the state of Michigan and surrounding areas, including northern Ohio and western Ontario, by testing the hardiness of plants from around the world in the twenty-six growing conditions found at the Arboretum. Whittemore continued to emphasize display grouping and massing of plants for artistic purposes and landscape gardens while also preserving and improving "habitat collections and natural associations of native trees, shrubs and wildflowers." He deeply appreciated the value of the Arboretum as "a place of quiet recreation and study for members of the University and other citizens and visitors." Whittemore sought to enlarge the Arboretum proper by the approximately thirty-six acres east of the Arboretum, which were later donated by Detroit Edison and are known today as Alex Dow Field. He also proposed adding the nearly seventeen acres across the Huron River (between the river and the Michigan Central Railroad tracks) that he noted were primarily used by "hobos" at the time.[11]

One dream held by both Tealdi and Whittemore was to create a series of bulletins on various aspects of plant use and landscape design. In May 1937, Whittemore published the first and only such bulletin, "Trees and Shrubs Hardy in Southern Michigan and Surrounding Territory,"[12] which featured the woody plants then found in Nichols Arboretum. In the introduction to the bulletin, Whittemore noted special groups of plants in the Arboretum, such as the French hybrid lilacs, Asiatic flowering crabapples, and the various azaleas, rhododendron, and broad-leaved evergreens grown in the area known as "Heathdale." He mentioned plans to establish a collection of "woody vines, honeysuckles, and other types" and "replicas of native habitats such as a Colorado mountain hillside."[13] Neither of these last two groups of plants have survived, and the collections of lilacs and crabapples are much reduced from what they must have been during Whittemore's tenure.

Among Whittemore's lasting contributions to the Arboretum was the acquisition of Alex Dow Field, which he saw as an opportunity for developing new collections, including additional Asiatic crabapples and flowering trees such as the Japanese cherries donated by U-M alumni in Tokyo. He also added the Inglis House and its nine-acre grounds, a gift to the university by James Inglis. During his tenure as director, over three miles of trails were added to the two miles of roads. The trails were constructed by a crew of thirty African American men working under the Works Progress Administration.[14] During this period, the Arboretum routinely employed some ten to twenty students working under the National Youth Administration program, a work study-type New Deal program that provided work experience and financial support for college and high school students.

When Professor Walt Chambers took over the directorship in 1958, he had dreams of further transforming the Arboretum into a dynamic center for plant display and landscape design, continuing with an emphasis on plants from around the world that were hardy in southern Michigan. By 1965, he proposed creating an ambitious "study center" that would include lecture facilities and meeting rooms, a conference area, a research laboratory, and greenhouses. In Chambers's mind, the best place for the study center would be Dow Field, or perhaps the magnolia collection area along the drive leading to the Arboretum's field office/superintendent's residence. He saw the Inglis House with its varied gardens as serving as a center for landscape design with displays of herbaceous plants. The natural areas of the Arboretum would provide demonstrations of well-designed nature trails and conservation/preservation of natural resources.

Very little of this vision came to fruition, however, as Chambers spent much of his tenure battling the university administration over what he saw as blatant neglect of the Arboretum and ignoring destructive recreational uses and vandalism. Like the directors before him, Chambers was opposed to keeping the Arboretum open for automobiles: "You can't look at plants from a car. You have to get out and touch them." The number of students and student-owned vehicles had grown dramatically by the early 1960s, and an increasing number of them drove off the main roads in the Arboretum, damaging plants and tearing up the ground. Chambers's assistant Arlie Braman observed that the lack of enforcement of regulations "has created a feeling of 'being able to get away with anything' among users. . . . At the present the university regards the arboretum as a recreation area for its students, as well as for other people. The fact that cars, motor-bikes, scooters, bicycles, horses, sleds and skiers are freely allowed to cover every square inch of the arboretum is plainly evident. . . . I would sincerely doubt that anyone wishing to visit the Arboretum for the purposes for which it was established could have derived much enjoyment from such a visit as they would have spent most of their time stepping over lovers and dodging golf balls, baseballs, footballs, volleyballs, and what have you."[15]

Fig. 123. Scene from *Two Gentlemen of Verona*, Shakespeare in the Arb, 2008. Botanical Gardens and Nichols Arboretum.

Chambers grew so frustrated that in March 1965 he wrote to Dr. Roger W. Heyns, vice president for academic affairs, and President Harlan Hatcher, questioning the university's support of the Arboretum and proposing a dramatic solution. The land given by the Nichols family should be returned to its heirs: "It seems apparent that under the circumstances this tract cannot meet the stipulations of the Nichols Gift of the Arboretum—under present administrative policies and practice. Since under the terms of the Nichols Gift, the Professor of Botany—later succeeded by me—is responsible for the oversight of the Arboretum, I would suggest that unless changes are made, legal efforts be initiated to return the tract to the heirs of the donor."[16] Fortunately, the university did not take Chambers up on his suggestion but instead responded with promises of increased police surveillance and rule enforcement.

When Charles W. Cares assumed the directorship in 1970, he brought a more tolerant perspective to balancing recreational uses with the Arboretum's mission to education, writing:

> A former director of the Arboretum [Chambers] made the following observation, ". . . the use was nothing more than what I would call a recreational park where there could be wild drinking parties, petting parties, all kinds of recreational uses, sledding and tobogganing in the winter and a continual abuse and malicious vandalism." This is a well-founded allegation for which much proof is available. Such uses make certain plant collections almost impossible to maintain. On the other hand, it is estimated that a minimum of 60,000 visitors use these facilities for all degrees of stress mediation and passive recreation including the study of living organisms. One cannot believe the Arboretum is a failure in its present use.[17]

Cares suggested a flexible policy for the Arboretum that recognized "park" functions but barred field sports, loud music, and uses disruptive to quiet relaxation. He recognized the appreciation of the Arboretum by ornithologists, mycologists, entomologists, and naturalists and supported an unstructured approach that avoided providing picnic facilities, tables, benches, directional signs, and toilets. Cares noted: "In a society which is necessarily more and more counted, regimented, punched, and codified, opportunities to make discoveries such as the Arboretum provides are valued as a contrast to the structured routine of life."

By the 1970s, the university's Matthaei Botanical Gardens was growing, and Cares carefully considered how this might impact the role of the Arboretum. He suggested that "information on all aspects of obtaining and cultivating trees and shrubs in this region might be handled by Matthaei Botanical Gardens." Plantings at the Arboretum might be limited to placing plants that enhance the natural beauty of the site and or provide functional solutions to particular needs. "Due consideration for introducing new species to expand the flora should be considered, but only within the general requirements of the scenic and qualitative development. Thus, the scenic and character requirements of the Arboretum should be given priority over expanding collections per se." As a result, fewer new plantings were added to the collections during Cares's tenure. However, Cares was the first to recognize the value of the native prairie flora of Dow Field and in the mid-1970s began letting the grasses and forbs grow all summer, mowing them down once in the fall. Later, under Director Harry Morton, Professor Bob Grese initiated a fire management program in 1989 to further restore the remnant prairie ecosystem.

Cares also saw the value of using the Arboretum for community events such as the annual Medieval Festival that was held in the Arb from the early 1970s through mid-1980s. These practices would continue under directors Harrison Morton (1986–99) and Robert Grese (1999–present) with Peony Garden parties, poetry walks, Shakespeare in the Arboretum, and various music festivals and concerts.

Morton initiated a number of key improvements including establishing a Friends support group in 1991 and relocating the historic Burnham House to the Arboretum to serve as a visitor center, classroom space, and offices. A strong supporter of students and the teaching use of the Arboretum, Morton routinely hired students as assistants on such projects as a complete inventory and update of the woody plants found in the Arboretum, an inventory of the Peony Garden, mapping of landscape ecosystems found in the Arboretum, and the rehabilitation of the Heathdale area, which had become known as "Appalachian Glen." He also initiated a series of master's projects that included the development of a master plan in 1988 and various other studies.[18]

One of the key changes was an administrative shift—moving the Arboretum from the School of Natural Resources and Environment to report directly to the provost. This move put the Arboretum on par with other "public goods" across the university such as the Museum of Art, Bentley Historical Library, Clements Library, and University Musical Society, and encouraged greater outreach to the university at large.

The changes made by Morton allowed for continued growth under Grese, who became director in 1999. Restoration activities were extended to other areas of the Arboretum, and invasive shrubs such as common buckthorn and shrub honeysuckle were removed in attempts to encourage a return of native wildflowers and other understory plants. Grese continued garden and collection renovation projects begun under Morton, such as renovating the Appalachian Glen/Heathdale (restoring the older name of Heathdale) and the Peony Garden. He established the Gateway Garden at Washington Heights, and he created the River Landing along the Huron River and a boardwalk through the sensitive wetland habitat near the center of the Arboretum. The historic shrub collection was extended into the Centennial Shrub Garden at the Geddes entrance. Grese also encouraged temporary art installations by art, architecture, or landscape architecture students, the most extensive of which was "Imagine/Align"—a half-mile line of twenty thousand daffodils—created by Susan Skarsgard, an MFA student, in 2004. Under Grese's leadership, efforts to rebuild infrastructure in the Arboretum became a priority as well, including correcting major drainage problems such as erosion in School Girls' Glen and Harvard Drain and rebuilding the major roads (2010–14).

In 2004, Nichols Arboretum combined with Matthaei Botanical Gardens as one administrative unit within the university. Grese was named as director of the combined unit and led the integration of the two staffs, volunteer base, and Friends organizations.

Today, Nichols Arboretum serves much of its original mission, providing a place within a fifteen-minute walk of campus for students to

Fig. 124. French peony cultivar "Exquisite" in the Peony Garden, photo by Michelle Yanga. Matthaei Botanical Gardens and Nichols Arboretum.

learn many of the lessons of vegetation. It is also much more. The Arb, as it is known, has served generations in the university and Ann Arbor community as a place of respite in the heart of the city, a valued refugia for birds and native flora, and a place to celebrate the seasons—from early spring wildflowers, to the glorious displays of azaleas, rhododendron and peonies, to the vibrant colors of fall foliage, to the quiet beauty of winter. It is often counted among the most treasured and scenic places in Ann Arbor.

Entry prepared by Robert E. Grese.

Nichols Arboretum is open every day from sunrise to sunset; entrances are on Geddes Avenue and Washington Heights across from the U-M Medical Center. http://www.lsa.umich.edu/mbg/see/NicholsArboretum.asp

Notes

1. Newcombe, 738.
2. Bartlett, 11.
3. "The Landscape Gardener Here and At Work."
4. *Proceedings of the Board of Regents* (1920–23), 715.
5. Tealdi, 714
6. Ibid., 716
7. University of Michigan, Board of Regents. *Proceedings of the Board of Regents* (1920–23), 576.
8. Tealdi, "Peony Garden," 3; Tealdi, "The Nichols Arboretum," 25.
9. Shaw, Wilfred B. Advisory Committee on University Arboretum, Meeting held at the Michigan Union, April 28, 1934. Matthaei Botanical Gardens and Nichols Arboretum Archive, University of Michigan.
10. F. N. Menefee to President M. L. Burton, March 26, 1921; M. L. Burton to F. N. Menefee, June 14, 1921; Aubrey Tealdi to President C. C. Little, September 22, 1927. Letters. Matthaei Botanical Gardens and Nichols Arboretum Archive, University of Michigan; see also Tealdi, "Letter to the Editor."
11. Shaw, Advisory Committee on University Arboretum, 1934 (see endnote 9).
12. Nichols Arboretum—Bulletin, 1937. Matthaei Botanical Gardens and Nichols Arboretum Archive.
13. Whittemore, "Trees and Shrubs," 3.
14. Whittemore, "The Founding of a Metropolitan Park System," 1–2.
15. Walter L. Chambers to R. R. Wilkinson, April 18, 1962, and Arlie B. Braman to Walter L. Chambers, September 10, 1964. Letters. Matthaei Botanical Gardens and Nichols Arboretum Archive, University of Michigan.
16. Walter L. Chambers to Roger W. Heyns, March 1, 1965. Letter. Matthaei Botanical Gardens and Nichols Arboretum Archive, University of Michigan.
17. Cares 1972.
18. University of Michigan master's projects and practicums between 1988 and 2000 include: Theresa Callery, Michael Canteberry, Deborah Havens, Chester B. Hill, Stephen I. Houseal, Sang-Oh Kim, and Michelle Youngquist. "Nichols Arboretum: A Master Plan" (master's project, 1988); Elizabeth M. Elling, "An Environmental Education and Communication Plan for Nichols Arboretum" (practicum, 1993); Tracy Boyle, Kristina Nebel, Michael Psarouthakis, Gail Quaderer, Susan Simon, and Martha Wyatt, "Nichols Arboretum: Improving User Experience" (master's project, 1996); Shalini Priyadarshini, "A Home for the Arboretum" (practicum, 1999); Zayn Al-Abideen, Nu Ehgotz, Christopher Kunkle, James Lempke, and Susan Letts, "A New Vision for the Nichols Arboretum" (master's project, 2000).

REFERENCES

Bartlett, Harley Harris. "The Botanical Garden at the University of Michigan." *Bartlettia* 2, no. 3 (1982): 11–12.

Cares, Charles W. May 1, 1972. Untitled Memo. From Matthaei Botanical Gardens and Nichols Arboretum, University of Michigan.

"The Landscape Gardener Here and At Work." *Ann Arbor Daily Times* (Ann Arbor, MI), August 17, 1906.

Newcombe, Frederick C. *Comments to the University of Michigan Board of Regents*. University of Michigan Board of Regents. *Proceedings of the Board of Regents*. Ann Arbor: University of Michigan (1901–1906).

Tealdi, Aubrey. "Letter to the Editor: Winter Sports at Arboretum Impossible." *Michigan Daily* (Ann Arbor, MI), March 15, 1922.

Tealdi, Aubrey. "The Nichols Arboretum." *American Peony Society Bulletin* 46 (1931): 26–28.

Tealdi, Aubrey. "Peony Garden at Ann Arbor." *American Peony Society Bulletin* 39 (1929): 3–7.

Tealdi, Aubrey. "The University Arboretum, a Beautiful Setting for the Study of Plant Life." *Michigan Alumnus* 28 (March 30, 1922): 713–17.

University of Michigan Board of Regents. *Proceedings of the Board of Regents*. Ann Arbor: University of Michigan (1920–23).

Weyland, Hank B. Hank B. Weyland to Walter L. Chambers, August 28, 1963. Letter. From Matthaei Botanical Gardens and Nichols Arboretum, University of Michigan.

Whittemore, Harlow O. "The Founding of a Metropolitan Park System." Harlow O. Whittemore Papers, Bentley Historical Library, University of Michigan.

Whittemore, Harlow O. "Trees and Shrubs Hardy in Southern Michigan and Surrounding Territory." *Bulletin of the Nichols Arboretum*. University of Michigan Official Publication 38 (May 15, 1937).

MATTHAEI BOTANICAL GARDENS

1906–2017

From its origin in the 1890s as a plant collection for interdisciplinary research and teaching among faculty and students in pharmacy and botany, the Matthaei Botanical Gardens now actively embraces more than sixty faculty in seven U-M colleges and schools and over 1,100 students in on-site course and fieldwork—while also attracting over 350,000 public visitors including at least 4,500 precollege students in educational programs. The living collections span four properties and range from exotic conservatory specimens to high-quality natural areas.

The Matthaei Botanical Gardens and Nichols Arboretum have been jointly administered since 2004, nearly a century after they were established. Although a university "Botanical Gardens" was first created as a single unit in 1906 on the land that is now the Nichols Arboretum, it quickly became riven by competing faculty needs and agendas (see entry, this volume). The resolution to cleave the institution into a separate Botanical Gardens and Arboretum by 1914 was a pragmatic decision that endured for ninety years. The 2004 determination to reunite the Botanical Gardens and Arboretum is thus both a return to the founding vision and a new creation of a successful and fundamentally well-grounded institution.

THE BOTANICAL GARDENS ON IROQUOIS STREET, 1914–61

The rugged topography of the bluffs and slopes on the Nichols farm between Geddes and the Huron River that was the gardens' founding site met many aesthetic and landscape-scale collection criteria. It was, however, considered entirely unsatisfactory by faculty in botany and pharmacy, who needed land for uniform experimental field-plots and extensive greenhouses. Their dissatisfaction led the regents to appoint a committee in January 1914, determined to "purchase whatever land seemed needed." Within months, the first twenty acres of land on Iroquois Street, then outside the city limits, were purchased for $4,000.[1]

With Dr. Henry Gleason as director, the trajectory of the "new" Botanical Gardens focused on faculty needs in pharmacy and botany and became increasingly independent of the programs at the "old" gardens, renamed in 1923 as the Nichols Arboretum. The "new" Botanical Gardens collections included the herbaceous plants dug up from the Arboretum site, and presumably also some plants from the faculty-initiated gardens begun in 1899 on the Diag. In its early years, much of the garden space was allocated to "an extensive project of growing drug plants. It was wartime, and there were serious shortages of several drug plants formerly imported from Europe."[2] Additional perennials, likely of medicinal and economic significance, were donated by the Beal Botanical Garden of Michigan State Agricultural College (now Michigan State University).

By February 1916, the Botanical Garden's first greenhouses opened. Within a year they held over six hundred instructional specimens of "nearly every important plant family,"[3] including a diverse array of tropical materials donated by the Missouri Botanical Garden as well as a cactus and succulent collection from faculty expeditions to Mexico. Previously, some of these plants had been grown in rented greenhouse space and in U-M building windows for year-round, if inconvenient, access.

Under Harley Bartlett's directorship (1919-1955), the School of Forestry's nurseries were relocated to the gardens in 1924; the greenhouses

Fig. 125. Conservatory under construction, 1964/5. Photo by A. G. Norman.

were expanded in 1930 just as the Great Depression began, though not enough to meet aspirations for significant decorative and ornamental displays.[4] Over the ensuing years, graduate student and faculty research increasingly focused on plant pathology, genetics, and taxonomy. Research collections were assembled of Chinese chestnuts, delphinium, evening primroses, ragweeds, tomatoes, and wild roses. Of special interest was the genetics of *Chrysanthemum*, and their ornamental value led the Botanical Gardens to host what became an annual Chrysanthemum Show. As early as 1922, the gardens exhibited a "huge collection of chrysanthemums"[5] at the State Flower Show in Barbour Gymnasium.

By midcentury, encroaching development made access increasingly difficult and the Botanical Gardens' property had been truncated to forty acres. Instructional space was always inadequate, even after the 1952 addition that accommodated the popular course taught by Dr. Elzada Clover, U-M's legendary first woman PhD in botany who led field expeditions for her research on the taxonomy of desert and tropical plants.[6] In addition, soil and frost limitation on the property precluded developing a rich and diverse outdoor plant collection. Further, by 1954, not only was Bartlett's thirty-six-year tenure as director ending but the context of botanical research had entered a period of unprecedented change. The discovery of DNA's role in genetics (1953) and the launch of Sputnik (1957) were symptomatic of the post-World War II-era's paradigm shifts in all fields of scientific research and required new institutional resources to support them. At the university, an internal assessment of the Iroquois facility succinctly described it as "obsolescent."[7]

In 1954, LSA Dean Odegaard appointed a committee to find a new, and permanent, home for the gardens that would meet the "requirements of a rapidly growing faculty and student group, and the changing character of the plant sciences."[8] While several sites were considered, a mosaic of land near the settlement of Dixboro, about six miles north of campus, was selected. The core property was generously donated in 1957 by Frederick and Mildred Matthaei.[9] With the initial construction phase of the new Botanical Gardens completed in 1959 and collections transitioning from the Iroquois site well underway in 1960, the severe hailstorm of May 1961 that destroyed the old greenhouses was a coup de grace. The university sold the Iroquois site in part to fund the new facility.[10]

THE MATTHAEI BOTANICAL GARDENS, 1957–2004

Director A. Geffrey Norman's mandate to establish an institution "second to none" would require both significant resources and time. With the founding property in hand,[11] an ambitious construction program began to build a unified classroom, lab, office, and conservatory complex designed by Alden Dow of Midland, Michigan (later Michigan's architect laureate). The estimated $1.6 million construction cost was too great to bear at once, so the building was constructed in three phases with funding from the university, a National Science Foundation matching grant, the sale of the Iroquois site, and an anonymous donor. President Harlan Hatcher attended the institution's dedication in 1962, where it was noted that "the development of the site is likely to extend over many years."[12]

Fig. 126. Azalea, invited bonsai exhibit by Dr. Melvyn C. Goldstein, U-M Class of 1959, 2015. Photo by Michele Yanga.

While the labs, offices, and research greenhouses were already in use by the time of the dedication, the grounds and outdoor gardens had not been established and the final and crowning construction phase—a conservatory and linked multipurpose university and public space—was imminent. When completed in 1964, it was the largest free-span university conservatory in the United States, measuring 201 feet long by 66 feet wide and rising 32 feet above the tropical plants.[13] The grounds north of the building complex were designed by the landscape architecture firm of Eichstadt and Grissom Associates of Detroit, with significant implementation in 1962–63. Parker Brook was diverted and the glacial hills were reconfigured to create a two-level pond system that enhanced the long entry drive's sweeping views and provided aquatic to upland plant collection sites. Major plantings were anticipated and planned, though never fully executed.[14] A set of formal gardens adjacent to the conservatory was proposed, and a trail system was developed to lead visitors from the building complex, through the incipient gardens, and into the substantial natural areas along Fleming Creek. The new facility allowed, for the first time, on-site classroom and research facilities and the ability to host and support public programming indoors and out. The multipurpose auditorium could serve up to a hundred people. Regent Matthaei's objective—a facility that "would compare with or surpass in excellence the finest of such facilities wherever now existing"[15]—was at hand.

Under Warren ("Herb") Wagner's directorship, the new Botanical Gardens was an ascendant institution. The conservatory soon held over twelve hundred nonhardy plants of economic, ecological, or taxonomic interest. Three key founding outdoor gardens or collections—herbs, medicinal garden, and native plants—noted as underway in the dedication were completed (and in all cases much expanded in 2015); with less success, staff also attempted to add ornamental roses and other collections to the grounds.[16]

Unlike the Nichols Arboretum or Iroquois

sites, the Dixboro site included extensive riparian, wetland, and upland native plant communities. The Matthaei family had managed the property for decades and established a range of native and exotic tree plantings, as well as fishery improvements in Fleming Creek—all in accord with the conservation ethos of the era.[17] The original groves of high-quality natural areas were the "crown jewels" of the property.

At the time the site was developed, the Fleming Creek floodplain was well known for its "Cathedral Elms," an extensive native expanse of majestic American elms (*Ulmus americana*) with additional stands of hundreds of tamarack (*Larix decidua*). However, beginning in the late 1960s, waves of invasive exotic pests annihilated first these canopy trees and then (starting in the 1990s) the extensive native ash (*Fraxinus*) stands that had sprung up to replace them. Only Radrick Forest (approximately eight acres not regularly open to the public) and the much reduced (but being restored) fens are still recognizable from the period documents. The Botanical Gardens' earliest ecological restoration project began in 1967 as a long-term demonstration and is now known as the Marilyn Bland Prairie in honor of the U-M graduate student who initiated it.

During Wagner's tenure, many popular courses were offered through U-M Extension Service, some becoming the core of the decades-long Adult Education program of staff member Ellen Weatherbee. Several botanical, environmental, and horticultural organizations began meeting at the gardens and supporting its programs or mission through their activities, a tradition that continues today.[18]

In 1969, the regents "took pleasure in renaming the institution as the Matthaei Botanical Gardens."[19] It was responsible for its facilities on Dixboro Road as well as the properties now known as Horner-Mclaughlin Woods and Mud Lake Bog—each with different high-quality natural areas totaling over three hundred additional acres.[20] The Botanical Gardens' living collections extended from individual plants collected on international faculty research expeditions to representative samples of important ecosystems of the Great Lakes region. Although not clearly articulated, the collections as a concept had thus expanded from individual, albeit important, specimens to embrace entire landscape-scale native habitats. Perhaps not coincidentally, Barnes and Wagner's *Michigan Trees*, was published in 1981 a contemporary version of former Botanical Garden Director Charles Otis's *Michigan Trees* published in 1913 (see Nichols Arboretum entry).[21]

Public outreach continued in new and vigorous ways. The "Friends of the Matthaei Botanical Gardens" was organized by Dr. Helen Smith during the directorship of Eric Steiner. From its initial membership of eighty-four in 1974, the group now has over twenty-four hundred members and serves as a bridge between the university and the public.

Several display gardens were under incremental development, spanning the directorates of Wagner, Steiner, and Benninghoff. Noteworthy are the Perennial Garden, which began as a formal rose garden at the north end of the conservatory; the Marie Azary Rock Garden (dedicated 1985); the Helen V. Smith Woodland Wildflower Garden (dedicated 1989); the herb garden, which was significantly expanded as the Alexandra Hicks Herb Knot Garden (dedicated 1994 as part of a major garden plan developed by HPK of Dexter, Michigan); and the initial work on the Gateway Gardens of New World Plants (dedicated 1996).[22]

The spring Ann Arbor Flower Show was begun in 1990 during the directorship of Steiner and ended ten seasons later during the tenure of Director James Teeri. During its run, it was actively promoted by Patricia Hopkinson and increasingly became the programmatic activity of the institution. The Flower Show was to benefit the Botanical Gardens, but as there was no facility on-site to host it and the hockey team's conference schedule precluded recurrent use of the U-M Yost Arena, the show became associated with the Farm Council Grounds near Saline. As a public program involving the Botanical Gardens, it was a popular herald of spring, but it did not pass an internal review and was discontinued in 2000.

Fig. 127. Visitor approaching the Alexandra Hicks Herb Knot Garden, 2015. Photo by MBGNA staff.

A history of the roles and challenges in integrating flower shows oriented to the general public with the academic mission of a university garden has yet to be written. At U-M, such shows serve as a valuable bridge among university and community partners and have been tremendously popular. However, it is costly and time-consuming to produce these displays, which, although visually appealing, are not tightly integrated into the Matthaei Gardens' larger educational and research mission. Their intermittent occurrence had begun with two major floral exhibits early in the twentieth century, held in conjunction with significant art exhibits in Alumni Hall. Other shows included the chrysanthemum show sponsored from the Iroquois site and the midcentury Ann Arbor Flower Show at Yost Arena. All were predecessors of the 1990–2000 revival. Today, the Botanical Gardens' facilities host the annual shows of several mission-related groups including the Ann Arbor Bonsai Club, Ann Arbor Orchid Show, and African Violet Society.

During the Teeri directorate, the Gardenscape project, funded by the Ford Motor Company Fund, provided for five experimental and temporary landscapes created by Michigan-based designers and dedicated between 1999 and 2003. Of these, the Urban Pocket Garden by K. C. Runciman Associates of Saline, Michigan, has been integrated into the long-term vision of the property.[23] The conservatory had its first major renovation, with substantial assistance from Robert and Marjorie Alpern. In addition to revitalizing the Temperate House collections, this work arrested rust to key structural elements and restored the main components of the Alden Dow path system.[24]

Outdoors—and equally significant for long-term collection development—was the establishment of what has become the Sam Graham Trees, a more than half-mile-long trail through a series of replanted native tree groves. This trail is targeted to schoolchildren and families so that they can learn the key trees and shrubs of Michigan.

Fig. 128. U-M field class in Radrick Fen, 2014, Photo by David Brenner.

Public programming aimed at underrepresented audiences included "Out of Africa" in conjunction with the College of Engineering and the Ypsilanti public schools (2000–2003).[25]

THE MATTHAEI BOTANICAL GARDENS AND NICHOLS ARBORETUM, 2004–17

In 2004, the Matthaei Botanical Gardens and the Nichols Arboretum were administratively reunited under Director Bob Grese and are refocused on serving all the university's academic audiences while building on the core strengths of the existing collections, gardens, and natural landscapes that historically serve the natural sciences. Diverse student and faculty research projects use the Matthaei Gardens site, mostly for the greenhouse space, field-plots, and natural areas.

Noteworthy additions to the gardens and collections, often featuring work by U-M landscape architecture students, include the Sue Reichert Trail (dedicated 2007), the Gaffield Children's Garden (dedicated 2009), continued plantings of the Sam Graham Trees (dedicated 2010), the Bonsai and Penjing Garden (dedicated 2013), and the Great Lakes Garden (dedicated 2013). In addition, student interests related to faculty activities in sustainable agricultural systems have led to the founding of the Campus Farm (2012) and an experimental permaculture plot, informally called the "Food Forest" (initiated 2014). The newest garden, developed with significant guidance from U-M faculty in pharmacy and medicine, returns to the founding theme of the institution: the Medicinal Garden (2015). Faculty in the humanities have substantially engaged the collections and landscapes—ranging from the temporary public display of the little-seen fourteenth-century Syrian cenotaph owned by the U-M Museum of Art (2007), through dance performances in the conservatory, to Checkov's "The Cherry Garden" in the perennial garden.

The Botanical Gardens has hosted several temporary experimental buildings made by faculty and students from U-M schools of architecture and engineering. These include the MiSo House (Michigan Solar House) and two that were both ephemeral and award winning: the *Shadow Pavilion* (2009) and Hair, Spikes, Cattail and Turkeyfoot (2010). Public sculpture and art include works by U-M faculty and other Michigan artists, notably the Jane LaRue Sundial by Rush (dedicated 1986), the *Great Lakes Habitats* mosaics by

Fig. 129. Potting House in the Botanical Gardens building on Iroquois Street, 1916. Bentley Historical Library.

Y. Hanachen and community volunteers (installed 2008), and *Guardian Angel* by J. Brose (1960s; installed 2013).[26]

Public outreach at the Botanical Gardens has grown significantly, with the Visitor Operations staff managing a facility and programs that operate seven days a week and all but four days a year. Precollege school groups, which at first drew only locally and occasionally, now come from the Detroit metropolitan region and must be booked months in advance. The docent program began with twenty community volunteers in 1976; it has grown to 115 volunteer ambassadors and docents in 2015. Public attendance is currently estimated at over 140,000 visitors a year.

Frederick Matthaei's vision of exceptional and integrated academic and public programs at a "facility that would compare with or surpass" any university botanical garden in this country has not been a static goal; rather, it has been a continuing quest. The current ambitious and ongoing vision of the Matthaei Botanical Gardens and Nichols Arboretum—coupled with the integration of their multiple properties, programs, staff, and unique strengths—sets the stage for a new era of service to the university and citizenry.

Entry prepared by David Michener.

Open Fall/Winter every day 10 am–4:30 pm; Wednesdays 10 am–8 pm; open Summer every day 10 am–8 pm; 1800 N. Dixboro Road. http://www.lsa.umich.edu/mbg

Notes

1. Bartlett *The Botanical Gardens*, 24; *The Botanical Garden at the University of Michigan*, 16.
2. The controversy over commercial-scale drug production with university resources is presented in the above two histories by Bartlett.
3. Flook and Gleason, *The University's Botanical Garden*, 7.
4. Ibid. The comment that "last year" [1916] the annual chrysanthemum show had to be abandoned for lack of greenhouse space in which to grow the plants is the earliest note yet seen of an institutionally supported recurrent flower show and its resource conflicts with academic priorities.
5. *Michigan Alumnus*, 138.
6. Several of Dr. Ezalda Clover's living research specimens are still on display at the Matthaei Botanical Gardens. Her papers are at the U-M Bentley Historical Library.
7. Norman, 30.
8. *University of Michigan Botanical Gardens Dedicatory Exercises*, June 14, 1962.
9. In 1960, Frederick Matthaei Sr. was elected regent and served until 1967; Fredrick Matthaei Jr. served as regent from 1967 to 1968.
10. Shackman, 2001.
11. The land of the Matthaei Botanical Gardens was assembled from several owners; only part became the Matthaei Botanical Gardens. Contiguous parcels became

part of the University of Michigan Radrick Farms Golf Course, the University of Michigan's Recreational Sports facility, and the Huron Valley Humane Society. The Matthaeis retained and developed a secluded perimeter parcel.

12. *University of Michigan Botanical Gardens Dedicatory Exercises*, June 14, 1962.

13. Several articles on the conservatory's history and significance are in the *Friends* newsletters of 2004–6.

14. Within the framework established by Eichstadt and Grissom, U-M professor of landscape architecture C. Cares developed detailed plans for the entry plantings and a conceptually innovative "Genetics Garden." Most of these plans were only partially implemented (Cares, pers. comm.).

15. *Proceedings of the Board of Regents*, September 1957, 113–14.

16. Many collection themes were tried but did not become integrated into the institution. Some were unable to deal with the alkaline soils site conditions. Others had too much potential to be sources of invasive exotic pest plants.

17. The extant plantings of European and Rocky Mountain conifer species on the north and west boundaries, the horse-chestnuts in the long-abandoned Cadillac gravel and railroad operation [as shown in pre-WWI land title maps] near Radrick Forest, and the upland stands of honey-locust were all planted by the Matthaei brothers (pers. comm. of Fredrick Matthaei Jr.). The fishery buildings and stone weirs on the Botanical Gardens' property were removed 1990 to circa 2010.

18. In 1969, six mission-related groups are listed as active on-site at least annual programs; in 2015 there are twenty-two.

19. *Proceedings of the Board of Regents*, April 1969, 1726.

20. The university's acquisition of natural areas in southeastern Michigan continued after World War II. One large tract with a significant intact bog is Mud Lake Bog. Separately, the Michigan Botanical Club purchased what is now Horner-McLaughlin Woods as a spring wildflower preserve. An IMLS-funded project has made the natural-area resources for all four properties available online at http://www.lsa.umich.edu/mbg/see/gis

21. Barnes and Wagner. *Michigan Trees* has been continually in print including a revision in 2004.

22. The initial work includes the Jean Cobb Gateway funded with the assistance of the U-M Alumnae Club Birthday Greeting Funds of 1989 and 1990.

23. The *Gardenscapes* were intended to bring the Flower Show "home" and to be long-term temporary. All five can be found searching the MBGNA web site.

24. Marge Alpern was deeply concerned that the conservatory and its collections' conditions at that time did not lead visitors to appreciate the divine, however any person might understand that relationship (pers. comm.). The dedicatory plaque reads, "As you explore the variety of plants and surround you, delighting in their fragrance and beauty, we hope you will experience a sense of wonder, joy, and appreciation for the miracle of life and the natural world. Much of what you see in the Conservatory is a gift from Marge and Robert Alpern. The University of Michigan is appreciative of the devotion and support."

25. "Out of Africa" resulted in several publications, including Michener and Klatt, 2002; Michener et al. 2002; and Michener and Schultz, 2002.

26. Each of the projects can be found within the U-M digital environments by keyword searching.

REFERENCES

Barnes, Burton V., and Warren H. Wagner. *Michigan Trees: A Guide to the Trees of Michigan and the Great Lakes Region*. Ann Arbor: University of Michigan Press, 1981.

Bartlett, Harley Harris. "The Botanical Gardens." *Bartlettia* 2, no. 3 (1982): 21–28.

Bartlett, Harley Harris. "The Botanical Garden at the University of Michigan." *Bartlettia* 2, no. 3 (1982): 9–20.

Flook, L. R., and Henry A. Gleason. "The University's Botanical Garden." *Bartlettia* 2, no. 3 (1982): 5–8.

Michener, David C., and Brian Klatt. "Peoples, Plants and Cultures at the University of Michigan Matthaei Botanical Gardens." *The Public Garden* (April 2002): 27–30.

Michener, David C., and Inger J. Schultz. "Through the Garden Gate: Objects and Informal Education for Environmental and Gardens." In *Perspectives on Object-Centered Learning in Museums*, ed. Scott G. Paris, 86–100. Mahwah: Lawrence Erlbaum Associates (2002).

Michener, David C., Karen Simpson, Beth Ramsey, Ellen Ransey, Debbie Taylor, Liz Elling, and Susan Gawecki. "Out of Africa Expands Horizons." *The Public Garden* (Summer 2002): 22–27.

Michigan Alumnus, vol. 29, p. 138. Ann Arbor: Alumni Association of the University of Michigan (1922).

Norman, Arthur G. "The Botanical Gardens 1940–1975." *Bartlettia* 2, no. 3 (1982): 29–36.

Shackman, Grace. "The Botanical Gardens on Iroquois." *The Ann Arbor Observer* (Ann Arbor, MI), May 2001. http://aaobserver.aadl.org/aaobserver/15253

University of Michigan Board of Regents. *Proceedings of the Board of Regents*. Ann Arbor: University of Michigan (1837–present). http://quod.lib.umich.edu/u/umregproc

University of Michigan Botanical Gardens Dedicatory Exercises. June 14, 1962. Unpublished brochure on file Matthaei Botanical Gardens.

STEARNS COLLECTION OF MUSICAL INSTRUMENTS

1898–2017

The Stearns Collection of Musical Instruments is part of the School of Music, Theatre and Dance (SMTD) and holds approximately twenty-three hundred musical instruments from all over the world. More than half the collection was acquired by Frederick Stearns (1831–1907), who donated about nine hundred instruments to the University of Michigan in 1899 and added around four hundred more in the years until his death. The collection also holds instruments obtained during expeditions by Joseph B. Steere (1871–75) and Carl E. Guthe (1922–25). Among the most historically important artifacts in the collection is the first commercially produced Moog synthesizer, the earliest surviving bass clarinet by the maker Adolphe Sax, and instruments from a Vietnamese orchestra used at the Paris Exposition in 1900. In addition to its historical artifacts, the collection maintains reproductions and modern ethnographic musical instruments for student use to support the academic endeavors of SMTD.

FREDERICK STEARNS: A BRIEF BIOGRAPHY

Frederick Augustus Stearns was a successful manufacturer of pharmaceuticals in Detroit who, in his later years, became a wealthy patron of the arts and sciences. He was born in Lockport, New York, on April 8, 1831. At the time, Lockport was a newly incorporated village on the Erie Canal. His family moved to Buffalo in 1842 and Stearns began his pharmaceutical career in 1846 when he apprenticed at the drugstore Ballard & Green. The firm was short-lived, and around 1848 Stearns was employed at A. I. Mathews, one of Buffalo's leading retail druggists. Eventually becoming a partner in the firm, Stearns sold his interests at the end of 1855 and moved to Detroit to establish his own business in partnership with Lewis E. Higby. In March 1856, Higby & Stearns opened as a wholesale and retail drugstore.

Stearns secured contracts with the state of Michigan to provide medical supplies to its regiments during the Civil War. At the end of 1863, Higby withdrew from the partnership, leaving Stearns as sole proprietor. For the rest of the decade, Stearns expanded the manufacturing side of his business, replacing cook stoves with steam power, milling machinery, and extraction equipment. In 1881, Stearns sold his retail drugstore and turned his attention solely toward manufacturing. The following year, the company was incorporated with $200,000 in capital stock. At the annual meeting in 1887, Stearns announced his retirement, naming his son Frederick K. Stearns as his successor.

Stearns traveled the world in his retirement, using his wealth to obtain specimens for creating extensive collections of musical instruments, gems, mollusk shells, Asian art, rare books, antiquities, and more. His collecting emphasized the customs of everyday life and methods of manufacture over aesthetic beauty, which Stearns attributed to his strong desire to supplement his moderate early education. In the 1890s, Stearns donated many of these collections to either the Detroit Museum of Art (now the Detroit Institute of Arts) or the University of Michigan. On January 13, 1907, Stearns died in Savannah, Georgia, while preparing to spend the winter in Egypt.

THE MAKING OF A COLLECTION AND DONATION TO UNIVERSITY

Stearns purchased the first instrument for his collection in 1881 at an antiquarian's shop in Prague, where he was intrigued by an instrument (Stearns 1130) that seemingly merged the shape of a harp with the functionality of a guitar. The collection grew at a modest pace over the next fourteen years, after which Stearns accelerated his acquisition rate. In 1895, an article in the *Detroit Free Press* reported an attic fire at Stearns's home and indicated that the collection numbered nearly two hundred instruments. When Stearns began discussing the possibility of donating the collection to the University of Michigan in September 1897, he said that he possessed about five hundred instruments. This figure was revised to eight hundred just two months later, and was increased again to more than nine hundred at the time of donation.

The donation was publicly announced on November 27, 1898. In accepting the gift, the university allocated the north side of the third floor of the University Museum for its display. Stearns provided additional funds for new exhibition cases, and over the course of the next year the collection was mounted with the help of Professor Albert Stanley, who along with Francis Kelsey worked with Stearns to secure the donation.

Stearns wanted to create a "complete" collection—one tracing the evolution of musical instruments from their most primitive to their most advanced forms—that would be the source for a comprehensive catalog he intended to publish with the university's imprint. To that end, he continued to add to the collection after transferring it to the university. In 1900–1901, Stearns traveled to Europe and North Africa, acquiring more than three hundred additional instruments. During this trip, he relied on dealers—who often took advantage of his lack of musical knowledge. Among these was Leopoldo Franciolini, a notorious Florentine dealer who sold his own fanciful creations to unsuspecting collectors. Stearns returned with nearly fifty forgeries and dozens of opera props, prompting the university to discourage him from making future acquisitions on their behalf without proper guidance. The university allocated additional space on the fourth floor of the University Museum to accommodate the new additions.

In 1907, following the death of Stearns and the appointment of Alexander Ruthven as curator of the museum later that year, the collection's presence within a natural history museum began to be questioned. The lack of space in the building had long been an issue—even an empty elevator shaft was used for storage—and it was evident that Ruthven had little interest in the musical instruments. Whereas Ruthven's successor, Charles Adams, included the collection and its activities in his annual report to the Board of Regents, Ruthven ignored it completely except for a single mention where he described his plans for the space the collection occupied. In 1912, the regents decided that the collection would be moved to Hill Auditorium, which was currently under construction, and displayed on the second floor. The move was overseen by Stanley and completed in April 1914.

Fig. 130. Lira-Chitarra, 1807, Naples, Italy. Stearns 1130.

The entire collection of nearly fifteen hundred instruments was mounted in just eighteen cases.

Though Stanley had been actively involved with the collection since 1899, he was officially made curator in 1916. His primary task was to fulfill Stearns's wish of completing a catalog. Stearns had drafted one prior to donating the collection, but Kelsey and Stanley determined that this version required significant reworking. In 1904, a new catalog was completed by Philip L. Schenk, a graduate student who from 1902 to 1903 had received a fellowship funded by Stearns to work on the project; however, questions surrounding its cost prevented its publication, and following Stearns's death the project was shelved. With the financial backing of Frederick K. Stearns, Stanley revised Schenk's manuscript and published the catalog in 1918.

1920–80

Stanley retired from the university at the end of the 1920–21 academic year; his final act with the collection was completing the catalog's second edition in 1921. For the next eighteen years, the collection remained on display in Hill Auditorium entirely unmanaged. Historical records from this period are almost nonexistent. Donations of instruments are occasionally referenced in the *Proceedings of the Board of Regents* during this time, but none of these can be positively identified in the collection's current holdings.

In 1940, the University Musical Society transferred control of the School of Music to the regents of the university, and steps were taken to address the staffing needs of the collection. The position of curator was reestablished and offered to William Lichtenwanger, who had just completed his master's thesis at U-M on the history of the clarinet and played a critical role in establishing an independent music library for the school. Lichtenwanger turned down the curatorial position, however, instead accepting a position at the Library of Congress, where he went on to have an illustrious career.

There was no immediate alternative to Lichtenwanger as curator of the collection, and the search for a curator was put on hold due to the outbreak of World War II, and remained suspended until the end of hostilities. In 1947, William Raymond Kendall was named curator of the collection. A year later, he resigned to become dean of the Department of Music at the University of Southern California.

There does not seem to have been a curator in the years immediately following Kendall's departure, and by the early 1950s the instrument displays were in a dilapidated state. Dry conditions inside Hill Auditorium caused many instruments to crack and deteriorate, and most were covered with dust and soot. A contemporary account described the site as "a set of dark and dirty forms, which inspired more ridicule than admiration."

By 1953, Professor Marion McArtor became involved with the collection, holding the title of "acting curator" according to a *University Record* article in February of that year. Eventually becoming the permanent curator, McArtor obtained a small budget for the collection and began the process of rehabilitating the original displays. Sadly, only one case was completed before his untimely death from a heart attack on January 13, 1956. He was forty-one.

Robert A. Warner was named curator in 1959 and under his leadership the collection enjoyed a level of stability not seen since the days of Albert Stanley. Warner resumed the rehabilitation process begun by his predecessor, and by 1960 most of the original exhibits were dismantled and replaced with new ones featuring fewer instruments. Instruments no longer displayed were stored in locations across campus until space was allocated in Burton Tower as a combination storage area and workroom. In 1964, Warner established the collection's presence inside the School of Music by securing two display cases in the lobby of the new Earl V. Moore building. Also around that time, Warner improved public accessibility of the collection's displays, previously open only during auditorium events, by establishing regular visiting hours and arranging for guided tours.

Warner recognized the importance of making the collection an academic resource to students and faculty. Toward that end, he formed a relationship with the school's early music ensemble (Collegium Musicum). The ensemble's modern reproductions were added to the collection, while some of the collection's historical instruments were made available to the ensemble.

Greater attention was paid to the collection's non-Western holdings, thanks to the ethnomusicologist William Malm, who joined the School of Music's faculty in 1960. In 1966, Malm acquired the Javanese gamelan *Kyai Telaga Madu* (Venerable Lake of Honey) for the collection. Two years later, it was used to establish the U-M Gamelan Ensemble, and it remains in use today. Malm also established the Japanese music group for undergraduate students, which used shamisens and other instruments from the collection.

It had long been recognized that the environmental conditions inside Hill Auditorium were detrimental to the collection. An assessment written around 1965 identified the lack of humidity control as the greatest problem facing the collection and suggested that the best solution would be to redisplay it in a new location. But it wasn't until the first Ann Arbor Hash Festival in 1972—when the audience of a Get Out the Vote rock concert in Hill Auditorium caused damage to exhibit cases and instruments—that serious efforts were made to move the collection to a new location. Later that year, the university purchased the former Zeta Beta Tau fraternity house on North Campus (now the Stearns Building) and the collection was entirely relocated by the end of 1973. Exhibits in the Curt Sachs room were completed in April 1974, just prior to hosting the third annual meeting of the American Musical Instrument Society.

Fig. 131. Baritone Oboe in C, mid nineteenth century, Paris, France. Stearns 0676.

1980–2015

Warner announced his retirement effective at the conclusion of the 1979–80 academic year and was succeeded by William Malm as director and School of Music faculty member and musicologist James Borders as curator. The following year, the School of Music faced a budgetary crisis that led to the removal of the collection from the Stearns Building; by the end of 1982, it was moved to the Argus II Building located west of downtown Ann Arbor. The facility offered little more than shelter, with environmental controls consisting solely of two heaters and a single humidifier drawing water from the boiler room. The collection remained almost entirely in storage until late 1985, when it was allocated two rooms and some hallway areas for exhibition space in the School of Music's new Towsley Wing.

Malm was very successful in expanding the collection's public programs and improving its financial condition. In 1986, he launched the collection's newsletter and established the 2+2+2 Lecture Series. This series was renamed to honor Virginia Martin Howard in 1992 after she established an endowment to support the collection's public programming. A year later, Malm's efforts helped establish the collection's general endowment.

Malm retired from the collection at the end of the 1993, though he continued his involvement for another ten years. Margo Halsted was named his permanent replacement in the beginning of 1995, though she stepped down prior to the start of the 1997–98 academic year, when Joseph Lam became director. The collection celebrated several important milestones during Lam's directorship, including its one hundredth anniversary and its return to Hill Auditorium following its renovations in 2004.

In late 2001, Christopher Dempsey was appointed as a graduate student curator. Dempsey oversaw several modernization efforts, including establishing the collection's web presence, digitally photographing its holdings, and converting its paper cataloging system into an easily searchable database. He brought the conditions of the collection's storage facility to the attention of the School of Music's senior administration, beginning the search for proper facilities. Additionally, he sought to reestablish the collection as an academic resource for students and faculty and to rebuild its professional reputation. To these ends, he introduced policies that facilitated loans of select instruments and spearheaded the return of the American Musical Instrument Society's annual conference in 2009.

Following departure of Lam and Dempsey from the collection's leadership in 2006, at the conclusion of the AMIS conference Steven Ball was appointed director. Under Ball, the collection grew substantially, with nearly two hundred instruments added from the collections of Leo Sarkisian and Keith Heiss. Discussions regarding the collection's storage concerns intensified, and the search began for a new storage facility with the proper environmental controls necessary to maintain the collection.

Ball left the university during the summer of

Fig. 132. Sho, date unknown, Japan. Stearns 0704.

2013, and the School of Music asked Dempsey to return to the collection on an interim basis. At that time, the school also appointed Arthur F. Thurnau Professor and ethnomusicologist Lester Monts as director designate. It was expected that Monts would take over the collection in the fall term of 2015 following his final year serving as the university's senior vice provost for academic affairs and a subsequent year of sabbatical. The collection faced two impending moves. Since construction was scheduled to begin in the summer of 2014 on an expansion and renovation project most of the collection's remaining Moore display areas were moved to the new displays cases on the Hill Auditorium balcony level lobby. The bulk of the collection was also scheduled to occupy a new climate contolled storage and work facility in the fall of 2014; however, this move was postponed when

Fig. 133. Hill Auditorium exhibit case, before 1973. Bentley Historical Library.

another, more suitable facility became available in the North Campus Research Complex.

THE FUTURE

As of the fall of 2015, the Stearns Collection's storage remains in the Argus II Building, with its move to the North Campus Research Center now scheduled for the spring of 2017. Once again, the collection finds its primary display areas in Hill Auditorium with public accessibility limited to scheduled events. Lester Monts has assumed leadership of the collection, and with the support of former School of Music, Theatre, and Dance dean Christopher Kendall (no relation to William Raymond Kendall), coordinator of communications and outreach, Carol Stepanchuk, and an advisory board drawn from across the university and community, planning for the next chapter of the collection, finally to be housed in appropriate storage and work facilities, is underway. The collections attributes in Western historical performance, in world music, and as a history of instrument technology position it ultimately to realize a role as an important resource for the University of Michigan.

Prepared by Stearns Collection of Musical Instruments Staff.

Contact information.

http://www.music.umich.edu/research/stearns_collection/

email: stearnsoutreach@umich.edu

REFERENCES

Stanley, Albert A. *Catalogue of the Stearns Collection of Musical Instruments.* 2nd ed. Ann Arbor: University of Michigan, 1921.

Stearns Collection of Musical Instruments Records, Bentley Historical Library, University of Michigan.

The Stearns Newsletter (Ann Arbor, Michigan), vols. 1–19, 21, 26 (1986–2006, 2008, 2013). Available online at http://www.music.umich.edu/research/stearns_collection/newsletters.htm

SINDECUSE MUSEUM OF DENTISTRY

1992–2017

Over the course of its history, the University of Michigan has boasted two distinct dentistry museums. While only a few specimens of the first nineteenth-century study collection and museum have survived, the contemporary Sindecuse Museum of Dentistry presents period-style exhibits of dental offices from the 1860s to 1940 and holds important collections pertaining to the history of dentistry and of the University of Michigan School of Dentistry.

Between the 1880s and 1928, the School of Dentistry (originally the College of Dentistry) housed a specimen collection of human crania and odontology created by medical faculty member Corydon L. Ford to teach anatomy to dental students. A year before his death in 1894, Ford formally donated his highly regarded specimen and library collections to the College of Dentistry where the Dental Museum or Ford Museum of Crania served as a study collection for dental students. With the addition of five hundred animal crania donated by dental alumni Dr. William Mitchell (1878) and Dr. L. J. Mitchell (1884),[1] the regents renamed the collection as the Ford-Mitchell Museum in 1904. When the Ruthven Museums Building opened in 1928, the remaining collections were transferred to the Museum of Paleontology[2] and continued to be used by dentistry faculty to teach anatomy.[3]

Today's Sindecuse Museum of Dentistry has a history separate from the earlier specimen-focused teaching collection. Rather than training dental students, its purpose is modeled on other history museums that address a broad audience interested in learning about history through artifacts and exhibitions. The Sindecuse's location within the School of Dentistry affords a built-in audience. Faculty, students, staff, and patients of the dentistry school, as well as members of the campus community at large, can visit the exhibits.

The beginnings of the museum date to 1990, when Dr. Gordon H. Sindecuse offered the School of Dentistry an unexpected and generous gift, worth approximately one million dollars, for the purpose of creating and operating a dental history museum. Dr. Sindecuse, a 1921 alumnus of the school, had retired after thirty years of practicing dentistry in Kalamazoo, Michigan, followed by ownership of a successful investment business. At the time, J. Bernard Machen, dean of the School of Dentistry, praised the new museum: "Through Dr. Sindecuse's extraordinary generosity, the education of our students will be enhanced with a unique historical perspective. The museum will . . . promote greater awareness of the history of the profession."[4] Dean Machen stressed that this "gift will foster a deeper understanding of the way science and technology evolve in accordance with the desire of the human spirit to minister to others and to heal."[5] In early 1992, he organized a museum advisory committee, which was tasked with selecting the first museum curator.[6]

In April 1992, the committee selected Jane E. Becker as a part-time curator. Becker had prior experience at two university museums in California: the A. W. Ward Museum of Dentistry at the University of the Pacific in San Francisco and the Lowie Museum of Anthropology[7] at UC Berkeley. Her first concerns were to acquire objects to display and to prepare for the museum's grand opening. Becker posted announcements about the museum's pending opening and the need for donations to the collection in several journals, including the *Journal of the Michigan Dental Association* and *DentalUM* (formerly *School of Dentistry Alumni Bulletin*).

Fig. 134. Ford-Mitchell Dentistry Collection from original museum. SMD 03090070c.

The School of Dentistry selected its Art Deco-period Kellogg Foundation Institute Building to house the museum and constructed four exhibit cases in the ground level lobby and two cases in the first floor lobby. A storeroom and curator's office were placed in the basement of the adjoining Dental School Building.

The Sindecuse Museum of Dentistry opened on September 18, 1992. Artifacts were displayed either on the shelves of the exhibit cases or in dental office settings. Museum advisory committee member Dr. Ronald Berris and his wife, Maggie, both alumni, loaned their historical dental advertising and antique dental office pieces for the museum's first displays.

To build its permanent collections, the museum accepted artifacts that had been acquired by dental school faculty. For example, Dr. Charles C. Kelsey had accumulated many photos, documents, scrapbooks, dental tools, and student mementos offered by alumni as he conducted research for articles that he published in the *School of Dentistry Alumni Bulletin* during the 1970s and early 1980s. He also donated dental instruments related to his area of teaching, prosthodontics. Museum staff continues to rely on his writing and research about the history of dentistry and the dental school.

Another faculty collector was Professor Albert G. Richards, head of dental radiology, who retired in 1981. An innovator in modern dental X-ray technology, Richards was keenly interested in and published on the history of X-ray technology in dentistry. While conducting research, he collected and tested antique dental X-ray units, tubes, and accessories, and gathered supplier photographs. His collection—including the associated correspondence with the original equipment owners and his own publications—form the nucleus of the museum's X-ray technology collection.

In addition to individual collections, other dentistry-related artifacts, photos, and documents originating from alumni or staff and faculty of the school had accumulated for decades, many having been saved from an earlier building and moved into the new Dental School Building in 1970. These objects had been stored and preserved by the Dental Library and were offered to the Sindecuse Museum in 1996. More than 650 items were transferred to the Sindecuse Museum. Included among them were correspondence from the dental school's first dean, Jonathan Taft; documents related to Willoughby D. Miller, a prominent nineteenth-century dental researcher who discovered the cause of dental caries; a large collection of tooth keys, one of the earliest types of tooth extractors; numerous early dental instruments; lecture slides; a St. Apollonia bronze

Fig. 135. Al G. Richards Collection. SMD 273.

sculpture created in 1981 by dentist/alumna Dr. Carmen Nolla; teaching specimens; scrapbooks made by dental alumni; and photographs and negatives. The dental librarian also transferred the School of Dentistry's archive of dental and hygiene class composite photographs to the museum. Both sets of a nearly complete run of the two series were digitized by the museum in 2009, with digital copies provided to the Bentley Historical Library.

Such collections were invaluable and gave the museum a central role in the preservation of the history of the U-M School of Dentistry. In addition, Becker continued to seek large dental artifacts for the dental "operatories" or dental office displays to augment the collections and serve the museum's larger historical mission. During this formative period, the museum purchased a large, historically significant, dental artifact collection from Dr. Jack Gottschalk, an Ohio dentist and collector.

To improve and expand exhibition space while continuing to build a more comprehensive collection, plans for expanding the museum's storage and displays were soon underway. Becker was vital to this process, working with planning staff to integrate conservation standards into exhibit and storage areas and communicating frequently by letter with Dr. Sindecuse as expansion plans developed. Sadly, on September 22, 1993, Dr. Sindecuse, age ninety-five, died in Sarasota, Florida, without having visited the museum named in his honor. Upon his death, 75 percent of his financial assets were given to the University of Michigan specifically to continue support of the Sindecuse Museum of Dentistry.

In 1998, renovations of the Kellog Dental Building—including the museum space—began. As one of the last projects of her six-year tenure, Becker supervised the packing and temporary removal of the collections to an off-campus storage space. While renovations were still underway, the School of Dentistry hired a new curator, Tammy Szatkowski-Reeves, who had museum conservation experience at the U-M Kelsey Museum of Archaeology. Szatkowski-Reeves, also working part-time, focused her efforts on preparing for the collection's return to expanded climate controlled storage and designing new exhibits for the renovated spaces. During this time and with the help of part-time student assistants, the museum's database was converted to a new platform and museum forms and accession records were updated to meet professional museum standards.

The new exhibitions reflected the strengths of the collection, including period-specific dental operatories, articulators used in prosthodontics, toothpowder and pastes, and dental X-ray

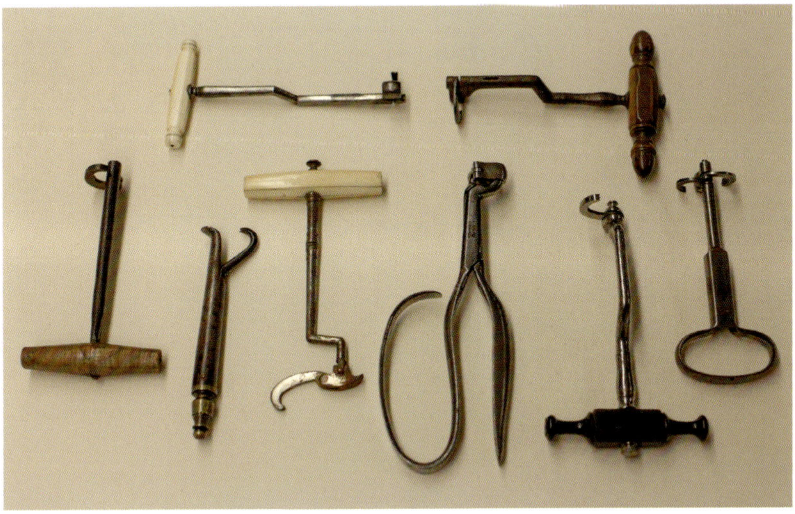

Fig. 136. Historic tooth-keys. SMD161.

technology. On September 8, 2000, during the 125th anniversary of the founding of the School of Dentistry, Dean William Kotowicz rededicated and reopened the museum and acknowledged the two-year effort of building improvements.

Most important of these improvements for the Sindecuse was the construction of a beautiful enclosed atrium that replaced a former outdoor courtyard located between the Kellogg Institute and the Dental School Building. Today, the Sindecuse Atrium is the primary museum gallery, with two large museum display cases running along its north and south walls. The atrium also doubles as event and reception space for the School of Dentistry. The building renovations also included the conversion of a former locker room to large museum storeroom and the construction of built-in wall cases in a nearby corridor. All of the remodeled spaces meet contemporary environmental standards; as a preservation measure, fiber optic lights were built into the new exhibit cases. In addition, the original 1992 exhibit cases remain in use, adding to the museum's total space for exhibitions.

In 2003, Szatkowski-Reeves successfully lobbied for an assistant curator position. Shannon O'Dell—an anthropologist and museum professional who had managed historical collections for a decorative arts museum and a county historical society in Ithaca, New York, as well as in the U-M Libraries Preservation Division—was chosen to fill that post in a full-time capacity. After the departure of Szatkowski-Reeves in 2004, O'Dell was promoted to full-time curator and the position of assistant curator was dissolved.

O'Dell focused efforts on building the collection and mounting interpretive exhibits. The generous endowment left by Dr. Sindecuse has allowed the museum to acquire dental merchandising artifacts, advertising, games and toys, and photographs that depict dental offices and workplaces. Digital imaging of the collection—all efforts started by Szatkowski-Reeves—continued. The museum's mission was defined and published on a new website in 2006,[8] along with a new logo to support the museum's promotional initiatives. The museum staff focused on correcting, organizing, and refining records and cataloging new accessions.

In 2010, guided by a professional exhibit team, exhibits were refreshed and enhanced. The dental office artifacts were set into period backgrounds and provided with expanded interpretive and chronological context. The result—*Inside the Dental Practice: 1860–1940*—was a new permanent exhibit located in the north wall of the atrium and in two large cases in the adjoining Kellogg Institute Building lobby. New museum-quality

Fig. 137. Permanent exhibition: "Inside the Dental Practice."

LED track lighting improved exhibit viewing and lowered maintenance costs.

Two years later, O'Dell, with guidance from a professional exhibit design team, curated *Women Dentists: Changing the Face of Dentistry*, which was installed in the atrium's south wall. Smaller temporary shows, such as *WD Miller, Scientific Pioneer of Dentistry* (2007) and *Dental Hygiene—A Century of Progress* (2013), were designed to create appealing exhibitions that highlight the collection strengths and that convey important stories in dentistry's history through a biographical lens.

In the last years, the museum has undertaken substantial digitization efforts. To address long-held goals to expand access to the growing collection of more than twenty thousand objects, in 2011 the museum hired a collections coordinator, Kathy Daniels. With Daniels's skills, the museum added storage for large objects at an off-site location and migrated the museum database to a new collection management system, Past Perfect. This conversion was completed in late 2014 and today the public can search and view collection records through its website, while staff continuously add more records and images.

As we celebrate the bicentennial of the University of Michigan and recognize the important role of institutional history, the Sindecuse Museum of Dentistry will continue to acquire, document, study, interpret, make accessible, exhibit, and care for historical collections that further the museum's mission. In our role as a university museum, we care about responsible stewardship in our collecting. Through collection development and exhibition initiatives, we hope to stimulate understanding and curiosity about the field of dentistry and its stories through relevant, inclusive, and interactive exhibitions. The Sindecuse Museum is committed to increasing intellectual control over the collection and to showcase its many treasures while balancing the need for adequate space and resources to protect the collection—now and well into the future.

Entry prepared by Shannon O'Dell.

1011 North University Avenue. Open Monday through Friday 8 am to 6 pm. http://dent.umich.edu/about-school/sindecuse-museum/sindecuse-museum-dentistry

Notes

1. *University of Michigan, An Encyclopedic Survey*, Vol. 3, 1331–32.

2. *University of Michigan, An Encyclopedic Survey*, Vol. 3, 1331–32.

3. The Ford Collection of crania was accessioned by

the Museum of Anthropological Archaeology (Museum of Anthropology) in 1996. The collection is currently used in biological anthropology teaching and is available online at http://quod.lib.umich.edu/c/crania1ic

4. Deborah Gilbert, "Dental School receives $1 Million for Museum," *The University Record* (Ann Arbor: University of Michigan), September 3, 1991, 1.

5. University of Michigan School of Dentistry, "Dr. Gordon H. Sindecuse Endows Dental Museum," *DentalUM*, Fall, 1991, 12.

6. This inaugural committee, led by Senior Associate Dean William Kotowicz, was made up of alumnus Dr. Ronald Berris; Susan Segar, head of the Dental Library; Grace DeShaw-Wilner, a staff representative from the Michigan Dental Association; Dr. Gerald Charbeneau, professor emeritus of dentistry; and Dr. Charles C. Kelsey, professor of dentistry, editor of the *School of Dentistry Alumni Bulletin*, and dental school historian.

7. Renamed the Phoebe A. Hearst Museum of Anthropology in 1991.

8. http://dent.umich.edu/about-school/sindecuse-museum/sindecuse-museum-dentistry

REFERENCES

Shaw, Wilfred B., ed., *The University of Michigan: An Encyclopedic Survey*. 4 vols. Ann Arbor: University of Michigan Press (1941).

Section 2

Libraries

THE UNIVERSITY LIBRARY'S SPECIAL COLLECTIONS

1899–2017

In 1899, to safely and conveniently store its books of exceptional value, the University of Michigan Library created a number of caged areas in the stacks—thus establishing one of the first rare book rooms in the country. By 1913, these caged areas had come to be known as the Rare Book Rooms, and they continued to provide protection for rare and valuable volumes as well as for other types of increasingly vulnerable collection material including manuscripts, photographs, maps, and musical scores. After assuming responsibility for the Papyrology Collection and the Joseph A. Labadie Collection in the 1960s, the unit was renamed the Department of Rare Books and Special Collections. Early in 1992, in recognition of both the breadth and the depth of the collections that had come under its care, the Department of Rare Books and Special Collections became the Special Collections Library. Today the Papyrology Collection is administered separately from the Special Collections Library, as are a number of other University Library units with important holdings of special collections material, including the Art, Architecture, and Engineering Library; the Askwith Media Library; the Stephen S. Clark Library for Maps, Government Information, and Data Services; and the Fine Arts, Museums, and Music libraries.

HISTORY AND INFLUENCES

On June 5, 1837, "the day appointed by the Governor for the first meeting of the Board of Regents of the University of Michigan," the regents began to demonstrate their intent to establish a library that would honor the institution whose organization and location had recently been entrusted to them. While the original act of the university's founding in 1817 had already called for the establishment of "libraries and museums," it was only after the university had moved from Detroit to Ann Arbor and embraced a new organizational structure, including a Board of Regents, that an official "Committee on Library, Philosophical Apparatus, and Cabinet of Natural History" was formed.[1] The regents declared the Reverend Henry Colclazer "duly elected" as the university's first librarian.[2]

Less than a year later, signaling their belief in the importance of acquiring books for a library that would eventually be housed at the center of the new campus in Ann Arbor, the regents ordered "one copy of Audubon's Ornithology for the use of the University," for which they rendered $970, a substantial amount of money at the time.[3] John James Audubon's *Birds of America*, which consists of eight "elephant folio" volumes containing 435 magnificent hand-colored plates, is to this day one of the great treasures of the University Library and by far the most valuable work in its collection.

It is worth noting that these volumes were intended to be actively used—which, apparently, they were. In 1875, Albert Ten Brook (librarian from 1864 to 1877) stated that in that year, "It is now thirty five years that the leaves of these ponderous volumes have been turned over constantly by students and visitors, excepting only a few years during which they were laid aside in the hope that by avoiding wear they might be transmitted as the property of the [L]ibrary to distant ages."[4] The pages were indeed showing signs of wear, for in 1942 William Warner Bishop (librarian from 1915 to 1941) recalled that when he entered the university as a student in 1899, Audubon's *Birds* were "exposed to inspection and handling by every visitor to the art gallery in the [L]ibrary."[5] Not long before Bishop became librarian, the set was transferred to the newly established Rare Book Room in which, in 1933, a "metal case with roller shelves was designed and built especially for these eight volumes, providing the maximum of protection for this precious work."[6] The same case is still in use to this day.

Soon after ordering the copy of Audubon's *Birds*, the regents passed a resolution commissioning Dr. Asa Gray, the recently appointed professor of botany who had requested a year abroad prior to taking up his teaching duties, to buy books:

> That Dr. Asa Gray, Professor of Botany . . . in his contemplated tour in Europe be requested to purchase a library for the University of Michigan, that the sum of five thousand dollars be appropriated and placed in his hands to carry that object into effect, . . . and that fifteen hundred dollars be advanced to Doctor Gray for defraying the expenses which he may incur in the execution of the commission hereby confided to him.[7]

Additional resolutions specified the types of books to be purchased: "Standard English Works in history, biography, voyages and travels, poetry, statistics, science, and one copy of the most approved British Annual Register from its commencement to the time when the purchase shall be made together with such Scientific Works enumerated in the catalogue furnished by him as he may deem most desirable."[8]

In an autobiographical essay, Gray conveys that while he was busy visiting the botanical gardens of Europe, he delegated the task of buying the books to the American book publisher and seller George P. Putnam. "George P. Putnam, of the firm of Wiley and Putnam, was then resident in London," he writes, "and through him I managed the expenditure of the money placed in my hands for the purchase of." The regents' library committee was satisfied with the result of this arrangement: having "examined with as much attention as the time would permit the Catalogue of Books purchased in Europe for the University," the committee expressed its opinion "that the trust committed to Professor Gray has been executed in a manner which reflects much credit upon his judgment and discrimination in the selection and cost."[9]

As a result of this commission, the earliest holdings of the University Library consisted of approximately thirty-four hundred volumes "embracing the various departments of history, philosophy, classical literature, sciences and arts, jurisprudence, etc.," including rare editions and works that could not be obtained in America. The library committee deemed it "fortunate for the interests of the Institution that such a basis has been laid for building up a Library which shall do honor to the Institution."[10] Although over the years some of the Asa Gray volumes have been lost, disposed of, or replaced, those still extant are shelved together in the Special Collections Library. This library traces its history to the early twentieth century and owes its establishment as one of the first rare book rooms in the country to a long-standing and deeply felt commitment on the part of the University Library to protect the rare and valuable materials in its care.

As early as 1841, the regents' committee on the library recognized the need to distinguish and safeguard "special collections" by placing a halt on their open circulation and making it a policy that they could be accessed only in the library.[11] More than fifty years later, Raymond C. Davis (librarian from 1877 to 1905) took further steps to protect the growing number of "rarities" in the collection. In his report for the year 1894–95, he wrote:

> We have for a long time needed some means of keeping separate from the great mass of books, and securely, those volumes that, are, for one reason or another, rare and valuable, and often frail. If suffered to stand on the shelves with other

> books, they come to harm in the hands of persons who are ignorant of their value. I am glad, therefore, to report, that during the winter several closed cases were constructed in the stock-room, in which our rarities can be safely placed.[12]

One year later, Davis included in his report an appeal for an expansion of the library:

> We already have a good deal of matter that is rare and valuable, and some that is unique. It is impossible under present conditions to guard this against indiscriminate handling . . . injury, or loss. There should be rooms . . . where this can be preserved from danger, and yet made accessible to any who may need to consult it. Having such conveniences for the safe keeping of especially valuable material would greatly increase our chances of securing more of it.[13]

Reporting for the 1899–1900 year, Davis described how the University Library had successfully created a number of caged areas in the stacks for valuable books as well as for "auxiliary matter that we are accumulating, such as engravings, photographs, maps, and musical scores."[14] The McMillan Shakespeare Library was held in one of these guarded places; in another, "a collection of rare and expensive books of a miscellaneous nature; and in yet another the Stearns Musical Collection."[15] These caged areas came to be known as the Rare Book Rooms into which, beginning in 1913, the library's most valuable acquisitions and holdings were regularly segregated.[16]

Fig. 138. "Blue Jay," detail of plate CII in *The birds of America, from original drawings* by John James Audubon.The first purchase made for the University of Michigan Library, by vote of the Regents in 1838, Audubon's masterpiece (1827–38) depicts more than four hundred avian species in hand-colored engravings with scientific detail and artistic mastery. Special Collections Library.

SPECIAL COLLECTIONS TODAY

Because rare and valuable volumes were pulled for safekeeping from all areas of the general collections, much of the original rare book collection quite naturally reflected existing collection strengths. Special collections have since been developed according to several guiding principles, resulting in a collection that is widely recognized for its breadth of coverage and equally noted for its pockets of strength. Among the practices that have influenced the growth of the University Library's special collections since at least the early twentieth century are a preference for purchasing to complement and add to strengths, an inclination against acquiring antiquarian materials in a wholesale way, and a tradition of sending to the Rare Book Room materials from the University Library that seem suitable candidates for inclusion and special safeguarding. In this way—by focusing on overall collection growth—the University Library has intentionally developed special collections as research collections related to the strengths of the general collections.

At the same time, the development of the general collections and of the special collections in particular was greatly affected by the emergence of several specialized libraries on campus. The William L. Clements Library, home to an exceptional collection of early Americana, opened its doors in 1923. In 1935, the Michigan Historical Collections (now the Bentley Historical Library) were established to serve as the archives of the university and as a repository for materials documenting the history of the state. Other libraries specializing in different subject areas, including music, fine arts, and the natural sciences, were built and developed; many of them had their own designated spaces for housing rare and valuable collection material.

The turn of the twenty-first century witnessed an evolution from rare book rooms intended primarily for providing protection for valuable books to dedicated settings for ensuring access to and enabling use of an ever-widening assemblage of special collections material. Similarly, twenty-first-century special collections are composed of considerably more than book rarities; they encompass a wide variety of materials broadly characterized by their artifactual or monetary value, physical format, uniqueness or rarity, and/or an institutional commitment to long-term preservation and access. These include books, pamphlets, newspapers, and other printed or published texts; government documents; archival and manuscript material; maps, atlases, and globes; posters, prints, photographs, and other visual material; sheet music; computer games; and audio and moving-image recordings.

Today most but certainly not all of the University Library's special collections are held in three repositories, all housed in the Hatcher Graduate Library on central campus: the Papyrology Collection; the Stephen S. Clark Library for Maps, Government Information, and Data Services; and the Special Collections Library. Other University Library units with important special collections include the Art, Architecture, and Engineering Library; the Askwith Media Library; the Fine Arts Library; the Museums Library; and the Music Library.

THE PAPYROLOGY COLLECTION

The University Library is home to the largest collection of ancient papyri in North America.

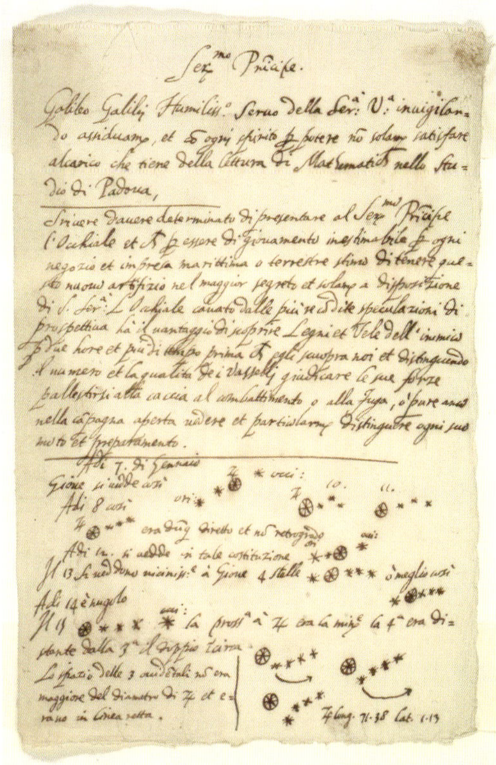

Fig. 139. The Galileo Manuscript. Draft of a letter to the Doge of Venice (August 1609) and notes on the moons of Jupiter (January 1610) by Galileo Galilei (1564–1642). Special Collections Library.

The documents span roughly two thousand years and contain not only important religious texts—including sixty pages of the oldest known copy of the Epistles of Paul—but also personal letters, school primers, sales contracts, wills, marriage agreements, and other records that paint a unique portrait of everyday life. Of the eighteen thousand pieces in the collection, about five thousand have not yet been studied or translated, and they continue to attract scholars from across the country and around the world.

The collection is the brainchild of U-M professor Francis Willey Kelsey (see Pedley, this volume), whose belief that students would benefit from directly studying historical objects led him to travel throughout Europe and the Middle East, purchasing papyri and other ancient artifacts for the university. From 1920 to 1936, the University

Fig. 140. Opening panel of a *murakkaa* (calligraphic album), executed by the master calligrapher Mahmud Celâleddin Efendi in 1791/92 in Istanbul, Turkey. Special Collections Library, Islamic Manuscripts Collection.

of Michigan regularly acquired batches of papyri, often by way of faculty members who were in Egypt on archaeological excavations. Since then there have been only occasional purchases, the most recent one occurring in 1982.

THE STEPHEN S. CLARK LIBRARY FOR MAPS, GOVERNMENT INFORMATION, AND DATA SERVICES

The Stephen S. Clark Library, which opened in 2011 and is named for U-M alumnus and donor Stephen S. Clark, holds more than 380,000 cartographic works dating from the fifteenth through the twenty-first centuries. One of the largest collections of its kind in an academic institution, its strengths include printed atlases; celestial charts; and rare and other maps depicting the Americas, Europe, the classical world, and Japan. Many of these have come to the university by way of gifts, including those of Colonel Thomas M. Spaulding (1882–1973), U-M regent Lucius L. Hubbard (1849–1933), U-M alumnus and regent William L. Clements (1860–1934), and many others. Since the founding of the Clements Library in 1922, the map collections of the University and Clements libraries have been developed largely in concert.

THE SPECIAL COLLECTIONS LIBRARY

In keeping with the University Library's mission to collect, describe, preserve, and make available the record of human knowledge, the Special Collections Library acquires, cares for, interprets, and promotes the use of important collections of unique, rare, primary source, and other material in all formats and in a variety of subject areas. Holdings consist of approximately three hundred thousand published volumes and more than seven thousand linear feet of archival and manuscript material, including the records of the university's Hopwood Awards Program and the papers of a number of distinguished authors, illustrators, poets, activists, and filmmakers. In addition to the notable collections and subject strengths described below, the Special Collections Library holds impressive collections of early printed books, including more than 350 incunabula; medieval and Renaissance manuscripts; European political and military history; English and American literature, including extensive collections of works by William Shakespeare and William Faulkner; Jewish life and customs; and artists' books.

History of Astronomy

The collection documents the early history of the

Fig. 141. "The Last Strike" Nedeljkovich, Brashich, and Kuharich / Industrial Workers of the World (1912). From a collection of more than two thousand political posters. Special Collections Library, Joseph A. Labadie Collection.

field in hundreds of pre-1800 publications including works by Nicolaus Copernicus (1473–1543) and Johannes Kepler (1571–1630). Also included in the collection are all four of the "big four" star atlases produced during Europe's "Golden Age" of celestial cartography: Johann Bayer's *Uranometria* (1603), Johannes Hevelius's *Firmamentum Sobiescianum sive Uranographia* (1690), John Flamsteed's *Atlas Coelestis* (1729), and Johann Elert Bode's *Uranographia* (1801). The highlight of the collection, and one of the "treasures" of the University Library, is a manuscript in Galileo's own hand that illustrates his discovery, sometime between 1609 and 1610, of the four moons of Jupiter.

History of Medicine

Consisting of approximately eighty-five hundred works of historical and scholarly significance, the collection includes fifty-two magical amulets dating from 100 to 500 CE; medical "fugitive sheets"—the Renaissance version of a pop-up book that allowed for the study of different layers of the body; early editions of medical classics, including several extremely rare early editions of the works of Galen in the original Greek, a first edition of Andreas Vesalius's *De Humani Corporis Fabrica Libri Septem* (1543), and a first edition of William Harvey's *Exercitatio anatomica de motv cordis et sangvinis in animalibvs* (1628); and one of the world's most comprehensive collections on the history and development of homeopathy.

Hubbard Imaginary Voyages Collection

The Hubbard Imaginary Voyages Collection was presented to the University Library in 1923 by Dr. Lucius L. Hubbard (1849–1933), who served as a regent of the university from 1911 until his death in 1933. Hubbard assembled one of the world's best collections of editions, translations, adaptations, and spin-offs of two novels that changed the course of literature: Daniel Defoe's *Robinson Crusoe* (first published in 1719) and Jonathan Swift's *Gulliver's Travels* (first published in 1726). Other imaginary voyages, chiefly in English, Dutch, and German, figure in the collection, which now consists of approximately thirty-five hundred volumes and includes editions, translations, and adaptations of Johann David Wyss's *Swiss Family Robinson*, first published in 1812.

Fig. 142. Map of the Americas appearing in Abraham Ortelius' *Theatrum Orbis Terrarum*, first published in Antwerp in 1570 and considered to be the first true modern atlas. Stephen S. Clark Library.

Islamic Manuscripts

The collection consists of 1,095 volumes (and a small number of single leaves) dating from the eighth to the twentieth century CE, containing roughly 1,795 texts chiefly in Arabic, Persian, and Ottoman Turkish. The collection ranks among the largest and most significant such collections in North America and offers a vast range of material for philologists and historians of various disciplines, including Islamic social history, knowledge transmission and acquisition, manuscript production and ownership, and the arts of the book. The subjects covered by the manuscripts—the bulk of which were acquired via purchase in the 1920s and 1930s—include the Qur'an (texts and commentaries); commentaries and other works of criticism; Islamic traditions, theology, and jurisprudence; and philology, philosophy, geography, history, mathematics, astronomy, and astrology.

Children's Literature

Composed of approximately thirty thousand published volumes and complemented by several archival and manuscript collections, this collection contains the artwork, correspondence, drafts, and other material created by a number of notable authors and illustrators of children's books. The collection is noted for a number of strengths, including pop-up and movable books, Newbery and Caldecott medal winners, works by Michigan authors and illustrators, and an extensive collection of limited and finely bound editions illustrated by Arthur Rackham. Among the many treasures in the collection are early editions of Lewis Carroll's *Alice's Adventures in Wonderland* and A. A. Milne's Winnie-the-Pooh books.

Janice Bluestein Longone Culinary Archive

Established in 2005 and transferred from the William L. Clements Library to the University Library in 2012, the Janice Bluestein Longone Culinary Archive is recognized around campus and across the country as a premier collection that both documents and defines the American culinary experience. Shaped by the donation of a rich assemblage of cookbooks, menus, and other material gathered over many years by Janice Bluestein Longone and Daniel T. Longone, the collection includes a diverse assortment of charity cookbooks and food-related advertising ephemera.

Joseph A. Labadie Collection

The Joseph A. Labadie Collection is one of the oldest, largest, and most comprehensive collections of its kind in the United States. It was named for Detroit labor organizer and anarchist Jo (Joseph Antoine) Labadie (1850–1933), who in 1911 donated to the University of Michigan the books, pamphlets, newspapers, magazines, manuscripts, and memorabilia he had assembled over the years. The collection includes material pertaining to the history of anarchism, anticolonialist movements, antiwar and pacifist movements, atheism and free thought, civil liberties and civil rights, labor and workers' rights, LGBTQ movements, prisons and prisoners, the New Left, the Spanish Civil War, and youth and student protest.

Philippine History

In 1914, Dean Conant Worcester (AB 1889) (1866–1924) donated his extensive assemblage of material related to Philippine history to the University of Michigan, where he had served as assistant professor of zoology and curator of the Museum of Zoology. The collection now consists of published works, manuscript items, and photographs documenting many aspects of Philippine history, with particular emphasis on the period between 1899 and 1913, when Worcester served as a member of the United States Philippine Commission and as secretary of the interior for the Insular Government of the Philippine Islands. The University Library's holdings in Philippine history are complemented by equally impressive collections held by the Bentley Historical Library and the Museum of Anthropological Archaeology (see Rice, this volume).

Screen Arts Mavericks & Makers

In 2005, with the acquisition of a significant collection of primary source material documenting the creative activities of Orson Welles during the last two decades of his life, the University of Michigan Library embarked on a collecting adventure that now distinguishes it as a major destination for research on maverick American filmmakers. Complementing a rich and extensive body of Welles material—which includes scripts, photographs, and personal as well as production-related

Fig. 143. One of thirty leaves from a papyrus codex containing the Epistles of St. Paul (excluding the Pastoral Letters) and the Epistle to the Hebrews. Dated to the late second or early third century AD, the leaves are the oldest surviving manuscripts of these works of Christian scripture. Papyrology Collection,

correspondence—are the papers of several notable independent filmmakers, including Robert Altman, Alan Rudolph, Nancy Savoca, and John Sayles.

Transportation History

The greatest strength of the Transportation History Collection is its breadth, as it provides in-depth coverage of various modes of transportation and the infrastructure that supports them. These include automobiles, ballooning and dirigibles, bicycles, bridges, canals, carriages, railroads and rolling stock, and roads and highways. The collection was established in 1923 by John S. Worley (1876–1956) when he came to the University of Michigan to head the newly created Department of Transportation and Railway Engineering. In 1964, the collection was placed in the Engineering

Library, and in 1991 the historical materials were transferred to the Special Collections Library. The collection contains over seventy thousand items, including pamphlets, monographs, serials, prints, photographs, manuscripts, brochures, timetables, maps, and guidebooks.

It is entirely fitting that today one of the eight volumes of Audubon's magnificent work is on display in a centrally located, purpose-built space bearing his name. Completed and dedicated in 2009, the Audubon Room represents a significant milestone not only in the history of the University Library but also toward its mission of sharing its collections with the widest possible audience. In the same space where a page of Audubon's *Birds* is turned weekly, visitors will find a regular rotation of themed exhibits that draw attention to both the well-known and the less-familiar treasures that so remarkably distinguish the library's special collections.

Entry prepared by Martha O'Hara Conway.

The Special Collections Library's Reading Room is located on the eighth floor of Hatcher Graduate Library. Open Monday–Friday 10 am–5 pm (Tuesday until 8 pm); the Audubon Room in the Hatcher Graduate Library is open Sunday 1 pm–7 pm, Monday–Friday 8:30 am–7 pm, and Saturday 1 pm–5 pm.

Notes

1. *Proceedings of the Board of Regents*, 1837–1864, 2.
2. Ibid., 7.
3. Ibid., 36 and 111.
4. Ten Brook, 166.
5. Bishop, 384.
6. Bidlack, 101; see also "Report of the General Library 1933–1934," 12.
7. *Proceedings of the Board of Regents*, 1837–1864, 56.
8. Ibid.
9. Ibid., 143.
10. Ibid.
11. Ibid., 169.
12. Davis 1894–95, 5.
13. Davis 1896–97, 6.
14. Davis 1899–1900, 6.
15. Ibid.
16. Bishop, 1376.

REFERENCES

Bidlack, Russell E. "The University of Michigan General Library, a History of Its Beginnings, 1837–1852." Thesis, University of Michigan, 1954. Photocopy of typescript. Ann Arbor, MI: University Microfilms, 1966.

Bishop, William W. "Rare Book Rooms in Libraries." *The Library Quarterly* 12, no. 3 (1942): 375–85.

Bishop, William W. "The University Library to 1941." In *The University of Michigan: An Encyclopedic Survey, Vol. IV, Part VIII*, edited by Walter A. Donnely, Wilfred B. Shaw, and Ruth W. Gjelsness, 1369–96. Ann Arbor: University of Michigan Press, 1958.

Davis, Raymond C. University of Michigan Librarian's Report, 1894–95. 1–7.

Davis, Raymond C. University of Michigan Librarian's Report, 1896–97. 1–9.

Davis, Raymond C. University of Michigan Librarian's Report, 1899–1900. 1–10.

Gray, Asa. *Letters of Asa Gray*, edited by Jane Loring Gray. Boston: Houghton Mifflin, 1893.

Report of the General Library Reprinted from the President's Report, University of Michigan, 1933–1934. Ann Arbor: University of Michigan.

Ten Brook, Andrew. *American State Universities, Their Origin and Progress: A History of Congressional University Land-Grants*. Cincinnati: Clarke, 1875.

University of Michigan Board of Regents. *Proceedings of the Board of Regents* (1837–1864). Ann Arbor: University of Michigan. (Available at http://quod.lib.umich.edu/u/umregproc)

BENTLEY HISTORICAL LIBRARY

1938–2017

The Bentley Historical Library (originally named the Michigan Historical Collections) is devoted to collecting, preserving, and making available archival material on the history of the state of Michigan and the University of Michigan. Collecting efforts began in 1935 with an emphasis on the state of Michigan. The regents formally named the program the Michigan Historical Collections in 1938 and mandated that it serve both as a historical collection for the state of Michigan and as the archives of the University of Michigan. A gift from the family of Alvin M. Bentley provided the library with dedicated facilities when it opened on North Campus in 1973.

THE LEWIS G. VANDER VELDE YEARS, 1935–60

University of Michigan associate professor of history Lewis G. Vander Velde (see Williams, this volume) had a new idea to collect local history when he applied for a Faculty Research Fund grant in 1934. The modest $700 grant awarded to Vander Velde allowed him to begin to locate and collect materials relating to the history of the state of Michigan, mainly for use in his graduate seminar on Michigan history. Vander Velde was able to hire students and seek out historic material scattered across the state. The first targets were the papers of former governors and other leading Michigan citizens from the 1800s. Letters inquiring about personal papers and historic records went out to family members in the hope that the material was still extant. The first papers to arrive were those of former governor John Bagley, accessioned on October 1, 1935. Thus, an important collecting theme—acquiring papers of elected officials—was established from the outset. A room in the William L. Clements Library was allotted to store, organize, and catalog the collected material.

University President Alexander G. Ruthven took note of Vander Velde's small collecting program. Ruthven was concerned about university records, and in the fall of 1935 appointed Vander Velde to serve as secretary of the Committee on University Archives, which was charged to collect and preserve manuscripts and printed material relating to the history of the University of Michigan. Letters seeking early university records and publications were sent to former regents and faculty, and their families, as well as to alumni; circulars went to all campus units inquiring about the nature and type of records maintained in each office. Early acquisitions included the papers of former presidents Erastus Otis Haven and James B. Angell. The original 1817 draft of the act to establish the "University of Michigania" eventually came to the archives along with the libraries of founders Gabriel Richard and John Monteith. Papers were also received from several former regents, including content from their time as regents and documentation on their personal activities and business ventures, which shed light on the history of the state.

In recognition of what Vander Velde and his small staff managed to accomplish in a brief amount of time, in February 1938 the regents formally designated "the collections of historical books and manuscripts" under Vander Velde's care as a research institution of the university, named him as director, and designated it as the Michigan Historical Collections (MHC). By this action, the dual mission of the new unit was ratified: to collect, preserve, and make available important historical records of the state of Michigan and

Fig. 144. Michigan Historical Collections Reading Room in Rackham ca. 1962.

to serve as the archives of the university. In June 1938, following completion of the Horace H. Rackham School of Graduate Studies Building, the university provided Vander Velde and his program with two offices and a large storage room in the basement of the new structure.

As MHC continued to acquire material, policies and collecting themes were established. A key policy was that material would be acquired entirely by donation rather than purchase. While budget constraints determined this policy from the onset, acquisition through donation remains a core principle.

Collecting themes continued to be developed. Church records were highlighted and their systematic collection began when several denominations placed their records in the archives. Papers of governors and public officials continued as collecting areas along with university records. Records detailing economic activity in Michigan became another theme, and record books of small general stores were collected along with business records from larger enterprises. Some family papers covered a number of areas—such as the Post family papers, which included early records of the Postum Company and the work of C. W. Post, tracing the simple origins of the General Foods Corporation.

Materials documenting the extraction of natural resources were eagerly sought out to document the state's economic and industrial history. Michigan's lumber industry is well represented in the extensive papers of early lumbermen such as Charles Mears and William B. Mershon and in the records of the Wisconsin Land and Lumber Company. While primarily collected for the information they provided on the economic impact of the lumber business, collections such as the Mershon papers are equally rich in telling the story of early conservation efforts in the state.

The diverse population of the state came to be reflected in significant holdings relating to ethnic groups and minorities, a priority identified by Vander Velde. Early collecting efforts centered on Native Americans in Michigan and on the Dutch, German, and Finnish groups that were among the state's early European settlers. Ethnic collections expanded to encompass other nationalities.

The response of Michigan citizens and organizations to national emergencies and wars was another early focus of collecting. Civil War letters and diaries were solicited for the insight they provided into the horrors of combat, camp life, the home front, national politics, and general observations on society. Content from subsequent conflicts continues to be sought out. Among the particularly notable World War I holdings are the letters, diaries, and photographs of the American soldiers dubbed the "Polar Bears" who fought a little-known campaign in northern Russia against the Bolsheviks. As World War II unfolded, the position of war historian was created to document the massive changes on campus due to the war and the many contributions of the university to the war effort. The lack of such information from the WWI period prompted a proactive effort to identify and preserve material from the latest conflict.

Printed works were actively collected, particularly county histories, atlases, city directories, and published annual reports from Michigan's

Fig. 145. Gerald R. Ford, Robert M. Warner, and Robben W. Fleming reviewing Ford's congressional papers at the Bentley Library in 1974.

governmental entities as well as other public and private enterprises throughout the state. Microfilm of content relating to Michigan but held in other repositories was another collecting priority. By 1940, MHC already housed more than three hundred thousand unbound manuscripts, five thousand bound manuscripts, and ten thousand imprints. New Deal programs during the Great Depression funded workers who helped create indexes, transcripts, and calendars of correspondence in the collections, making them more accessible to researchers.

The success of the collecting program soon filled up the allotted space in the Rackham Building. Records of former Michigan governors and University of Michigan presidents and regents—along with business records, church records, diaries, and scrapbooks—soon overwhelmed the space, which doubled as offices, reading room, storage, and processing space. Large collections added to the space demands when university admissions records dating back to the 1850s arrived along with voluminous business records of Michigan railroads, most of which consisted of enormous ledgers and accounting volumes. Adequate space for the expanding collections became a pressing need.

The professional staff also grew under Vander Velde's leadership. F. Clever Bald, who had served as war historian, was named assistant director in 1947 to help cover for Vander Velde, who had taken on additional responsibilities as chair of the Department of History. A full-time librarian for printed works was added in 1951 and a dedicated field representative was hired in 1958. To give the growing enterprise more oversight, an executive committee consisting of five members was appointed by the regents in 1955. The committee oversaw staffing and the formulation of policies and procedures. Ralph A. Sawyer, dean of the Rackham Graduate School, was the committee's first chair. The president and other executive officers have been members of the committee over the years.

F. CLEVER BALD, 1961–66

When Vander Velde reached mandatory retirement age in 1960, he was succeeded by Bald. Just as the size of the holdings continued to increase, research visits were also on the rise. More than fourteen hundred research visits were recorded in 1962, signaling the need for a position devoted to reference. In 1965, the position of reference archivist was established, leading the way for the standardization of reference services while bringing the size of the professional staff to six. Being a relatively new field, there were limited opportunities to recruit trained archivists. As a result, the library developed its own training program, pulling in graduate students from the history department and the Department of Library Science who learned on-the-job. MHC became a model program for the administration of archives. Its best practices and experiences were compiled in the 1966 book *The Modern Manuscript Library* authored by MHC staff members, which became a standard textbook for a generation of archivists and manuscript curators.

Under Bald, political collections remained a priority. The papers of G. Mennen Williams, Michigan's governor from 1948 to 1960, became the first modern political collection to arrive. It was also the largest collection then received, comprising some 750 linear feet of records. In 1964, the library also began collecting the papers of Congressman Gerald R. Ford, then serving as US representative from Michigan's Fifth Congressional District. Yearly installments were received from Ford's office.

ROBERT M. WARNER, 1966–80

Robert M. Warner, who started working for MHC as a graduate student in 1954 while pursuing a doctorate degree in history, was appointed field representative prior to becoming assistant director in 1961. Following Bald's retirement in 1966, Warner became MHC's third director. Among his first priorities was securing badly needed space for the collections. Additional space assigned in Rackham over the years had long been filled, forcing the majority of the collections to be stored offsite in poor conditions at the Willow Run Airport. Research visits, which had topped two thousand, were challenging given the seventeen-mile distance between campus and Willow Run.

Warner worked aggressively to raise funds for a new building. With the assistance of many donors and a generous gift from Mrs. Arvella Bentley, a structure designed to house the riches of Michigan and University of Michigan history became a reality. At their meeting in December 1971, the regents accepted the Bentley gift for a building to house the collections, naming it in honor of the late Alvin M. Bentley, former congressman and regent. Ground was broken in 1972 for the Bentley Historical Library on a site on the U-M North Campus. The new building provided considerable space for the archives to grow and made new program initiatives possible when it was occupied in 1973.

The programmatic reach of the archives grew significantly during Warner's administration. In 1971, he inaugurated a course in archival administration using the methods developed and refined by the staff. By 1976 it developed into a sequence of courses offered to graduate students in history and library science. A conservation laboratory was added in 1974 when Jim Craven, who had been with the U-M bindery for more than twenty years, came to the Bentley Library.

Warner aligned the Bentley with the larger archival community. In the 1960s he was one of the founders of the Michigan Archival Association. He served as executive director of the Society of American Archivists from 1970 to 1974, operating the professional archival association from his office. Warner also launched ambitious efforts to secure external funding to preserve and make the holdings more accessible. In 1972, the library received its first federal funding, a grant from the National Historical Publications and Records Commission to microfilm the Detroit Urban League papers. From 1975 to 1978, a comprehensive microfilm edition of papers relating to temperance and prohibition was undertaken in conjunction with the Ohio Historical Society and the Women's Christian Temperance Union. Other grants followed.

On August 9, 1974, when Gerald Ford became president of the United States, his papers took on new importance. Warner planned for appropriate housing of the Ford papers under the terms of the Presidential Libraries Act. The designs for a separate building next door to the Bentley Library were approved in 1978. Coincidentally, Warner left his post at the Bentley Library in 1980 to become archivist of the United States just in time to accept the new Gerald R. Ford Library as part of the National Archives in 1981.

Fig. 146. Francis X. Blouin and Jennifer Granholm in the archive's stacks in 2012, shortly after Granholm transferred her gubernatorial papers to the Bentley.

FRANCIS X. BLOUIN JR., 1981–2013

When Francis X. Blouin Jr. was named director in 1981, he took over a library holding more than 20,000 linear feet of material, representing more than 6,500 collections, 500,000 photographs, 45,000 printed works, and 2,500 reels of microfilm. Research visits also exceeded 3,000 for the first time.

During Blouin's tenure, the Bentley continued to enhance its reputation as a premier archival program. Blouin succeeded in securing funding from the Andrew W. Mellon Foundation for a grant program that offered fellowships to archival professionals to pursue research projects in Ann Arbor. From 1983 through 1997, the program funded a total of 122 fellows, resulting in more than seventy articles and monographs relating to problems in the selection, use, understanding, and administration of contemporary records.

The Bentley developed an international reputation based largely on a series of exchanges that engaged the archival community. Among the first was the Vatican Archives project that began in 1984. Archivists from the Bentley were given unprecedented access to the Vatican Archives, which allowed them to enhance access to those archives and to produce a database of collection-level descriptions and an award-winning book. Archival exchange programs were initiated with institutions in Denmark, France, Russia, and China, opening up the exchange of ideas and the sharing of archival practices. Bentley staff was also active with the International Council on Archives.

By the early 1980s, as the Bentley Library Building was quickly filling up with new collections, the staff undertook a thorough collection analysis to assess the state of the collections in terms of size, number of collections, and subject areas represented in the collections, as well as anticipating future collecting areas. The analysis tightened the collecting focus and set priorities. In 1985, the library began to devote its collecting energy toward important collections in several targeted subject areas that built on traditional strengths but added new depth to areas long collected. It prioritized to acquire the archives of African American churches to complement earlier collecting of mainline Christian denominations. The library also expanded its collecting efforts to new areas with strong research potential, such as the environment and gay and lesbian issues, along with newer ethnic populations, including Arab Americans and Latinos in Michigan.

Fig. 147. Bentley Library reading room, 2014.

Despite being in a building less than a decade old, the stacks were already close to capacity. Two off-site buildings were assigned, temporarily solving storage needs. By 2004, the need for space and additional offices led to an expansion of the Bentley Library. A 36,000-square-foot addition included new offices, a meeting room, a new conservation lab, and two new storage cubes, which more than doubled the existing storage capacity.

Technological advances continually improved access to the collections. The typed card catalog in the reading room gradually gave way to machine-readable catalog records when Bentley records were entered into the Research Libraries Information Network starting in 1985. Beginning in 1994, Bentley records were entered into MIRLYN, making collection-level description of the holdings accessible to all. Even greater strides occurred when the Bentley became an early adopter of Encoded Archival Description, making the text of finding aids fully searchable online. By 2014, more than twenty-eight hundred of the Bentley's finding aids were encoded and available online. The advent of the digital age also meant archiving email and acquiring records in a variety of digital formats ranging from preserving websites to capturing social media feeds from notable events. In 2011, a division of digital curation was created to respond to the pressure of digitally created content as well as to manage analog content that had been digitized.

Annual research visits routinely exceeded four thousand by the late 2000s, including yearly visits by more than four hundred students representing thirty different university courses across a variety of disciplines. The large numbers of courses holding sessions at the Bentley Library were the result of actively engaging professors and students to promote new ways of incorporating archival material into classroom instruction. In recent years the library has become an important center for research into midcentury modernism in architectural design. The acquisition of designs and drawings by several prominent Michigan-based architects has resulted in new users discovering the library and several exhibitions, including a joint exhibition with the University of Michigan Museum of Art.

Furthermore, the university's athletic records are a major draw for students and researchers. In 1990, a unique partnership with the Athletic Department transferred its photographs, films, videos, and records to the archives along with partially underwriting an archivist's position. Bentley staff supports internal university requests while also responding to hundreds of external requests from The Big Ten Network, ESPN, *Sports Illustrated*, as well as former athletes, their relatives, and general fans. Dozens of books, films, and documentaries on Michigan athletics have been produced, drawing upon the library's extensive holdings on the history of sports and athletic competition.

TERRENCE J. MCDONALD, 2013–PRESENT

Historian and former dean of the College of Literature, Science and the Arts, Terrence J. McDonald, became the director in 2013, succeeding Blouin who stepped down as director after serving in that capacity for thirty-two years. McDonald has worked to position the Bentley to develop greater digitization capacities and to prepare the library for a central role in the bicentennial celebrations planned for 2017. He introduced a course on the history of the university taught from the Bentley's holdings, and secured one of

Fig. 148. Bentley Library, main entrance, 2001.

the university's Transforming Learning for a Third Century grants designed to foster innovative new partnerships and understandings of teaching with primary sources. Several digitization initiatives are underway, among them an online fully searchable edition covering all 125 years of *The Michigan Daily*, and a digital edition of the papers of James B. Angell. Both projects will be available in time for the bicentennial celebration.

Directors of the Bentley Historical Library

Lewis G. Vander Velde, 1935–60
F. Clever Bald, 1960–66
Robert M. Warner, 1966–80
Richard M. Doolen, 1980–81 (acting)
Francis X. Blouin Jr., 1981–2013
Terrence J. McDonald, 2013–present

Entry prepared by Brian Williams.

Contact Information and Opening Hours:

Open Monday–Friday 9 am–5 pm; Saturday 9 am–1 pm; 1150 Beal Avenue.

http://www.bentley.umich.edu

REFERENCES

The Bentley Historical Library: Its History and Purpose. Ann Arbor: Bentley Historical Library, University of Michigan, 2004.

An Enduring Gift: The Bentley Historical Library, Keeper of Michigan's History. Ann Arbor: Michigan Historical Collections, Bentley Historical Library, University of Michigan, 1983 (MHC Bulletin no. 32).

The History and Purpose of the Michigan Historical Collections/The Bentley Historical Library, The University of Michigan. Ann Arbor: Michigan Historical Collections, Bentley Historical Library, University of Michigan, 1983 (MHC Bulletin no. 33).

History of the Michigan Historical Collections. Ann Arbor: Michigan Historical Collections, Bentley Historical Library, University of Michigan, 1974 (MHC Bulletin no. 23).

The Michigan Historical Collections of the University of Michigan. Ann Arbor, 1948, Updated 1956 (MHC Bulletin no. 3).

The Michigan Historical Collections: The Vander Velde Years: Twenty-five Years of Leadership. Ann Arbor: University of Michigan, 1961 (MHC Bulletin no. 11).

WILLIAM L. CLEMENTS LIBRARY

1923–2017

The William L. Clements Library is a special collections repository that acquires, preserves, and provides access to primary source materials documenting the history of early America. University of Michigan alumnus and regent William Lawrence Clements founded the eponymous library in 1923 with the donation of his personal collection of rare Americana and a building in which to house it. The library's current holdings of 88,000 rare books, 30,000 maps, over 4,000 volumes of newspapers, 2,600 manuscript collections, and upward of 100,000 prints, photographs, and other visual materials make the University of Michigan a premier destination for scholars of American life between 1492 and 1900.

William L. Clements (U-M class of 1882) made his fortune as president of the Bay City (Michigan) Industrial Works, a firm that manufactured cranes and other heavy machinery for large-scale construction projects. He served on the U-M Board of Regents from 1909 to 1933. Mr. Clements had an abiding interest in American history and a love for the primary source materials that described it. His founding donation included upward of twenty thousand titles on the period of European exploration and discovery of the New World, European encounters with indigenous Americans, seventeenth- and eighteenth-century colonial America, the American Revolution, the formative years of the United States, and the early Republic. Approximately two thousand volumes of early newspapers, several hundred maps, and one significant manuscripts group accompanied the donation of his book collection in 1923.

THE FOUNDING DONATION, 1923

Mr. Clements began assembling his Americana book collection in 1903 and acquired many rare titles over the next twenty years. These begin with the first printed account of the New World, Christopher Columbus's brief report to the king and queen of Spain (*Epistola Christofori Colom*, Rome, 1493). Martin Waldseemüller's geographical work *Cosmographiae Introdvctio* (1507) includes his proposal to name the new-found land after Americus Vespucius. Spanish exploration of what is now Central and South America is represented by works such as Xerez's account of Pizarro's conquest of Peru (*Libro Primo de la Conqvista del Perv*, 1535) and tracts by Bartolomé de las Casas relating atrocities committed against indigenous peoples. Rare printed accounts of voyages by the French include such significant works as the English translation of Jacques Cartier's travels into the St. Lawrence River in 1534 and 1535 (*A Shorte and Brief Narration*, 1580) and Samuel de Champlain's *Les Voyages* (Paris, 1613), describing New France. Clements also compiled an impressive collection of the Jesuits' reports of missionary activities in New France between 1632 and 1672. Dutch explorations and the history of New Netherlands may be found in Emanuel van Meteren's *Nederlantsche Historien* (1612), David Pietersz. de Vries's *Korte Historiael* (1655), and Daniel Denton's *A Brief Description of New-York* (1670).

Printed works pertinent to the English colonization of North America begin with the exceptionally rare account of the ill-fated Roanoke colony, Thomas Hariot's *A briefe and true report of the new found land of Virginia* (1588) and subsequent editions illustrated by Theodorus de Bry. Captain John Smith's narrative of the first lasting English settlement at Jamestown may be found in his *A Trve Relation* (1608). Mr. Clements's original gift included thousands of tracts, sermons, polemics, and other printed works related to colonists'

Fig. 149. The Clements Library façade in wintertime.

interactions with Native Americans, religious discourse, the first Great Awakening, and struggles between colonial powers in seventeenth- and eighteenth-century America.

Over two thousand of the political, philosophical, and economic pamphlets that stimulated public debates leading to the 1776 separation of the American colonies from Great Britain came to U-M with the Clements collection. These include first and subsequent printings of works by James Otis, John Dickinson, John Hancock, Samuel Adams, Thomas Paine, Thomas Jefferson, Lord Grenville, William Pitt the Elder, Junius, Charles J. Fox, Edmund Burke, Samuel Johnson, and others.

Although Clements acquired several small groups of manuscripts over the years, it was not until he purchased the papers of William Petty, second Earl of Shelburne, from Sotheby's auction house in 1921 that he began to seek out manuscripts in earnest. The 170 volumes of Lord Shelburne's papers document British foreign, colonial, and domestic affairs throughout much of the eighteenth century.

Clements finalized his gift agreement with the University of Michigan in 1922. The regents hired Detroit architect Albert Kahn to design the Clements Library Building. His plans, inspired by a sixteenth-century casino at the Villa Farnese in northern Italy, were executed by master contractor Henry L. Van der Horst of Kalamazoo, Michigan. The library was formally dedicated on June 15, 1923. In his gift agreement, Clements stipulated that his library would focus on collecting and preserving primary sources for early American history, that it would be independent from the university's library system, and that it would be available only to advanced scholars. The Clements Library was to be governed by a Committee of Management, selected by the regents and headed by the university president. The trustees were to include the library's director, the head of the University Library, a professor from the Department of History, and two members at large.

DIRECTOR RANDOLPH G. ADAMS, 1923–51

Clements recommended Randolph Greenfield Adams (PhD, University of Pennsylvania, 1920), assistant professor of history at Trinity College, North Carolina, to be the first director of the Clements Library. Adams shared Clements's appreciation for American history and his love of rare books. As an administrator, Adams championed the safety and security of the collection over any concerns for access by faculty, students, or the public. He maintained that special collections

Fig. 150. Avenir Foundation Room, Clements Library. Photo by Philip Dattilo.

were not for general and indiscriminate use, but treasures that needed to be preserved and guarded from those who did not have a reverence for the objects themselves. When Sir John Forsdyke, director of the British Museum, visited in the mid-1940s, he remarked that "the function of the Clements Library is to collect the archetype." Dr. Adams expounded, "we are not satisfied with facsimiles, photographs, or microfilms. We strive to acquire the originals from which these useful copies may be made."[1]

Increased tax pressures following the First World War made it necessary for many English families to sell the contents of their libraries and muniment rooms. With the able assistance of Adams, Clements purchased important manuscript collections from descendants of prominent figures in Revolutionary War-era history. He obtained the papers of commanders in chief of the British Army in America, Thomas Gage (1763–75) and Henry Clinton (1778–82); colonial secretary Lord George Sackville Germain; undersecretary William Knox; secretary of war Thomas Townshend, first viscount Sydney; solicitor general John Lee; attorney general Alexander Wedderburn; and Hesse-Cassel minister of war Freiherr von Jungkenn. The most significant collection acquired for the American side of the conflict was the largest extant group of Major General Nathanael Greene's correspondence and documents. Clements acquired a number of these collections personally and they did not arrive in Ann Arbor until the settlement of his estate in 1937.

The pre-Revolutionary War period is highlighted in the collection by the papers of New York colonial governor George Clinton (1743–53); Sir Peter Warren, the naval commander who triumphed at Louisbourg in 1745; Sir William Mildmay, commissioner of claims following the Treaty of Aix-la-Chapelle, 1748; and lord lieutenant of Ireland George, first Marquis Townshend (1767–72). These and other acquisitions in the 1920s and early 1930s form the cornerstone of the manuscripts collection and remain the most utilized resources at the Clements Library.

By 1923, Mr. Clements had acquired several

notable atlases and some exceedingly rare maps bound into books. However, it was the fortuitous purchase in 1923–24 of the entire fifty thousand-volume library of Henry Vignaud, an American expatriate scholar of the Discovery period, that established the Clements Library as an outstanding cartographic repository. The Vignaud Collection included almost all editions of the great atlases of the fifteenth through eighteenth centuries by Italian, Dutch, German, English, and French mapmakers. Simultaneously, Mr. Clements purchased en bloc the unsurpassed collection of printed maps of the American Revolution assembled by the London bookselling firm of Henry Stevens. The great Revolutionary War manuscripts acquisitions of the 1920s and 1930s also included hundreds of individual manuscripts and printed maps of the American colonies, fortifications, and battles from the French and Indian and Revolutionary wars. By the mid-1930s, the Clements Library had become the preeminent collection of eighteenth-century American maps and was among the finest general resources for the history of cartography in the world.

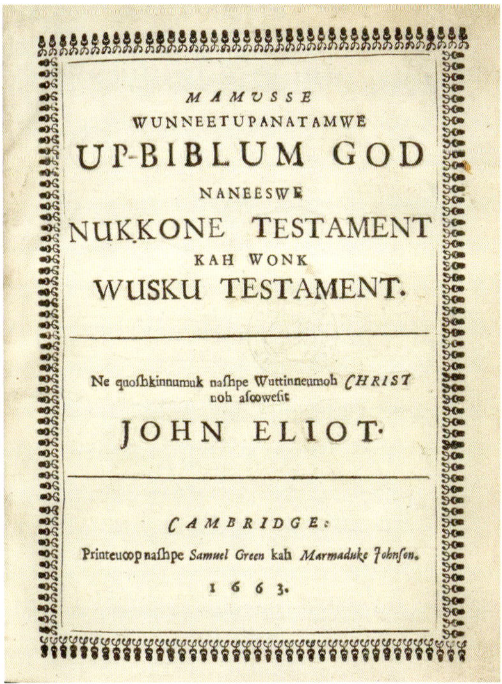
MAMVSSE
WUNNEETUPANATAMWE
UP-BIBLUM GOD
NANEESWE
NUKKONE TESTAMENT
KAH WONK
WUSKU TESTAMENT.

Ne quoshkinnumuk nashpe Wuttinneumoh *CHRIST* noh asoowesit

JOHN ELIOT.

CAMBRIDGE:
Printeuoop nashpe *Samuel Green* kah *Marmaduke Johnson*.
1663.

Fig. 151. Eliot Bible. The first Bible printed in what is now the United States; this bilingual English-Massachusett volume was printed in 1663 by Indian missionary John Eliot (1604–90).

Director Adams divided the library into three separate divisions in 1931—Books, Maps, and Manuscripts—and appointed a curator to head each. Following Clements's death in 1934, Adams modestly expanded the subject matter and chronological scope of the library's holdings. Clements had amassed British statesmens' papers up to the War of 1812 but collected few later items. In 1939, with the help of Professor Dwight L. Dumond, the library acquired the papers of antislavery advocates James G. Birney and the Grimké and Weld families. The same year, Albert H. Greenly of Hoboken, New Jersey, donated over one thousand titles comprising his collection of Abraham Lincoln rarities. A 1944 endowment in memory of Herbert C. Ely helped build a significant collection of eighteen- and nineteenth-century works of American music and drama. Subject matter pertinent to Westward Expansion, the Mexican War, the Civil War, and other defining events of the nineteenth century slowly began to make an appearance at the Clements Library. Despite the broader scope of the holdings, Adams maintained an emphasis on rare Americana, bringing to the Clements, for example, a first edition of John Eliot's bilingual English-Massachuset *Mamvsse Wunneetupanatamwe* (1663), the first Bible printed in America.

In October 1947, the library formed an organization of friends, the Clements Library Associates (CLA), that supports the library with acquisitions funds. CLA dues and contributions have helped the library purchase many resources that might not otherwise have been within budget. Examples of acquisitions by the CLA in the first years of its existence include manuscript maps of the American Revolution by Edouard Colbert, comte de Maulévrier, 1781; issues of *The Royal American Magazine*, illustrated by Paul Revere, 1774–75; diaries of Isaac Bonsall, Quaker missionary to Native American settlements, 1803–23; and many other rare items.

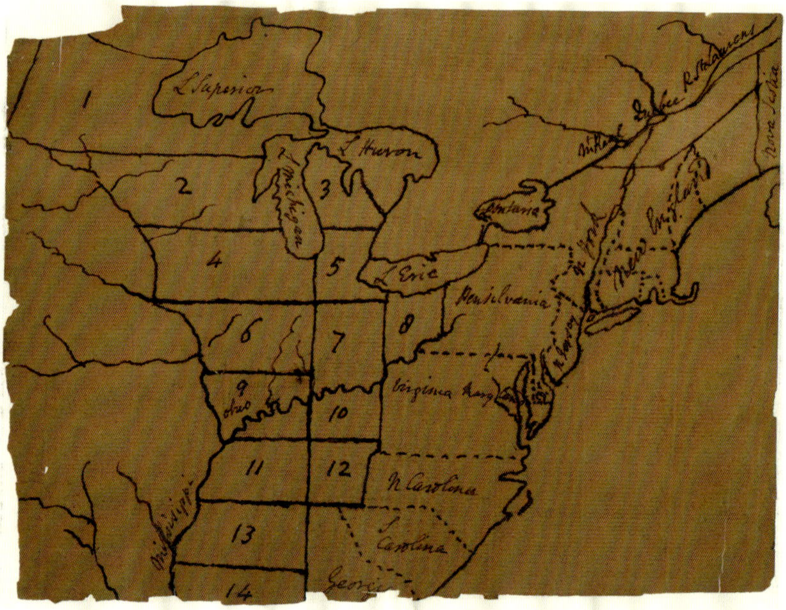

Fig. 152. "Jefferson-Hartley" map of 1784, showing Thomas Jefferson's original concept for dividing the lands between the Appalachian Mountains and the Mississippi River into states. Both known manuscript copies are held by the Clements Library. Clements Library, David Hartley Papers.

DIRECTOR HOWARD H. PECKHAM, 1953–77

In 1953, the U-M regents appointed Howard H. Peckham director of the Clements Library. As curator of manuscripts from 1936 to 1945, Peckham had not only arranged and described the magnificent Revolutionary War acquisitions, but he had a firm grounding in Adams's philosophy of the institution and an understanding of its collections, professional standards, and historical librarianship. Over the course of his service as director, Peckham made modest but significant steps toward shifting the library from an exclusive vault of treasures to a research-centered facility that balanced preservation needs with user access. Increased use of the collections by graduate students and scholars produced many more acknowledgments of the library in scholarly journals, doctoral dissertations, and books. History and geography classes began to meet regularly in the library. These and other efforts helped to improve awareness of the value of discriminating book collecting to historical scholarship.

The Clements Library's collections did not have a clear chronological end date during the Adams years. Peckham was the first to commit to paper a collections development policy. He continued to emphasize rare works of bibliographic and historical significance. Acquisition of "new" materials that had a direct relationship to items already in the collection fleshed out areas of strength, but there was also expansion into other subjects, with broader holdings on slavery and antislavery movements, architecture and furniture design, the development of political parties, education, Westward Expansion, the sciences, and religious movements. The focus of the library under Peckham was 1492 to just before the Civil War.

Adams and Peckham both stressed the importance of maintaining respectful and collegial relationships with dealers and collectors. As a result, the library increasingly attracted sales offers and donation opportunities. Mrs. Hubert S. Smith of Bay City, Michigan, donated her late husband's extensive collection of maritime and naval books and manuscripts in 1953–54. The library was a considerable force at the sales of prominent Americana collector Thomas W. Streeter in the mid-1960s. With the support of the CLA, Peckham successfully purchased Antonio Pigafetta's account of Magellan's voyage around

Fig. 153. Family papers of Angelina and Sarah Grimke, prominent abolitionists and women's equality advocates in the early mid-nineteenth century.

the world, *Le Voyage et Nauigation* (Paris, 1525), the highest-valued book in the 1966 Streeter sale. The CLA assisted with the acquisition of papers of Major General Anthony Wayne; seventeenth- through nineteenth-century imprints; printed and manuscript Revolutionary War and Seven Years' War maps; manuscripts of British abolitionist Thomas Clarkson; the family papers of Episcopal Bishop John Henry Hopkins (1800–1932); and other important assets.

A most significant expansion of the library's collecting scope occurred in 1974 when Committee of Management member James S. Schoff donated his massive holdings of American Civil War books, manuscripts, photographs, and engravings. Although his collection included important materials of prominent individuals (Lincoln, Sherman, Hooker, Lee, and others), its overwhelming emphasis was on the experiences of ordinary soldiers. The James S. Schoff Civil War Collection extended the chronological holdings of the library firmly through the 1860s.

DIRECTOR JOHN C. DANN, 1977–2007

John C. Dann (PhD, College of William and Mary, 1975) worked under Howard Peckham as the Clements Library's curator of manuscripts from 1972 to 1977 before beginning his thirty-year directorship. Dr. Dann's administration was marked by extraordinary growth of collections, increased access to the library's holdings by U-M and other students, and an entry into the world of electronic cataloging and description.

Like Peckham, Dann was a shrewd and knowledgeable collector. Unlike his predecessor, the new director was far more aggressive about expanding into new areas of inquiry. He heartily embraced the plunge into Civil War material in the wake of James Schoff's gift. He firmly believed that archivists and librarians had a responsibility to anticipate future trends in historical inquiry and to build supporting collections before the subject matter became popular and expensive. He led the library into such genres as county atlases, visual materials of all sorts (especially photography), crime, religion and faith, and the history of ideas. In 2002, Dann established a Graphics Division with its own curator. Subjects such as education, women's history, travel, business and trade, labor, religion, and Westward Expansion became more heavily emphasized. Dann believed as well in bolstering the older collections when opportunity permitted.

Dann's thirty years of collecting rapidly

increased the library's holdings. Manuscript collections came in from every direction. Wonders such as the Lyttelton Papers, Tailyour Family Papers, Tousard Papers, Eyre Coote's archive, and the 1698 Hacke or "Pirate" Atlas attracted attention to the library's resources. In addition, the library acquired hundreds of Civil War soldiers' archives and war-related correspondence as recent as the Vietnam War. By the turn of the century, this large-scale collecting, though enriching the library's resources, had exhausted its space. Lesser-used items were transferred to an off-site facility. An effort to incorporate culinary history as a fifth division of the library further strained its resources. The 1923 Kahn building was showing its age and needed both renovation and expansion.

DIRECTOR J. KEVIN GRAFFAGNINO, 2008–PRESENT

J. Kevin Graffagnino (PhD, University of Massachusetts, 1993) was the first director not to have learned the ins and outs of the Clements Library from his predecessor. He is also only the institution's fourth director in its eighty-five-year history. Graffagnino arrived at an exciting time for the library. Planning for the much-needed renovation of the 1923 Kahn building had just begun. The university not only allotted considerable funds for this project but also enhanced staff support and provided some extra positions, the most significant of which was a director of development—the first time a development officer had ever been assigned to the Clements. The availability of development staff has greatly increased the outside support available for the library, including a $6,000,000 gift to assist in renovating the Kahn building and a $4,000,000 endowment to support nineteenth-century collecting and studies, and substantially increased giving by both long-time and new friends of the library.

In addition to an increased commitment to collecting nineteenth-century material, Graffagnino has more clearly defined the library's collecting policy, establishing a terminal date of 1900 for most materials. He has also committed the library to continuing its efforts to attract U-M students to use the collection and to increasing fellowship support for scholars from outside the university.

The Clements Library has also continued to collect within its areas of strength. The years since 2008 have seen major acquisitions of manuscripts and the receipt of gifts of substantial additions to existing collections, such as the Oliver Hazard Perry, David Bates Douglass, and Weld-Grimké Family Papers. Major acquisitions have included the Henry Strachey, Henry Burbeck, and Samuel Latham Mitchill Papers and gifts from collectors Keith Hook and David P. Harris. The photography collection has been massively expanded by the gradual receipt of the David V. Tinder Collection of Michigan Photography. Similar growth has occurred in the book and map collections as well.

The future of the Clements Library appears bright and includes increasing digitization efforts, expanded services for U-M students and classes, and continued healthy growth of collections.

Entry prepared by Cheney J. Schopieray.

Open for research Mondays through Thursday, and to the general public on Friday, 10 am to 4 pm; 909 South University Avenue

http://clements.umich.edu

Note

1. Peckham, 1408.

REFERENCES

Peckham, Howard H. "The William Clements Library of American History." In *The University of Michigan: An Encyclopedic Survey Part VIII*, edited by Walter Arthur Donnely, Wilfred Byron Shaw, and R. W. Gjelsness, 1402–10. Ann Arbor: University of Michigan Press, 1956.

GERALD R. FORD PRESIDENTIAL LIBRARY AND MUSEUM

1981–2017

The Gerald R. Ford Presidential Library and Museum is unique among the presidential libraries across the country in having two sites, located 130 miles apart. The library—holding the papers and audiovisual materials—is located in Ann Arbor on the University of Michigan's North Campus, and the museum—holding the artifacts and memorabilia—is located in Grand Rapids, Michigan. The two are run as a single institution under the auspices of the Office of Presidential Libraries of the National Archives and Records Administration, a federal agency.

THE BEGINNINGS

The story of why and how these two sites came to be is much more complicated than the site selection processes for other presidential libraries. Gerald Ford's hometown was Grand Rapids. He attended the University of Michigan in Ann Arbor from 1931 to 1935, where he was voted MVP of the football team in 1934. Although he subsequently earned his law degree at Yale University, Ford maintained a keen interest in the University of Michigan, even after he established a law practice in Grand Rapids. After naval service in World War II, he returned to Grand Rapids and was elected to Congress in 1948. That victory was the first of thirteen elections that would send Ford to represent Grand Rapids in the US House of Representatives.

The genesis for this split library/museum operation developed after President Ford began donating his congressional papers to the University of Michigan in the mid-1960s—long before he would become vice president or president. In March 1964, Congressman Ford was approached by staff of the University of Michigan's Michigan Historical Collections (MHC, now the Bentley Historical Library) about donating his papers. Ford admitted he had been "systematically destroying all their files when they became two years old, because they had no space in which to keep them. . . . Ford was a little taken aback—he didn't see himself as 'history'—a little flattered, and a little pleased that he had a practical solution to his storage problem."[1] At the end of the year, Ford sent his first contribution of nine boxes of files from 1963, shipped just one week after he was elected House Minority Leader. He continued to send files each year thereafter.

In October 1973, when Ford was selected by Richard Nixon to replace Spiro Agnew as vice president, his papers were moved into the temperature- and humidity-controlled stacks of the brand new Bentley Historical Library on the north campus of the university.

At this time, both Ford and the university assumed that Ford's career would wrap up at the close of his term as vice president and that he would retire as planned. However, as the Nixon presidency became increasingly clouded due to the Watergate revelations, the possibility that Ford might become president came up in a meeting that Robert Warner, then-director of the Bentley, had with staff of the Office of Presidential Libraries at the National Archives.

When Ford was sworn in as president on August 9, 1974, it "drastically change[d] the rules of the game for handling his archives. Now we had presidential papers on our hands and the prospect of having a presidential library in Ann Arbor was very real."[2] While the National Archives now was a major player as the legal repository for presidential papers, according to Warner, the decision

Fig. 154. President Gerald R. Ford, accompanied by Dr. Harry A. Towsley, prepares to place mortar on the Ford Library's cornerstone during the cornerstone ceremony, June 19, 1979. Courtesy of the Gerald R. Ford Presidential Library.

about where to locate the papers was "largely in the hands of President Ford and the University of Michigan."[3]

As Ford's time in office came to a close, Warner approached Ford, a very loyal U-M alumnus, about creating a presidential library on campus and consolidating his congressional, vice-presidential, and presidential papers there. According to Warner, "there was never any discussion about the papers not being in Ann Arbor, but the Grand Rapids people really wanted the museum. So the possibility arose of having a split facility with the museum in Grand Rapids and the library in Ann Arbor. The university's position was that they would be happy having the library and the museum or the library alone. . . . Everyone thought this approach was fine except some National Archives administrators who thought that separate buildings in separate cities would complicate the role of the director."[4] And so it does.

As was typical of President Ford, he made a decision that reflected his personal feelings about

Fig. 155. President Gerald R. Ford meets with University of Michigan President Robben Fleming and other University officials regarding the Gerald R. Ford Library Project in the Hussey Room at the Michigan League, December 19, 1976. Courtesy of the Gerald R. Ford Presidential Library.

both Grand Rapids and the University of Michigan while also respecting the interests of both communities. Thus, the only presidential library and museum with two sites was born.

Planning for the eventual library and museum remained on hold as Ford and his staff focused their entire energy on the election in fall 1976. Only when the election results were announced did plans move forward. On December 13, 1976, President Ford signed a deed donating the prepresidential and presidential papers to the federal government. The deed stipulated that all papers were to be housed at the University of Michigan. James B. Rhoades, archivist of the United States, and Robben W. Fleming, president of the University of Michigan both supported this arrangement.

A month later, Ford's papers, then fifteen million pages in all, arrived in Ann Arbor in a nine-truck caravan. Temporary storage for the papers was found in low-rise warehouses in the shadow of the U-M stadium where Ford had played football years before. The initial housing for classified materials was a paint locker fitted out with thirty heavy Moseler safes, each with a different combination. Presidential artifacts, artwork, and other memorabilia were sent to temporary storage at a National Archives facility in Laguna Niguel, California.

At the same time, committees for fund-raising, site selection, and building planning were formed in Grand Rapids and Ann Arbor. President Ford played an active role in the fund-raising initiatives for both sites and some funds raised in Grand Rapids were even transferred to support the U-M library project.

CREATING THE LIBRARY

For the library, the university established a university building committee, chaired by Warner, that began meeting at the end of January, 1977. For the first time, the committee included a representative from the National Archives, a practice that continues for planning new presidential libraries to this day.

The site selected for the library was located just south of the new Bentley Historical Library and the same architectural firm that designed the Bentley Library building—Jickling, Lyman and Powell—was also chosen for the new presidential library. The university "decided that it would retain ownership of the land on which the building was constructed instead of turning it over to the Federal Government."[5]

Plans for the library were presented to President Ford in April, 1978. "About the only change Ford suggested was to reduce the size of the presidential suite because he knew he would not use it very much."[6] Warner referred to the library design as "modest yet dignified, straightforward and functional . . . designed to be unostentatious . . . and efficient, while fitting harmoniously with the surrounding landscape."[7]

Groundbreaking for the library took place on January 15, 1979, with the cornerstone laid on June 19, 1979. The staff moved into the building in 1980, and the dedication ceremony was held on April 27, 1981. Several hundred attendees mingled with former cabinet members and senior members of the Ford Administration, as well as Michigan governor William Milliken and University of Michigan president Harold Shapiro.

The library, with 50,760 square feet set on 4.2 acres, cost $4.3 million to build. The two-story brick-and-glass building features a full two-story lobby from which there is a beautiful view onto a garden that showcases a metal kinetic sculpture by George Rickey. Large west-facing windows provide the research room and staff offices with an open and airy feeling and great workspace that is cherished by staff, researchers, and visitors alike.

CREATING THE MUSEUM

For the museum, planning for site selection, design, and construction took place on a parallel track. Local leaders selected the site as part of their vision for the revitalization of downtown Grand Rapids; the architect was Marvin DeWinter and groundbreaking took place on June 13, 1979. Two years later, the dedication ceremony on September 17, 1981, was an important community event that celebrated a reenergized downtown and riverfront. More than 250,000 people attended to see American and international dignitaries including President Ronald Reagan, Governor William Milliken, numerous international leaders, members of Ford's cabinet, and senior administration officials.

During the dedication, Ford described his vision for the museum: "The past can instruct the present not only in the art of leadership, but also in the opportunities of citizenship."[8] His view of presidential libraries was that "unlike the pyramids, their function is not to deify the dead but rather to distill from the past the essence of experience that may illuminate the dim path into the future."[9] Sixteen years later, at the rededication of the museum in April 1997, President Ford went on to state that the museum "is not a monument to any one man or any one presidency. Rather, it is a classroom of American democracy, a place where school kids as well as scholars will enjoy privileged access to the innermost workings of their government."[10]

The museum, with 55,200 square feet set on 4.6 acres, cost seven million dollars to build. The two-story triangular building of steel and concrete has 11,000 square feet devoted to ten galleries of permanent exhibits and 4,500 square feet for feature exhibits. The building also includes 4,200 square feet for offices of the Ford Presidential Foundation. From the lobby, visitors view the Betty Ford Garden and fountain, sparkling river rapids, and the skyline of the city that the president loved so much.

For several years, planning has been underway for the total renovation of the museum's permanent exhibits, incorporating new scholarship on Ford's life and career as well as the latest technologies for displays and interactives. The addition of a new two-story DeVos Learning Center of 7,300

DALLAS TRIP

June 7, 1964

Saturday, June 6, 1964:

Flew from Kent County Airport to Dallas, Texas (Love Field) with Air Force plane from SAM fleet. Left at 11:45 p.m. Grand Rapids time and arrived two hours later. Met at airport by Secret Service agent who drove me to Sheraton-Hilton Hotel for the night.

Sunday, June 7, 1964:

Met at Statler-Hilton by Secret Service Agent Wolferd at 8:00 a.m. who drove me to Hensley Field (N.A.S.) where we met Chief Justice Warren, Rankin, Ball, Spector.

Drove to Texas School Book Depositary building where the group was met by Mr. Truly and several Secret Service personnel. We were shown the TSBD from top to bottom. Chief Justice and I with stop watch in hand followed Oswald's course from 6th floor window to exit down and out the front door. At 6th floor window sighted motorcade course with rifle. Went to 5th floor and knelt in window under O. while shells were ejected above. Followed course of Jarman and others on 5th floor after assassination. Also took course of Truly and Officer Baker from front of TSBD to 2nd floor and lunchroom scene.

By this time the "press" had assembled outside TSBD. ChJ. and I went out rear exit and got in car as "press" tried to question and take pictures. Driven by Secret Service to Sheriff Decker's headquarters across the street where he greeted us while we waited for a court reporter.

11:45 AM: Our party went to an upper floor in the police headquarters where juries are kept during a criminal proceeding. The room set aside for the Ruby interrogation was apparently a small kitchen and eating quarters for the jurors. It is about 10 ft. x 18 ft. Very austere surroundings. Table in center of room about 3 ft. x 8 ft. Chair set up for R. at one end with his back to the sink and kitchen facilities.

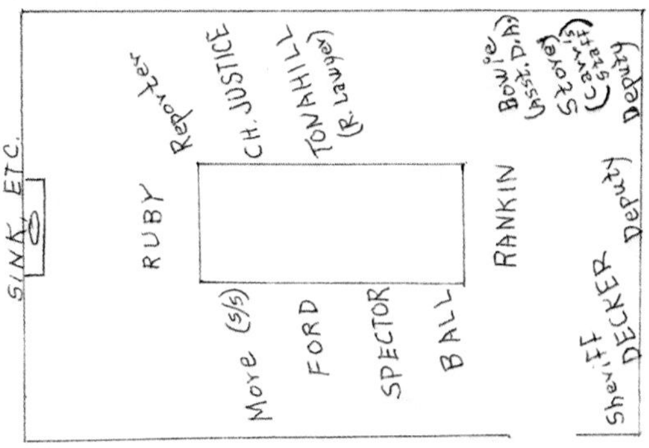

Fig. 156. Gerald R. Ford's Notes from his trip to Dallas, Texas, to interview Jack Ruby as part of the Warren Commission, June 7, 1964. Courtesy of the Gerald R. Ford Presidential Library.

square feet will greatly enhance the museum's educational programming, with special emphasis on the character and integrity that Ford brought to public service. Both the exhibits and addition are scheduled to open in June 2016.

THE LIBRARY AND MUSEUM TODAY

Overall, the Gerald R. Ford Presidential Library and Museum collections consist of twenty-five million pages of documents; five hundred thousand audiovisual materials; and more than nineteen thousand artifacts, the latter housed at the museum. Together, they draw several hundred thousand visitors each year, including school and tour groups, college and university classes, families, the general public, and researchers from around the world. In addition, the Ford website (www.fordlibrarymuseum.gov) consistently draws nearly 1.3 million visitors annually.

A major new initiative of the Ford Library and Museum is the digitization of documents and photos and imaging of artifacts, which makes them available to remote users around the globe via the website. Both the digitization and imaging proj-

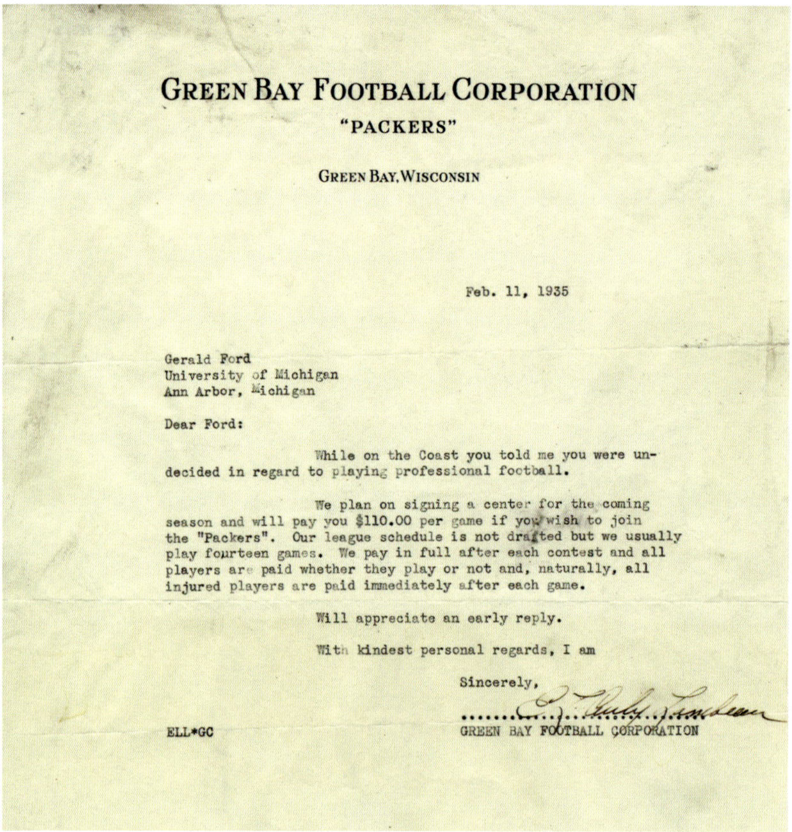
GREEN BAY FOOTBALL CORPORATION
"PACKERS"
GREEN BAY, WISCONSIN

Feb. 11, 1935

Gerald Ford
University of Michigan
Ann Arbor, Michigan

Dear Ford:

While on the Coast you told me you were undecided in regard to playing professional football.

We plan on signing a center for the coming season and will pay you $110.00 per game if you wish to join the "Packers". Our league schedule is not drafted but we usually play fourteen games. We pay in full after each contest and all players are paid whether they play or not and, naturally, all injured players are paid immediately after each game.

Will appreciate an early reply.

With kindest personal regards, I am

Sincerely,

....
GREEN BAY FOOTBALL CORPORATION

ELL*GC

Fig. 157. Letter from the Green Bay Football Corporation to Gerald Ford offering a professional football contract, February 11, 1935. Courtesy of the Gerald R. Ford Presidential Library.

ects are made possible by the generous support of the Ford Presidential Foundation.

The library draws researchers from across the country and around the globe each year, many of whom are in residence for several weeks. Some of these scholars are supported by Ford Presidential Foundation Research Travel Grants. The library staff also responds to several thousand email and phone inquiries each year, along with working with University of Michigan classes and others on specific archival projects.

Due to the unusual nature of the dual locations, the library has made a major effort to provide exhibits and information about President and Mrs. Ford in the library lobby, so that people who are unable to visit the museum in Grand Rapids, or arrive unwittingly in Ann Arbor expecting to see exhibitions, are not disappointed. The lobby offers an introductory film about the president and permanent and temporary exhibitions. The former includes a timeline of the president's life and career and the "Wolverine Corner," anchored by a full-size sculpture of a wolverine, the university's fierce mascot. This area highlights the president's time at the University of Michigan, and includes his number 48 football jersey and signed helmets. There is also a Betty Ford Corner that examines the life and accomplishments of the first lady. Nearby, the president's office features the furniture from his congressional office. Rotating lobby exhibit cases feature thematic exhibitions that draw on documents, artifacts, and photographs in the library and museum collections.

The Ford Library's collections cover US domestic issues, foreign relations, and political affairs during the Cold War era. The core elements of the library's collections are the 1974–77 presidential papers of Ford and his White House staff. In addition, the library has his vice presidential and congressional papers, along with the papers of more than four hundred former administration officials. Highlights of the collection include materials related to the Nixon pardon, Ford's service on the Warren Commission, the Conference on Security and Cooperation in Europe (Helsinki Accords), bicentennial celebrations, the end of US involvement in Vietnam, the Equal Rights Amendment, the role of Betty Ford, and many other topics.

At the Ford Museum, the permanent exhibits consist of ten galleries spanning Ford's childhood, his service in the US Navy during World War II, the congressional years and service on the Warren Commission, his selection for the vice presidency and then presidency, his pardon of Richard Nixon, domestic issues and international challenges, head of state gifts, the bicentennial and election in 1976, the post-presidency, and his funeral in 2007. Exhibits include a nearly full-size replica of the Cabinet Room and the Oval Office; this display of actual artifacts used by Ford in the White House is unique among the presidential libraries.

Of special note is the exhibit of the staircase from the US Embassy in Saigon, which was used to help evacuate thousands of US personnel and South Vietnamese during the final days of the Vietnam War in April 1975. Despite the reservations of others, Ford personally insisted that the staircase be included in the exhibit, dubbing it "the staircase to freedom."

While set apart at two sites, the Ford Library and Museum staff work closely together to preserve the records and historical artifacts of the Ford Administration, with the goal of providing access to these important materials to students, researchers, and the general public, thus enhancing civic literacy and the understanding of democracy. The completion of the updated and highly interactive exhibits and the new DeVos Learning Center at the museum—coupled with the library's vigorous efforts to process, open, and digitize materials for research—will continue to fulfill President Ford's vision for both institutions.

Entry prepared by Elaine K. Didier.

Notes

1. Ruth Borden in Warner, "A Presidential Library Comes to Ann Arbor," 7.
2. Robert Warner, 19.
3. Warner, 19.
4. Warner, 20.
5. Warner, 23.
6. Warner, 23.
7. Warner, 23.
8. "Ford Museum Dedication," Box Q11, 9/18/81, Post-Presidential Office File, Gerald Ford Presidential Library.
9. Ibid.
10. Ibid.

REFERENCES

Warner, Robert M., "A Presidential Library Comes to Ann Arbor," *Bentley Historical Library Bulletin*, no. 55, pp. 1–24. Ann Arbor: Bentley Historical Library, University of Michigan, 2007.

LAW LIBRARY

1859–2017

The University of Michigan Law Library is a comprehensive legal research library with one of the finest collections of works on domestic, foreign, comparative, and international law in the United States.

Collection areas of special depth include foreign and comparative law; public and private international law; collections on the legal status of Native Americans and other indigenous peoples; pre-Soviet Russia; early American session laws and court reports; current and superseded annotated statutes for all US states and the federal government; and Roman law. The Law Library became the first depository for European Union documents in an American university in 1957. It is also a selective depository for US government publications and collects law-related documents of many international organizations such as the United Nations, World Trade Organization, and Organization of American States.

Rare law books are found in the Jackier Rare Book Room. The oldest book is a first edition of Johannes Nider's *De contractibus mercatorum*, circa 1468, and the oldest manuscript is Johannes Calderinus's *De ecclesiastico interdicto: distinctiones decretalium (also a series of Questions disputed by Calderinus at Bologna between 1330 and 1346 and similar disputations of Paolo Liazari, canonist at Bologna).*

As of July 1, 2015, the Law Library's collection held 1,036,578 volumes in print and 517,968 digital titles. Collections consist of a mix of print, microform, and electronic formats; the entire print collection is housed in the closed stacks of the ten-story Legal Research Building and the three-story Allen and Alene Smith Addition, popularly known as the "Underground Law Library." The library's print collection is noncirculating.

BUILDING THE COLLECTION: HISTORICAL OVERVIEW[1]

The Law Library was established in 1859 as part of the Law Department and continues to be "maintained and administered as a part of the instruction and research operation of the Law School."[2]

The library has been considered the "apparatus" of the Law Department[3] and "the lawyer's laboratory."[4] Indeed, this underlying view led the library to build a comprehensive collection that would provide "the means necessary for original investigation"[5] and "permit scholars to do research work in any field of law, regardless of country or period."[6] The collection development policy—to collect primary sources of law: statutes, civil law codes, court opinions, regulations, as well as books and treatises about the law for all countries of the world in the original, official languages—remains consistent with this view.

The development of the Law Library from a small collection of 350 volumes housed in one room maintained by a law student[7] to a major global collection of over a million volumes took place in distinct historical phases.

The Early Years

Between 1859 and 1924, the dean and faculty of the Law School were deeply involved in decisions about the collection[8] and they focused on building a comprehensive collection of US state and federal court reports, statutes and administrative regulations, as well as on expanding the library's holdings of major legal treatises and textbooks.[9] The first foreign law materials were added to the collection in the 1890s but were limited to court reports and legislation for major common law jurisdictions—Great Britain, Canada, Ireland, and

Fig. 158. Legal Research Building. Photo by Law Library staff.

Scotland.[10] The library began adding public and private international law materials to the collection in 1919 followed by the earliest Roman and comparative law acquisitions in 1923.[11]

As faculty interest in other areas of law developed and the curriculum of the Law School broadened during this period, the collection expanded to support new areas of study and instruction.

The Foreign Law Years

The next phase of the Law Library's collection building took place between 1925 and 1960 through the collaborative work of Henry Moore Bates—who served as Law School dean from 1910 through 1939—and Hobart R. Coffey—who joined the Law School in 1924 as a law professor and served as law librarian (1926–44) and director of the Law Library (1944–64).[12] Bates is credited with the vision that led to the development of the Law Library collection into one of the foremost foreign, comparative, and international law collections in the United States.[13] Coffey turned Dean Bates's vision into reality.[14]

> [Hobart Coffey's] first and primary concern was with building the collection, and the Michigan Law Library today stands as a memorial to his towering achievement. The figures tell some of the story: about 80,000 volumes in 1928 when he came—a modest, primarily American collection; 350,000 in 1965 when he retired, comprising one of the great legal research collections of the world.[15]

During his tenure, Coffey, who was fluent in six languages,[16] worked tirelessly to identify and acquire foreign, comparative, and international law books; primary sources; and the documents of international organizations. His "tools" included advice and assistance from the Law Department faculty, the work of Law Library staff, travel abroad, advice and assistance from foreign book dealers, gifts and exchanges, and his own research.[17]

1960s to 1980s

By 1960, Coffey's major retrospective building of the Law Library collection had been completed and the main responsibility for selecting materials for the collection was transferred to reference librarians who were assigned jurisdictional selection responsibilities[18] (e.g., Anglo-American law, civil law and mixed jurisdictions, public international law). Collection development continued to be handled by reference librarians with law degrees and expertise in different jurisdictions and areas of law.

By the early 1970s, as collection building was

routinized, the Law Library turned its attention to considering alternative formats such as microform to preserve the collection.[19] By 1990, the Law Library had embarked on its "Preservation in Context" initiative under the leadership of Director Margaret Leary to both preserve the collection and stabilize its size to avoid the need for offsite storage.[20] During the ensuing two decades, the library replaced print volumes with microfiche and succeeded in opening up much needed shelving for new primary and secondary material that could not be so replaced.

The Digital Turn

The Law Library was an early subscriber to sophisticated full-text electronic legal research tools. In 1970, the Law School participated in an experiment initiated by the Mead Data Central Corporation, the creator of the LexisNexis database, to compare the use of computers with the use of standard print law digests to conduct case law research.[21] The library eagerly embraced this new technology as databases, notably Lexis and Westlaw, became available by license in the late 1970s to provide both legal research training to law students and invaluable legal research tools to librarians and the law faculty. The Law Library continued to add new electronic finding aids and electronic resources to its collection throughout the 1980s and 1990s, including the European Union's Celex database.

The advent of the World Wide Web in the 1990s and its dramatic growth as a source of legal materials triggered the next stage of the library's digital collection building. Today, the library provides web-based access to more than one thousand electronic databases and thousands of electronic journals and books.

While electronic resources have become a major part of the collection, the Law Library continues to collect primary sources of law, treatises, and journals in tangible formats to enable legal scholars to meet the requirement of citation to the official sources of law; to archive official versions; to ensure preservation of the laws of jurisdictions experiencing political or other instability or where digital preservation is not assured; and to provide access to materials not yet available electronically. This is especially important for foreign law materials, as a vast number of publications and sources are not yet available in stable, citable digital formats, and many are not yet available in any electronic form.

While digital sources have not replaced tangible formats, the library continues to add both free and for-fee web-based resources to the collection. These selections include foreign and domestic governmental websites that publish court opinions, legislation, and documents; websites of international organizations and associations that publish law and law-related documents and studies; full-text journal collections; historical full-text treatise collections; modern e-books; electronic journal indexes; and new electronic resources as developed. Reference librarians identify, vet, and select electronic materials.

The Law Library's comprehensive collection continues to attract students, lawyers, researchers, and scholars from all over the world, and its continuing work to enhance and maintain the collection in support of the Law School's educational and research mission remains paramount.

DEVELOPMENT AND GROWTH OF LAW LIBRARY FACULTY AND OTHER SERVICES

While collection building was an ongoing priority, the Law Library was also in the forefront of developing services for Law School faculty. It was the first academic law library in the United States to provide faculty document delivery services in 1966. This service was enhanced in 1984 by Director Margaret Leary who added a comprehensive research component to the service to increase support for faculty research and teaching.[22] This innovative research service began with two law students supervised by a dual-degreed reference librarian. Over the next fifteen years, the service grew and expanded so that today, a full-time reference and faculty services librarian with the

Fig. 159. Reading Room, Legal Research Building. Photo by Law Library staff.

assistance of six law students support the Law faculty via customized, expert legal and nonlaw research.

The library's dual-degreed[23] librarians and law students also provide legal reference services to researchers during the library's open hours in person, online, and via email. The librarians, moreover, offer individualized research consultations and group legal research training, as well as online research guides.

Other service initiatives include the in-house creation and maintenance of web-based resources such as the website listing *Publications of the University of Michigan Law School Faculty* since 1859[24] and the more recent creation of the *University of Michigan Law School Scholarship Repository* [25] that provides global access to Law faculty scholarship, the Law School's eight student-edited journals, and other Law School publications.

LIBRARY BUILDINGS

The need for space to house the library's growing collection has had a significant impact on the construction of Law Library buildings. The first Law Library in 1859 was a room in "the old South Wing of University Hall."[26] In 1863, the Law Library moved into the new Law Building (later renamed Haven Hall) near the corner of State Street and North University Avenue, where it remained until 1931. At this point the Law Library found a new home in the iconic Legal Research Building that was part of the recently constructed Law Quadrangle. The new building was constructed with money donated by William W. Cook to specifically hold the law library.[27] By the late 1970s, the collection had outgrown its space in the Legal Research Building, a development that led to the building's expansion in 1981 through the construction of the Allen and Alene Smith Addition.

The Law Library has long supported and accommodated the changing study and research needs of law students, law faculty, and scholars. When the Legal Research Building opened in 1931, it included eight "consultation rooms" to make it possible for law students to study together and engage in discussions without disturbing the quiet study and research atmosphere in the library.[28] As years passed, these rooms gradually fell into disuse and were subsequently converted into staff offices. However, by the early 2000s, demand for collaborative study space for law students once again led to the creation of consultation rooms—now called group study rooms. The

Fig. 160. Allan and Alene Smith addition. Photo by Barbara Garavaglia.

provision of quiet study and research space in the Law Library remains a critical priority.

EARLY LAW LIBRARY STAFF AND LIBRARY DIRECTORS[29]

Until 1882, only law students who served in a custodial capacity staffed the Law Library. Two law students who held the position of "law librarian" were Isaac Marston (1861–63) and Levi L. Barbour (1865–67). As the collection grew in size and sophistication the Law School recognized the need for expert services, hiring its first professional librarian in 1883. The first librarian to hold the position of Law Library director at Michigan was Hobart R. Coffey. The library's professional staffing needs changed throughout its history based on the increasing diversity and depth of the collection, format changes, and the development and addition of new services—including catalogers, librarians with extensive foreign language skills, dual-degreed reference librarians, electronic resources librarians, metadata librarians, and scholarly publishing librarians to name just a few.

Director	Years
Hobart R. Coffey	1944–64
Beverley J. Pooley	1964–84
Margaret A. Leary	1984–2011
Barbara H. Garavaglia	2011–16

CONCLUSION

The Law Library continues to evolve and change in response to and in anticipation of ongoing, rapid changes in legal education and the legal information landscape. Although formats continue to change, the need for effective access to legal information and instruction and assistance in its use remain critical, not only for law students and professors but also for a society that values the rule of law. Access to foreign and international law continues to increase in importance in response to globalization, making the early vision of Bates and Coffey to develop a comprehensive global collection truly prescient. Technological innovation will undoubtedly create new challenges and opportunities relating to law; the library is committed to moving forward into the next two centuries with the same vision and commitment to excellence exercised by those who built this great law library.

Entry prepared by Barbara H. Garavaglia.

Opening hours: Reading Room 8 am–2 am, Smith Addition 8 am–midnight; hours vary during semester breaks. Email: askalawlibrarian@umich.edu

https://www.law.umich.edu/library/info/Pages/default.aspx

Notes

1. This section is based almost entirely on the comprehensive history of the Law School and Law Library by Elizabeth Gaspar Brown, *Legal Education at Michigan, 1859–1959*, 359–87, 920–30. It also borrows liberally from Margaret A. Leary's "Building a Foreign Law Collection at the University of Michigan Law Library, 1910–1960." The book and the article are based on extensive research in the Bentley Historical Archive, including comprehensive study of reports, proceedings, minutes, and correspondence relating to the Law collection and the role played by Law School deans and faculty in building the Law Library collection.
2. Regents of the University of Michigan, Bylaw 12.02.
3. Henry S. Frieze, 1871–1872 President's Report. Reprinted in Brown, 364
4. *Law Students' Handbook* (1957), 15. Reprinted in Brown, 382.
5. Brown, 368.
6. Nathan Fred, ed., *Book of the Law Quadrangle at the University of Michigan* (1934), 43. Excerpt reprinted in Brown, 387.
7. Brown, 926–27.
8. Ibid., 360, 370.
9. Ibid., 370.
10. Ibid., 368.
11. Ibid., 371
12. Ibid., 927.
13. Leary, "Building a Foreign Law Collection," 397.
14. Leary, "Building a Foreign Law Collection," 403–9.
15. Pooley, 363.
16. Ibid., 364.
17. Leary, "Building a Foreign Law Collection," 409.
18. Ibid., 423.
19. University of Michigan Law Library.
20. Leary and Snow, 2.
21. "Computers and the Law: New Legal Research Methods," 3.
22. Leary, "Supporting Faculty Research," 1; and "Library Support for Faculty Research," 193.
23. Reference librarians at Michigan hold both the JD degree and the MLS/MLIS/MI degree. Many are also licensed attorneys.
24. Available at http://www.law.umich.edu/library/guests/pubsfaculty/Pages/default.aspx
25. Available at http://repository.law.umich.edu
26. Brown, 359.
27. Brown, 319; Leary, *Giving It All Away*, 169.
28. "World's Finest," 465, 467.
29. All information in this section is from Brown, 926–27.

REFERENCES

Brown, Elizabeth Gaspar. *Legal Education at Michigan, 1859–1959*. Ann Arbor: University of Michigan Law School, 1959.

"Computers and the Law: New Legal Research Methods." *Law Quadrangle Notes* 15, no. 2 (1971): 3.

Leary, Margaret A. "Building a Foreign Law Collection at the University of Michigan Law Library, 1910–1960." *Law Library Journal* 94, no. 3 (2002): 395–425.

Leary, Margaret A. *Giving It All Away: The Story of William W. Cook and His Michigan Law Quadrangle*. Ann Arbor: University of Michigan Press, 2011.

Leary, Margaret A. "Library Support for Faculty Research." *Journal of Legal Education* 53, no. 2 (2003): 192–98.

Leary, Margaret A. "Supporting Faculty Research: A Direct Role for the Library." *Trends in Law Library Management and Technology* 3, no. 1 (1989): 1–2.

Leary, Margaret A., and Barbara J. Snow. "Preservation in Context: The University of Michigan Law Library Beyond 2000." A Report to Dean Lee Bollinger, July 1989, revised January 1990.

Pooley, Beverley J. "In Memory of Hobart Coffey." *Law Library Journal* 63, no. 3. (1970): 363–64.

Regents of the University of Michigan. *Bylaws* (available at http://www.regents.umich.edu/bylaws)

University of Michigan Law Library. "The Law Library to the Year 2000: Storing and Preserving the Collections" (ca. 1972).

"World's Finest Educational Building: William W. Cook Legal Research Library Nears Completion as Unit of Law Quadrangle." *Michigan Alumnus* 37, no. 24 (1931): 465–67, 472–73.

Section 3

Other Art Collections

ROSS SCHOOL OF BUSINESS ART COLLECTION

1975–2017

THE BEGINNINGS, 1975–99

The art collection at the University of Michigan Business School (UMBS) was initiated in 1999 as the school celebrated its seventy-fifth anniversary. As one commemoration of this milestone, Dean Joseph White chose to establish a fine art collection. His goals were twofold: to bring an aesthetic to the buildings and to help the community develop an understanding of contemporary art. UMBS already had the seeds of a strong collection of contemporary art in a small number of works donated between 1975 and 1997.

EXPANSION OF THE COLLECTION, 2000–2009

The dean asked alumnus and Adjunct Professor Thomas Porter to chair a committee for the purpose of purchasing significant works of art for the school. Professor Porter engaged the services of Alice Simsar, an experienced and knowledgeable Ann Arbor art dealer. With funding from Porter, J. D. Williamson, and other alumni—and with their exceptional expertise—the committee was able to purchase approximately forty works, focusing primarily on works on paper. The collection included works by many prominent, contemporary artists, including Frank Stella, Robert Rauschenberg, Sol Lewitt, Richard Estes, Anni Albers, Joseph Beuys, Dorothea Rockburne, Ed Ruscha, and Dale Chihuly.

In 2001, Professor Robert J. Dolan was appointed dean of the U-M Business School. In 2002, Kathleen Splaine Dolan, a consultant in contemporary art, joined the art committee, with the goal of bringing additional art to the dated buildings to help make the environment lively and aesthetically pleasing for students, faculty, and staff. At that time, several alumni had donated additional funds to purchase more art. Initially, the collection was hung in public lobbies and maintained its initial focus of works on paper and abstract images. After 2002, the committee chose to diversify the collection by buying more creations by women and minorities as well as more works of realism. Artworks were placed in additional public spaces such as the admission offices and student lounges.

New art was hung monthly and the school's population was excited and challenged by the constant stream of artworks. The art committee believed it had a unique opportunity for art education. Every piece had a detailed didactic that explained the artist and the artwork. In addition, a printed list of works by location was produced for viewers to take for a reference. As part of the community education process, the Business School began to acquire a book on each artist for its library and offered tours of the collection to students, faculty, and staff, at times with the assistance of a docent from the university's Museum of Art.

In 2003, inspired by the collection, a member of the Business School's Advisory Board, an officer at JPMorgan Chase, suggested that there might be a possibility of a loan from their esteemed art collection to the U-M Business School. Several mem-

Fig. 161. *With the Current* (l) and *Forgetting the Other* (r), Deborah Butterfield, 2002, cast bronze.

bers of the committee met with the curator of the JPMorgan Chase collection, who agreed to loan significant and very large paintings to the school's executive education building. These included works by Frank Stella, Donald Baechelder, Mike Didlo, Richard Artschwager, Roy DeForest, Joel Shapiro, Terry Winters, Fred Braithwaite, Yvonne Jacquette, and Leon Golub, among many others. This loan was arranged and paintings changed annually at the discretion of JPMorgan Chase.

Building a New Campus at the University of Michigan Business School

In 2004, Dean Dolan was involved in planning and funding a new campus for the Business School. His main collaborator in this was Stephen M. Ross, an alumnus to the school and a New York real estate developer, who made a $100 million dollar gift to the effort. In recognition of his generosity, the school was renamed the Stephen M. Ross School of Business. Over the next two years, planning of the new building took place. The JPMorgan paintings were returned and the process of moving works in the school's art collection began, involving many challenges due to the dimensions and palette of the work. Finally, the art was moved and the few remaining large works were stored until the new building was ready.

Art for the New Building

Now, the elaborate process of choosing and purchasing art for the new building began. The first recommendations came from the architectural firm of Kohn, Pedersen and Fox. During the planning stage, the architects designated key spaces in the building where large commissioned

Fig. 162. *Suite of Six Etchings*, Ed Ruscha, 1980, and partial view of *Reunion*, Robert Rauschenberg, 2000.

works of art could be hung. These spaces gave direction to the art committee for new acquisitions and for planning the rehanging of over a hundred works that the school then owned. It was a daunting challenge to determine artwork, location, and funding. To help with identifying new acquisitions, the school was fortunate to have the volunteer services of a retired art consultant, Gail Young. In addition, Alice Simsar and the curator for the JPMorgan Chase collection actively looked for work and suggested artists who could be commissioned to produce new works to hang in designated spaces.

Commissioned Art Work

The initial task for the committee was to identify sites in the new structure that called for large works of art. The main entrance—a space requiring a large rectangular work—was their first focus. Sculptor Gene Flores (originally from California and now living in Massachusetts) was commissioned to fabricate a steel structure. He proposed a rectangular work in rusted steel, a representation of the Huron River. Its completed size is 60 feet long x 34.8 inches high. Entitled *Crossing the Huron at Broadway*, it is a representation of the actual space on the university campus.

Howard Ben Tre, an internationally known glass sculptor, conceived of the massive glass, gold leaf, copper, and patina walls, measuring 25 x 78 x 3 feet, in the atrium. They provide a subtle but strong backdrop and divider for the high-ceilinged space. Miguel Rocha, a sculptor and painter (Lisbon, Portugal, and New York City), was commissioned to make a spectacular sculpture for the lower atrium. This work, in wood and painted acrylic, spans the entire wall, measuring 8 x 30 x 8 feet. Rocha moved to Ann Arbor to complete this project; he worked in a studio on campus to cut and mold the wood and installed this large, complex sculpture over a period of several months.

Sculptor Katy Stone (Seattle, Washington) was chosen to complete a monumental sculpture for the colloquium room on the top floor of the new building. She is known for using nontraditional elements. Her work, *Grasswave*, is made of hundreds of hand-painted laser-cut aluminum elements, creating a precisely made work that appears spontaneous. This sculpture fills a large wall in the room and its "waving grasses" echo the green roof that can be seen outside the space.

Although not a commissioned work, the formidable life-size horse sculptures by Deborah Butterfield donated by Jeanne and Sanford Roberston add a breathtaking dimension to the glass-enclosed space outside the Blau Auditorium.

Forgetting the Other and *With the Current*, made of found welded steel, are statement works that provide a powerful backdrop to the many events held in this space.

Another important area for art is the corridor outside the colloquium space. Funded by the Alumni Board of Governors, this group includes works by many prominent artists such as Darren Waterston, Hung Liu, Jocelyn Krystosek, and Lynn McCarty. The collection also has a floor dedicated to contemporary photography, including works by Nikki S. Lee, Jeff Chien Hsing Liao, and Lorna Simpson.

An additional work purchased for the new building was by a young, local artist, John Schwarz. Titled *Reluctant Hero*, it is made of 90 percent recycled materials; some of the objects the artist used dated from when he was in high school. This piece, rising seven feet tall with skis for wings, is a favorite with many students. *Reluctant Hero* was also a favorite of Dean Dolan. He later expressed his thoughts about the sculpture: "I spotted the work in 2007 and unconsciously thought: oh that's a new student getting ready to take off on an adventure with us. Up on the tip toes ready to soar!"[1]

Fig. 163. *Proposal for a Colossal Monument in Downtown NYC*, Claes Oldenburg, etching, 1993.

THE COLLECTION TODAY, 2009–17

Between 2000 and 2009, the Ross School Art Collection grew dramatically. With the support of generous alumni donors, more than 230 artworks were added during that period, and while the phase of rapid growth slowed once the building was opened, the school continues to add important artworks, including a sculpture by Alyson Shotz, which had been commissioned by the art committee for the new building, and, more recently, photographs by Bruce Gendelman. Kathleen Splaine Dolan continued to oversee the collection until 2011.

Today, the Ross School is undertaking new building projects as it continues to expand and improve its facilities—dramatically renovating the Kresge Library (renamed Kresge Hall in recognition of its expanding uses) and adding a new academic building (Jeff T. Blau Hall). Under the supervision of collection manager Wendy MacGaw, some artworks have been relocated. MacGaw is working with the School on plans to rearrange and integrate existing collections in the expanded spaces and to identify new public spaces for which art works can be commissioned. The collection remains accessible to students, staff, and visitors to the Ross School of Business, and can be viewed online at http://www.bus.umich.edu/rossart/default.htm.

Entry prepared by Kathleen Splaine Dolan.

The Stephen M. Ross School of Business Art Collection is on view throughout the school buildings, and can be visited on line at http://www.bus.umich.edu/rossart/default.htm

Note

1. Cited at http://www.bus.umich.edu/rossart/collection/Artwork.aspx?artwork_ord=208

SCHOOL OF SOCIAL WORK, COLLECTION OF ARTWORKS

1998–2017

During the planning of the current School of Social Work (SSW) building in the mid-1990s, the school initiated an acquisition program to bring important works of contemporary art into its new space. The impetus for this ambitious program grew from a desire to enhance the professional identity of the school's groundbreaking new facility. Made possible by the generous contributions of donors, gift funds were set aside and an endowment was established at the School of Social Work by the estate of Clinton E. and Dorothy Purdy and earmarked for the new building, its classrooms and furnishings, and works of art. To date, the School of Social Work art collection features fifty-four artworks, including three major commissioned works, and art by Gerome Kamrowski, Keith Haring, and William Kentridge.

A special art committee was assembled by then Dean Paula Allen-Meares in consultation with the faculty executive committee. This art committee sought to include works by artists with University of Michigan connections—and they resolved to acquire works that reflected the school's philosophy and curriculum objectives. The committee established several themes to guide its selection. These themes speak to important and long-standing social issues of our time:

> *Multiculturalism:* which fosters community spirit and a humane respect for such differences as gender, race, ethnicity, and social class;
>
> *Family:* traditional and expanded notions that affect children, youth, adults, and the elderly; and
>
> *Protection of the vulnerable*: especially the poor, sick, and infirm.

As Allen-Meares later explained, "The art in our collection reflects the values of the profession and the mission of the U-M School of Social Work—to improve the wellbeing of the economically and socially disadvantaged and other vulnerable populations by using research and an innovative curriculum to advance a more just, healthy, and productive society."[1]

DEVELOPMENT OF THE COLLECTION

The first phase of the program was completed with the acquisition of forty works—including photography, prints, textiles, and sculptures—that

Fig. 164. *PS*,. Ross Bleckner. lithograph, 1997. SSW 98.24. © Ross Bleckner, courtesy of Mary Boone Gallery, New York.

Fig. 165. *In the Street*, Garry Winogrand, vintage gelatin silver print, 1969. SSW 98.20. © The Estate of Garry Winogrand, courtesy Fraenkel Gallery, San Francisco.

were installed in time for the building's dedication ceremony in September 1998. A centerpiece of the original collection is *The Real Blue*, a four-piece installation by the internationally renowned American artist Sam Gilliam, which was commissioned for the school's lower level. Gilliam is a color field artist known for his use of saturated color on large, unprimed canvas that is then stretched, wrapped, or draped. He was the first artist to introduce the idea of a painted canvas hanging without stretcher bars, a major contribution to the color field school.

Noted sculptor Bill Barrett was commissioned by the school to create *Kindred* (2001), a ten-foot bronze sculpture that was dedicated on September 20, 2002, and stands at the entrance of the school's Educational Conference Center. Barrett, whose work is included in public and private collections throughout the world, earned his bachelor's and master's degrees in design from U-M, as well as his masters of fine arts. He says that the fundamental style of his sculpture was developed during his years at U-M while working with his mentor, Tom McClure. The purchase of *Kindred* was facilitated in part through Barrett's generosity.

During the early 2000s, new works by important artists continued to be added to the collection. In the spirit of earlier acquisitions, they address social issues of family, work, ethnicity, and multiculturalism. Among the artists represented is Ed West, a photographer who currently teaches at U-M School of Art & Design. The two photographs acquired were first seen in a major exhibition of West's work at the University of Michigan Museum of Art (UMMA) in late 2000 and early 2001. Janet Taylor Pickett, a graduate of U-M School of Art & Design, works in various media and formats, including drawing, prints, installations, and fabric constructions. Each of Pickett's fabric constructions in the SSW collection plays upon the traditions and associations of the American quilt. Hung Liu, who came of age during the Cultural Revolution, emigrated to the United States from the People's Republic of China in 1984. Her paintings and prints address the

Fig. 166. *Kindred*, Bill Barrett, bronze sculpture, 2002.

"cultural collisions" that she experienced in China and has since encountered in the West.

Other important pieces in the collection include works by renowned artists Gerome Kamrowski, Keith Haring, and William Kentridge. Kamrowski taught at the School of Art & Design from 1946 until his retirement in 1982. A contemporary of Jackson Pollock, he was an important figure in the American art world's transition from abstract surrealism to abstract expressionism. Haring's graffiti-inspired imagery mirrors 1980s street culture, expressing themes of birth, death, sexuality, war, and social harmony through simple, cartoon-like figures that appeared in NYC subway stations and large public murals. His work continues to be exhibited in museums around the world. One of South Africa's most celebrated contemporary artists, Kentridge creates animated films, books, prints, collages, and sculptures that explore the themes of apartheid, colonialism, and totalitarianism.

Two lithographs included in the collection, Robert Rauschenberg's *Untitled* (1996) and Jaune Quick-to-See Smith's *Indian Heart* (1994), also present collage works of social critique. Rauschenberg had begun to assemble "found" mass produced photographs in the 1960s. Smith brings this collage technique to bear on Native American ethnic identity and culture, juxtaposing the stereotypical image of a Sioux chief in advertisement with other insignia of national and indigenous culture. With *Untitled*, a work to benefit the humanitarian organization Doctors of the World, Rauschenberg probe the politics of compassion by combining natural imagery, a wrought-iron gate, and images of children whose eyes meet the viewer.

The School of Social Work's innovative art program has benefitted the university community in significant ways. The collection affirms the school's identity by giving students and faculty day-to-day experiences with art that mirrors and celebrates the professional mission of this nationally distinguished School of Social Work. It also serves an educational function by exposing viewers to important contemporary art. And finally, the collection furnishes the campus with a university-wide aesthetic resource.

At this time, the school is not actively growing its collection. As the school looks at additional space to house its expanding academic programs it anticipates the possibility of bringing new works of art to the collection.

Entry prepared by Lisa Raycraft.

School of Social Work Building, 1080 S. University Ave.

http://ssw.umich.edu/art-at-ssw

Note

1. Dean Allen-Meares spoke at the September 2002 dedication of Barrett's sculpture; for her quote see "*Kindered* Installation," 2.

REFERENCES

"*Kindred* Installation." In *Ongoing*, University of Michigan School of Social Work, Winter/Spring 2003, 2.

UNIVERSITY OF MICHIGAN HEALTH SYSTEM, GIFTS OF ART

1986–2017

Gifts of Art is one of the country's earliest and most comprehensive arts in healthcare programs. It is dedicated to bringing works of art, exhibitions, performances, bedside programs, and art and music-making opportunities to patients, visitors, and staff of the University of Michigan Health System. Through permanent installations of artworks and temporary exhibitions displayed in nine galleries, Gifts of Art enhances the system's environment of care by drawing on the power of art to heal, nourish, and comfort.

BACKGROUND

Art has had a presence in places of healing throughout history. Depictions of the ancient temples of Asclepius or the monasteries for the sick of the Middle Ages contain references to art, restful aesthetic spaces, and peaceful fragrant gardens. Anthropologist Ellen Dissanayake, who takes an ethological and bioevolutionary approach in her book *What Is Art For?,* has suggested that art has selective or survival value for humans: "An ethological view presumes that art contributes something essential to the human being who makes or responds to it—not in the usual sense of being good for his soul or pleasurable for his mind or spirit (though these benefits are not denied), but beneficial to his biological fitness."[1]

It is no surprise that in the second half of the twentieth century, when medical care seemed more concerned with pathology than with the patient, art should emerge in an intentional way in the increasingly sterile and technical world of healthcare. By the time the current University of Michigan Hospital was being designed in the early 1980s, a handful of hospital arts programs were taking shape around the country. The National Endowment of the Arts (NEA) was an early partner in promoting these new programs through its Office of Accessibility, which supports access to the arts for underserved populations—in this case, patients in hospitals.

A grant from the National Endowment for the Arts to Project Art at the University of Iowa Hospitals and Clinics set in motion what was to become the Gifts of Art program at the University of Michigan. Planned for the new replacement hospital, the original Gifts of Art program was modeled on Project Art to include both visual and performing arts. A key difference between Project Art and Gifts of Art was that Iowa had—and still has—a state-wide percent for art programs (that is, a percent of the project budget is set aside for the purchase of original art for public buildings). With no percent for art funds, Gifts of Art instead focused on rotating exhibit galleries as the heart of its visual arts program.

Despite no government mandate, a modest amount of the construction budget was earmarked for original art in public areas of the new U-M Hospital that opened in 1986. The focus was on locally sourced artwork to underscore the hospital's position as a state-supported institution. Considered interior design, the intent was not on building a collection, nor on any future appreciation of the works of art, but instead on ornamenting hospital spaces. An art committee was convened that included the director of the U-M Museum of Art and art history faculty, among others. They commissioned a signature piece, the mobile *Aerial*, by U-M professor emeritus Gerome Kamrowski for the main entrance atrium lobby.

Fig. 167. *Hippocrates: Medicine Becomes a Science*, Robert Thom, oil on canvas, c. 1952. Photo by Steven Demos.

THE COLLECTION IN THE CONTEXT OF HEALTHCARE

Many early arts directors in healthcare were recruited from museums, art schools, or community arts organizations and they turned to those worlds when designing operating policies and standards. But when it came to defining the mission and vision for the art collection, such individuals set their gaze firmly on the mission and vision of hospitals and healthcare organizations. Hospital art collections are intended to calm, comfort, support human dignity, enhance healing, revitalize the spirit, engage, and distract. They are designed to provide landmarks to help patients navigate the large facilities (wayfinding). They also serve to underscore values such as excellence of care, patient and family focus, and diversity. No less important, the art assists with place-making by creating inviting public spaces that feel familiar by making reference to local ecology, industry, and architecture.

The mission for art at the U-M Hospitals reflects all the above considerations. In addition, it is mindful of the Hippocratic Oath to "first do no harm." To that end, Gifts of Art created criteria for art on display in public areas: "We take into consideration the medical and emotional sensitivities of our viewers. Art with overt sexual content, violence and nudity will not be considered. We will also avoid pieces that are confrontational, chaotic, obviously depressing or contain negative medical connotations."

Hospitals are inherently dangerous places for artwork. Besides being open 24/7, people are distracted and corridors are pathways for all manner of large equipment. To minimize damage, artwork is installed with museum-quality security systems. Patient safety is also a critical consideration. As a result, fire safety as well as infection control issues not only inform what art is installed, but how.

Hospitals are places in which the drama of life and death play out. Many who are touched by either tragedy or joy wish to memorialize an outcome or a loved one. Two important additions to the U-M Hospital collection celebrated lives cut short by tragedy. A Harry Bertoia sound sculpture was purchased to mark the life of a physician tragically taken while at work. Artist Doug Hollis was selected to create the installation piece *Rotations* to celebrate six lives lost while on a transplant and survival flight mission.

Some art in hospitals is made by patients and staff. *Dragon of Wishes Hopes and Dreams* was created in partnership with faculty and students at the U-M Penny W. Stamps School of Art and Design. The sixteen-foot paper dragon contains messages crafted by two thousand patients, visitors, and staff and folded within its colorful scales. Other collaborative art works have been installed throughout the facility. They honor everything from institutional milestones to the loss and grief that challenge caregivers.

Donations of art must meet institutional criteria. The most significant donation came from Pfizer, Inc. In 2007, the company donated the complete forty-five-piece series *Great Moments in Medicine* by Detroit artist Robert Thom, which had been commissioned by Park, Davis and Company. Placement, maintenance, and image usage requests are handled in partnership with the U-M Museum of Art. Patient sensitivity played a large part in the placement of these forty-five pieces. Paintings showing surgery or graphic procedures were installed in the Medical School, where students and faculty but not patients would see them. Research-themed pieces were installed in predominately staff areas. Paintings of gentle care grace public areas. The most famous medical names from antiquity have a place of honor in the amphitheater adjacent to the main lecture hall to inspire physicians and clinicians.

Fig. 168. *Rotations*, Doug Hollis, stainless steel, 2009. Collection of the University of Michigan Health System.

Fig. 169. *Dragon of Wishes, Hopes and Dreams*, community art project, 2006. Photo by Martin Vloet, collection of the University of Michigan Health System.

Fig. 170. *Sister Loon*, Betsy Youngquist, mixed media. Photo by Candace Meyers, collection of the University of Michigan Health System.

THE COLLECTION AND THE GROWTH OF THE HEALTH SYSTEM

Like its peers, the U-M Health System experienced significant growth in the late twentieth and early twenty-first centuries. By the time the Frankel Cardiovascular Center opened in 2007, the inclusion of art was no longer an afterthought. Healthcare architects had begun planning for the placement of art early in the design process. For the first time at U-M Hospitals, the art was planned as a site-specific collection intended to appreciate in value. Although some funding came from the project, philanthropy played a significant role in purchasing the artwork. A signature sculpture was commissioned that reflects the main focus of the hospital—the heart. The piece is called *Your Heart Is in Our Hands*, by Jane DeDecker, and is located at the main entrance.

The new C. S. Mott Children's and Van Voigtlander Women's Hospitals opened in 2011 with the expectation that philanthropy dollars would entirely cover the cost of art. This was in part because children's hospitals have a special ability to attract donor funds, somewhat of a universal phenomenon. In addition, significant donors to Mott Children's Hospital were also donors to the U-M Museum of Art, leading to a new collaboration with the intention of building a contemporary collection with guidance from the museum. An art committee was formed that included staff, faculty, donors, and local art specialists as well as the then-museum director and the curator of contemporary art. The committee's charge was to select art for the main public spaces that would appeal to patients and visitors of all ages and create a museum feel.

HOSPITAL OR MUSEUM?

Today's hospitals contain art that may be seen by more people than in museums. Indeed, some visit hospitals just to see the art. A collateral benefit is that the hospital may feel more familiar and less frightening when those same visitors return for medical care. Not surprisingly, museums and hospitals are entering into collaborative partnerships. Museums loan artwork to hospitals and aid in building and maintaining collections. They host gallery programs for special populations such as the "Meet Me at MOMA" for dementia patients. They produce video tours of their collections for hospitalized patients. Museums send staff artists in residence to work at the bedsides of patients in hospitals. These and other hospital/museum partnerships have grown from supporting individual healing to creating healthy communities, working with the power of art to transform lives.

Entry prepared by Elaine Sims.

Artworks in the Gifts of Art Collection are on display throughout the U-M Hospital System. For information visit http://www.med.umich.edu/goa

Note

1. Dissanayake, 8.

REFERENCES

Dissanayake, Ellen. *What Is Art For?* Seattle and London: University of Washington Press, 1988.

UNIVERSITY OF MICHIGAN PUBLIC ART

1846 2017

The University of Michigan is distinguished by a rich and engaging campus atmosphere where members of the academic community can be at home in spaces that are uniquely "Michigan."

Public art plays a significant role in contributing to this environment. The President's Advisory Committee on Public Art was established to actively curate public art for the university's shared environment with the goal of reflecting and enabling the expression of the university's academic aspirations.

A BRIEF HISTORY

Public art at the University of Michigan has been acquired and displayed almost since the university's inception. Sculptural works were the first recorded pieces of collected art—the first interior piece being Randolph Rogers's *Nydia, the Blind Flower Girl of Pompeii* and the first recorded outdoor piece *Professors' Monument* by William Peters. Constructed in 1846 of limestone and marble, *Professors' Monument* is now located on Central Campus on the southeast corner of Hatcher Graduate Library. It was erected by the regents following the death of Professor Joseph Whiting (1845), with panels added after the deaths of Professors Douglass Houghton (1845), Charles Fox (1854), and Samuel Denton (1860). Originally it was placed in the designated "University Burying Ground" on the east side of campus. Because the burying ground was never utilized, the monument was moved several times, and has been in its present location since 1918.

Over the years, the university acquired artwork that was inventoried, cataloged, and displayed in various museum spaces. As these collections grew, so too did the collection of public art on campus. Class gifts as well as statues, memorials, and sculptures were donated; numerous paintings, drawings, prints, and photographs accumulated in buildings across campus. In addition, plaques and tablets in honor of donors, key university figures, and distinguished alumni began to adorn interior walls of campus buildings. These items, located outside the physical space of any museum, were generally not inventoried among museum holdings.

In 1980, the passage of Michigan Public Act 105 required that all public agencies of state government inventory all objects of art in their possession and create a mechanism to update the inventory annually. In 1988, a team from the Detroit office of the Commission on Art in Public Places undertook the first phase of the inventory at U-M, entering some 363 listings into a database. University staff followed up with the next phase in the same year, producing 325 additional entries from 32 buildings. Elisabeth Thoburn completed the third phase in January 1990, which resulted in 389 new entries covering 115 buildings. After these three phases, the inventory was not systematically updated. Some inventory activity occurred briefly in 1997 and 1998 when the university's risk management staff undertook a survey of fine art primarily for insurance purposes.

Historically, public art efforts fell under the purview of the university's Exterior Elements Design Review Committee (EEDRC), which advises the university planner on outdoor design details but without the mandate or expertise to engage in public art review. In response, the university planner established the Public Art Review Group (PARG) to review individual proposals for new public artworks from units. PARG facilitated the placement of various works of public art on

Fig. 171. *Wavefields*, Maya Lin, earth, grass,1995.

campus and provided an important review mechanism, but it was not charged with proactively planning for or pursuing works of art with an overall aesthetic vision.

Thus the President's Advisory Committee on Public Art was created in 2006 to replace PARG and to complement the work of EEDRC, while also adopting a more strategic and long-term approach to public art on campus. The committee's charge was, and continues to be, to advise the president about public art on campus as well as to identify sites for new installations, review proposals for new works, and help facilitate the donation of such works. The committee is composed of faculty from a range of units and disciplines on campus and is chaired by the director of the Museum of Art. Since the time the committee was established, it has stewarded numerous new public art installations on campus.

The committee evaluates the significance of the artist or artists and each work's aesthetic value, uniqueness, ethical position (including consideration of its provenance), and the potential contribution it can make both to the university's

Fig. 172. *Two Open Triangles Up, Gyratory II*, George W. Rickey, stainless steel, 1982

educational mission and as part of the existing collection of public art. The committee works to ensure that public art complements a shared visual landscape.

THE COLLECTION

The variety of arts and media on our campus is emblematic of the diverse nature of the community, and artists range from local artisans to internationally renowned sculptors, painters, ironworkers, digital artists, functional artists, and more. Collected pieces are located in outdoor spaces, on the facades of university buildings, and in public lobbies of university buildings; they include both historical and contemporary works as well as permanent and temporary installations. A catalog of the works that fall under the committee is available online.[1] This entry features only a small sample of these works. The collection is ongoing with proposals for new works encouraged and welcomed by the committee.

The University of Michigan's size, complexity, and academic strength; its impressive array of resources and opportunities; the quality of its faculty and research institutes—all these elements contribute to the rich environment in which students learn and challenge themselves as they come into contact with people, cultures, and ideas from all over the world. The university's public art collection is a reflection of that strength and diversity, an important enabler of a creative and welcoming environment, and a signifier of the university's broad vision and stature.

Fig. 173. Lorch Column, marble and steel, 2007.

Public Art at U-M takes many forms. North Campus's *Wave Field* is located on the southeast side of the Bagnoud Building and was designed in 1995 by artist Maya Lin, who is best known for designing the Vietnam Veterans Memorial in Washington, DC, and the Civil Rights Memorial in Montgomery, Alabama. Commissioned by the Association FXB in memory of Francois-Xavier Bagnoud (1982, Aerospace Engineering), and gift of his mother, Countess Albina du Boisrouvray, *Wave Field* connects art to the landscape in a poetic way. It is an earth and grass sculpture that occupies a square space of ninety by ninety feet and represents a naturally occurring wave pattern.

Located on Central Campus is the bright orange *Orion* by Mark di Suvero. Created in 2009, the steel sculpture sits on the west lawn of Alumni Memorial Hall near the Frankel Family Wing of the Museum of Art. Like many of di Suvero's works, *Orion*—named for the Greek mythological hunt-

Fig. 174. *Impressions*, Jun Kaneko, glazed ceramics, 2013.

er—is a bright orange-red and makes a stunning contrast against the sky and the surrounding area.

Two Open Triangles Up, Gyratory II by George W. Rickey is a stainless steel sculpture located on North Campus in the courtyard on the east side of the Ford Library; it was commissioned for this site. A gift of Verna and Kelly Goss, the sculpture was dedicated by former president Gerald R. Ford in November 1982 (Mr. Goss and President Ford were fraternity brothers when both attended U-M). Rickey created this piece so that the triangles move elegantly and quietly in the breeze.

The Lorch Column, created in 1930 of steel and marble, is an architectural artifact located on North Campus on the north side of the Art & Architecture Building. Originally created for the Home Office Building of the Mutual Benefit Life Insurance Company of Newark, New Jersey, it was never installed. It was a gift of Colonel William Starrett (class of 1897) of New York. The column was first installed in the courtyard of Lorch Hall, which was the original home of the College of Architecture. In 2007, it was moved to North Campus as part of the centennial celebration of the founding of the department that became the Taubman College of Architecture + Urban Planning. It has informally been referred to as the Lorch Column in honor of Emil Lorch, first dean of the college and the person responsible for obtaining it.

Requiem by Erwin Binder (2004) is a bronze sculpture on Central Campus. It is in the east court between Alumni Memorial Hall and the Frankel Wing of the Museum of Art. It is part of the collections of the University of Michigan Museum of Art and was a gift of Jack A. and Noreen Rounick.

Impressions by Jun Kaneko was installed in 2013. The painted bronze and stainless steel sculpture is located on North Campus in front of the Earl V. Moore Building. The sculpture was a gift of Clarence L. (1951) and Ruth M. Roy. Artist Jun Kaneko began the heads series in the late 1990s; over the years, he has increased their size and had them cast in bronze.

Entry prepared by Kristy Demas and Liz Barry.

Note

1. University of Michigan, President's Advisory Committee on Public Art, "The Collection," http://www.public-art.umich.edu/the_collection

Contributors

Kerstin Barndt is Associate Professor of German Studies at the University of Michigan, and affiliated faculty member in the Museum Studies Program. Her research and teaching focus on museum, media, and gender history, literary modernism and the European avant-gardes. She is the author of *Sentiment and Sobriety. The Novel of the New Woman in Weimar Germany* (2003) and co-editor of *Weimar Subjects/Weimar Publics. Rethinking the Political Culture of Germany in the 1920s* (2010).

Richard Barnes is an internationally known New York-based photographer and installation artist. His photographs are part of public and private collections, including the Metropolitan Museum of Art, the Philadelphia Museum of Art, the San Francisco Museum of Modern Art, and the Hirshhorn Museum and Sculpture Garden. Barnes is the author of *Animal Logic* (2009) and the recipient of a number grants and fellowships including the Prix de Rome Fellowship from the American Academy in Rome and the National Endowment for the Arts.

Liz M. Barry is Special Counsel to the President of the University of Michigan. She oversees a variety of ongoing initiatives and projects in the Office of the President, including the President's Advisory Committee on Public Art. Prior to her role as Special Counsel, she served as Managing Director of the university's Life Science Institute and as the university's Associate Vice President and Deputy General Counsel.

Francis X. Blouin is Professor of Information in the University of Michigan School of Information and Professor of History in the College of Literature, Science and the Arts. He served as Director of the Bentley Historical Library for thirty years. His research focuses on archives, documentation, and institutions of social memory, and United States social history. He is the author of *Processing the Past* (2012), *Archives, Documentation and Institutions of Social Memory* (2006), and numerous articles on information, memory, archives, and social history.

Mackenzie Caple graduated from the University of Michigan in 2013, and is currently the project coordinator for the NSF-CSBR *Joint Museums Project*, which aims to digitally reunite groups of specimens, which, although they came from a single expedition, are housed in several different university museums. Her research interests include ecological and plant physiological responses to human impact, especially related to global climate change and invasive species.

David Choberka is the Mellon Manager of Academic Outreach and Teaching at the University of Michigan Museum of Art. He is primarily responsible for facilitating use of the art museum's collection for university teaching and research. He completed his PhD in German Studies and Graduate Certificate in Museum Studies at the University of Michigan and also holds an MA in History from the University of Tennessee.

Martha O'Hara Conway is Director of the Special Collections Library at the University of Michigan. Martha previously worked for the Newberry Library, Yale University Library, and the Library of Congress. She has published on archival collections assessment and is an active member of the Rare Books and Manuscripts Section of the Association of College and Research Libraries.

Kristy Demas serves as Correspondence and Communications Coordinator at the Office of the President at the University of Michigan.

Christopher W. Dick is Professor for Ecology and Evolutionary Biology, Associate Chair for Museum Collections, U-M Herbarium and Director of the E. S. George Reserve. His research focuses on tropical biodiversity, molecular ecology, and plant evolution. The recipient of a great number of important research grants, he is the co-author of *Tropical Rain Forests: Past, Present and Future* (2005) and *Michigan Shrubs and Vines* (2016).

Lori Ann Dick holds the position of Marketing and Communications Manager at U-M's Museum of Natural History. She also serves as the Communications Manager for the Carl Rinne Lewy Body Dementia Initiative, a program of the Michigan Alzheimer's Disease Center at the University of Michigan.

Elaine K. Didier is the Director of the Gerald R. Ford Presidential Library and Museum. A leader in library information technology and administration, she served as associate dean of the University of Michigan's Rackham School of Graduate Studies, dean of Kresge Library at Oakland University, and as president of the Michigan Library Association.

Daniel C. Fischer is the Claude W. Hibbard Collegiate Professor of Paleontology and Director of the Museum of Paleontology at the University of Michigan. His research focuses on the paleobiology and extinction of mastodons and mammoths. He is the recipient of a great number of important research grants, including a recent NSF grant in support of U-M's collections and a consultant for national museums of natural history. His research has been published in *Nature* and *Paleobiology*, among other journals.

Richard I. Ford is Professor Emeritus of Anthropology and Curator Emeritus in the Museum of Anthropological Archaeology at the University of Michigan. His ethnographic and archaeological research focuses in the U.S. Southwest, where he has conducted ethnobotanical research on how Native peoples used plants as food, medicines, and cultural symbols, and studied rock art and sacred landscapes of northern New Mexico.

Barbara H. Garavaglia is the Director of the University of Michigan Law Library. An attorney and legal research librarian by training, she has worked for the Law Library for over twenty-five years and served as head of the reference department and assistant director, building and maintaining an enduring infrastructure to support faculty and student research.

Philip D. Gingerich is Professor Emeritus of Earth and Environmental Sciences and Curator Emeritus in the Museum of Paleontology, which he directed from 1981 to 2011. He holds numerous honors and awards including membership in the American Philosophical Society and the American Academy of Arts and Sciences. His research examines the origin of mammals and rates of evolution, with particular focus on the evolution of whales and primates.

Martha Girsch graduated from the University of Michigan in 2013 with a BA in the History of Art and has since worked as an editor and researcher at the U-M Medical School's Department of Cell and Developmental Biology, which in its early history encompassed the Department of Anatomy. She edited and contributed to a 2015 publication about the history of the Department of Cell and Developmental Biology in celebration of its 150th anniversary.

J. Kevin Graffagnino is Director of the Clements Library at the University of Michigan. He came to the university in 2008, after serving in curatorial and administrative positions at the University of Vermont, State Historical Society of Wisconsin, Kentucky Historical Society, and Vermont Historical Society. He has published widely on aspects of

American history, libraries, historical cartography, and bibliomania.

Robert E. Grese is Professor and Theodore Roosevelt Chair of Ecosystem Management and serves as Director of the University of Michigan's Matthaei Botanical Gardens and Nichols Arboretum. His teaching and research involve ecologically based landscape design and management that respects the cultural and natural history of a region. He is the author of *The Native Landscape Reader* (2011).

Amy Harris serves as Director of the University of Michigan Museum of Natural History. In this role she supervises exhibitions and public programming. Currently, she is working with her team to envision and create the newest iteration of the museum that will be housed in the Biological Sciences Building.

Julie Herrada is the Curator of the Joseph A. Labadie Collection at the University of Michigan Library. In 2002 Herrada was one of the first honorees selected as a Mover and Shaker by *Library Journal.* The award is bestowed on those in the library profession who are "innovative, creative and making a difference." In 2011 Herrada received the Distinguished Alumna Award from Wayne State University's School of Library and Information Science.

Amanda Krugliak is a visual artist, performance artist and arts curator who has served as the University of Michigan's Institute for the Humanities arts curator since 2007. Her paintings have been exhibited and collected throughout California and the Midwest. She has performed her original monologues nationally and curated many exhibitions, most notably *State of Exception* (together with Richard Barnes and Jason de Leon).

Peter M. McIsaac is Associate Professor of German Studies and Museum Studies at the University of Michigan. He works at the intersections of German literary culture and museums; on popular anatomy exhibitions; and with digital humanities approaches to mainstream periodicals. His publications include *Museums of the Mind: German Modernity and the Dynamics of Collecting* (2007) and the co-edited volume *Exhibiting the German Past: Museums, Film, and Musealization* (2015).

Carole McNamara is the Curator Emerita of Western Art, at the University of Michigan Museum of Art (UMMA). She has held a variety of positions at UMMA, including Registrar, Collections Manager, Senior Curator, and Interim Director. She has curated more than fifty exhibitions ranging from the graphic work of James McNeill Whistler, to international loan exhibitions. McNamara has more than fifteen publications to her credit including gallery guides and scholarly catalogues.

David C. Michener is the Associate Curator at the U-M Matthaei Botanical Gardens and Nichols Arboretum. His current publications focus on ornamental peony domestication and he is the lead co-author of the in-press *Plant Lover's Guide to Peonies* (Timber Press). He is a faculty associate in the Museum Studies Program, the Program in the Environment, and a summer-instructor in the Michigan Math and Science Scholars program.

Rachel Miller is a PhD candidate in the History Department of the University of Michigan and a Museum Studies Certificate student. Her research pertains to nineteenth century US cultural history, visual and material culture, and theories of race and ethnicity. Her dissertation considers US theater and entertainment history, 1860–1930.

Shannon O'Dell directs and curates the Sindecuse Museum of Dentistry at the University of Michigan. She supervises all museum functions including acquisition and development, research and exhibitions, and facility maintenance.

John Pedley is Professor Emeritus of Classical Archaeology and Greek at the University of Michigan, where he taught from 1965 until his retirement in 2002. He served as Director of the university's Kelsey Museum of Archaeology from 1973 to 1986. He is the author of more than a dozen books and many dozens of articles and reviews, including *The Life and Work of Francis Willey Kelsey: Archaeology, Antiquity and the Arts* (2011).

Christopher Ratté is Director of the University of Michigan's Kelsey Museum of Archaeology, curator of Greek and Hellenistic collections, and Professor of Classical Archaeology in the Departments of Classical Studies and History of Art. His research focuses on the archaeology of Asia Minor and he is the author of *Lydian Architecture* (2012) and co-editor of *Aphrodisias Regional Survey* (2012) and *Archeology and the Cities of Asia Minor in Late Antiquity* (2011).

Lisa Raycraft is Communications and Public Relations Manager at the University of Michigan's School of Social Work.

Pamela Reister is Curator for Museum Teaching and Learning at the University of Michigan Museum of Art. She organizes the museum's K–12 programs and museum tours and helps to select, train, and oversee its volunteer docents.

Anton A. Reznicek is Curator of Vascular Plants at the University of Michigan Herbarium. His research interests include the conservation of the Great Lakes region flora, and the history of botanical exploration in the region. He also studies the biogeography of the northeastern North American flora, and the systematics and evolution of sedges, with a major present focus in the neotropics, especially in Mexico. He is the co-author of *Field Manual of Michigan Flora* and the continuously updated *Michigan Flora Online*.

Mark Rice is Professor of American Studies at St. John Fisher College, in Rochester, New York. His research focuses on the public role of photography in American society. He is the author of *Dean Worcester's Fantasy Islands: Photography, Film, and the Colonial Philippines* (2014). His current research project considers Worcester's work as one of two members of the Menage Zoological Expedition to the Philippines (1890–93).

Cheney J. Schopieray serves as Curator of Manuscripts at the William L. Clements Library. Trained at U-M's School of Information, he is responsible for collection development, grant writing, reader assistance, and class presentations about the library's primary source materials.

Elaine Sims is the Director of the University of Michigan Health System's Gifts of Art program, which is one of the first and most comprehensive arts in healthcare programs in the country. An early pioneer in the field, Sims has developed model programs in the visual and performing arts, healing gardens, caring for the caregiver initiatives, and other arts in healthcare offerings in public spaces, at the bedside, in classrooms and the community.

Carla M. Sinopoli is Professor of Anthropology, Curator of Asian Archaeology and Ethnology in the Museum of Anthropological Archaeology, and Director of the Museum Studies Program at the University of Michigan. Sinopoli has published on the archaeology of empires, political economy of craft production, archaeological ceramics, South Asian archaeology, and the history of anthropological collecting in museums.

Kathleen Splaine Dolan is an art curator and heads Splaine Dolan Fine Art consulting services. At the University of Michigan, she developed and curated the Ross Art Collection at the Stephen M. Ross School of Business.

Emily Talbot is a PhD candidate in History of Art at the University of Michigan. She specializes in nineteenth-century European art, and is currently completing a dissertation that considers the impact of photography on the production and conception of pictorial art at the end of the nineteenth century.

Priscilla Tucker is Professor of Ecology and Evolutionary Biology, Curator of Mammals, and Director of the University of Michigan Museum of Zoology. Her research focuses on biodiversity, genetics, and the molecular evolution and systematics of mammals and speciation. She has published a great number of papers based on her research, which have appeared, among others, in *Genetics*, *Nature*, and in *Molecular Biology and Evolution*.

Brian A. Williams serves as archivist at the University of Michigan's Bentley Historical Library and heads the Bentley Library's Bicentennial Initiatives. Prior to working at the Bentley Library he worked as archivist at Oberlin College. He has curated several exhibitions based on the Bentley's holdings and published *Michigan on the March: The University of Michigan in World War II* (1995), and *Thought and Action. John Dewey at the University of Michigan* (1998) in the Bulletin of the Bentley Historical Library.

Acknowledgments

Near the end of a University of Michigan committee meeting in the early fall of 2013, the two editors of this book each observed that we were interested in launching a project that would both critically examine and celebrate the University of Michigan's museums and collections during the period of the 2017 U-M bicentennial. As sometimes happens in such serendipitous conversations, we determined then and there to partner in these efforts. Among the results of our partnership—nearly four years after we began—is the present book. It is both larger and more ambitious than our vague early ideas, as we brought together our skills, interests, and energies into a bigger project than either of us had originally imagined, and as we added archives and library collections to our original focus on museums. Along the way, we have found enthusiastic partners, received funding from many sources, and have had the good fortune to receive assistance from curators and collection managers in whose care the collections reside, archivists and staff working in the libraries, and editors. We are pleased to acknowledge them here.

Our most important debt is to the many colleagues who authored contributions. We thank all of them for sharing their knowledge and insights and for their timeliness and patience with the long editorial process. We have learned a tremendous amount from reading your work. We especially acknowledge artist Richard Barnes, who visited Ann Arbor four times to make photographs of the collections and museum spaces. A small portion of his remarkable images appear in the photo essay that separates Sections 2 and 3 of this volume. Coming from our perspectives as a cultural historian and an archaeologist, respectively, it has been fascinating to view the collections through Richard's eyes and learn to see the unexpected beauty and fascinating juxtapositions that his work explores. We also thank Amanda Krugliak of the U-M Institute for the Humanities for helping to facilitate our partnership with Richard and for authoring the essay that accompanies his photos.

A large number of university programs and units have provided funding for this project. We gratefully acknowledge awards from the U-M Bicentennial Committee, the M-Cubed initiative, the Institute for the Humanities, the Provost's Collection Council, the College of Literature, Science, and the Arts, and the U-M Office of Research. In addition, the following libraries, museums, and programs contributed subvention support that allowed a portion of this book to be printed in color: Clements Library, Herbarium, Kelsey Museum of Archaeology, Matthaei Botanical Garden and Nichols Arboretum, Museum of Art, Museum of Anthropological Archaeology, Museum of Natural History, Museum of Paleontology, Museum of Zoology, U-M Health System Gifts of Art, and Museum Studies Program. The Bentley Library made its contributions "in kind" and generously provided high resolution scans of many of its remarkable archival photographs at no cost, which made it possible to include more than 175 photographs in the volume. We thank the directors of all of the above for their generous support.

Many other individuals also helped to make this volume a success. We are grateful to all of the collection managers and professional staff in our many museums, archives and collections for granting us access to their spaces and introducing us to the collections. In the Museum of Zoology, we particularly thank Janet Hinshaw, Mark O'Brien, Doug Nelson, and Cody Thompson. Rich

Rabelar and Mackenzie Caple generously hosted our many visits to the University Herbarium; and Dan Miller and Adam Rountrey did the same in the Museum of Paleontology. In the Museum of Anthropological Archaeology, we acknowledge the assistance of Kay Clahassey, Karen O'Brien Dively, Lauren Fuka, and Kerri Wilhelm; Sebastian Encina allowed us into the storerooms and closed spaces of the Kelsey Museum. Director Shannon O'Dell and Collections Coordinator Kathy Daniels welcomed us behind the scenes in the Sindecuse Museum of Dentistry. In the Bentley Library, we particularly thank Malgosia Myc, Karen Wight, and Brian Williams for their willing assistance in locating the "perfect" image or record for our project. Pablo Alvarez made us aware of treasures residing in the Special Collections Library pertaining to Renaissance cabinets of natural history, and Timothy Utter located historical maps in the Clark Library that guided our research as well.

Our gratitude also extends to administrators, research assistants and editors who helped us to not lose sight of the many details a project of this scale entails. Amy Smola of the Museum Studies Program ably managed and distributed project funds; in the German Department, Sheri Sytsema-Geiger supported Kerstin Barndt in securing and administrating grants relevant to *Object Lessons*. Emma Kinery and Kathryn Zoller, students in the Undergraduate Research Opportunities Program assisted Barndt in researching the early history of the University museums and compiling their chronology. Michael Abrahamson, graduate student in Architecture, created a fabulous timeline that visualized the differentiation of museums and collections over the course of two centuries. Graduate students Kathryn Holihan, Hannah McMurray, and Mary Hennessy provided research and editorial support as we prepared the manuscript, while Jill Rheinheimer expertly edited the initial drafts of all contributions as we prepared them for submission to the University of Michigan Press. Johannes von Moltke read and provided helpful comments on early drafts of the book proposal and has been a source of encouragement and insight throughout.

At U-M Press we acknowledge the encouragement and support of Associate Acquiring Editor Scott Decker Ham and Editorial Assistant Danielle Coty, who took charge of acquiring permissions for the many images reproduced in *Object Lessons*. We also thank Production Editor Mary Hashman for ushering the manuscript through the last phase of the publication process and the U-M Press design staff.

Four years after we began, we remain excited by this project and view this volume as only a first step into the study of the first two centuries of the University of Michigan's manifold museums, archives, and libraries. Our appreciation of the breadth, depth, and scholarly potential of the collections has only grown over this time. We hope this work provokes further study and future publications, exhibitions, and events that explore, critique, and make visible these extraordinary collections.

Kerstin Barndt
Carla Sinopoli
November 29, 2016

Photo Credits

The source of each photograph is given in the caption. If no source is noted, the photograph is part of the museum collection whose description it accompanies.

Photographs reproduced from the Bentley Historical Library, University of Michigan, are part of specific collections listed below.

Alexander G. Ruthven papers 69, 71
Alumni Association
  (University of Michigan) records 74
Bentley Historical Library
  (University of Michigan) records 75, 144–48
Building and Grounds Department
  (University of Michigan) records 129
Corydon L. Ford papers 41
Detroit Observatory
  (University of Michigan) records 82–84
Douglas Houghton papers 9, 11, 13, 14
Francis Willey Kelsey papers 52, 53
George Robert Swain photographs and papers 24
Henry Philip Tappan papers 7
Jasper Francis Cropsey visual materials 6, 80
Joseph Beal Steere papers 44, 45
Matthaei Botanical Gardens
  (University of Michigan) records 2, 120
Medical School
  (University of Michigan) records 25–28, 42, 43
Museum of Art
  (University of Michigan) records 31, 32
Museum of Zoology
  (University of Michigan) records 35, 66, 70
Museums (University of Michigan) records 1
Sesquicentennial Office
  (University of Michigan) records 39
Stearns Collection of Musical
  Instruments records 133
University of Michigan Architectural Drawings 23
University of Michigan at Ann Arbor:
  photo-gravures 22
University of Michigan News and
  Information Services Photographs 5, 8
University of Michigan Photographs
  Vertical File 3, 15–16, 19–21, 30, 34, 36, 86
University of Michigan Scientific Expedition
  to Lake Superior photograph collection 17
Walter Koelz papers 76–79
W. B. Hinsdale papers 48

Index

Note: Page numbers in italics refer to illustrations and tables.